Michael Struwe · Variational Methods

Michael Struwe

Variational Methods

Applications to
Nonlinear Partial Differential Equations
and Hamiltonian Systems

With 16 Figures

Springer-Verlag Berlin Heidelberg New York
London Paris Tokyo Hong Kong Barcelona

Prof. Dr. Michael Struwe
Mathematik, ETH Zürich
ETH-Zentrum
CH-8092 Zürich, Switzerland

Mathematics Subject Classification (1980):
34–02, 35–02, 35 A 15, 49–02, 58–02, 58 E XX

ISBN 3-540-52022-8 Springer-Verlag Berlin Heidelberg New York
ISBN 0-387-52022-8 Springer-Verlag New York Berlin Heidelberg

Library of Congress Cataloging-in-Publication Data
Struwe, Michael, 1955– Variational methods: applications to nonlinear partial differ-
ential equations and Hamiltonian systems / Michael Struwe. p. cm.
Includes bibliographical references and index.
ISBN 3-540-52022-8 (Springer-Verlag Berlin Heidelberg New York)
ISBN 0-387-52022-8 (Springer-Verlag New York Berlin Heidelberg)
1. Calculus of variations. 2. Differential equations, Nonlinear. 3. Hamiltonian
systems. I. Title. QA316.S77 1990 515′.64 – dc20 90-9965 CIP

© Springer-Verlag Berlin Heidelberg 1990
Printed in the United States of America

2141/3140-543210 – Printed on acid-free paper

Preface

It would be hopeless to attempt to give a complete account of the history of the calculus of variations. The interest of Greek philosophers in isoperimetric problems underscores the importance of "optimal form" in ancient cultures, see Hildebrandt-Tromba [1] for a beautiful treatise of this subject. While variational problems thus are part of our classical cultural heritage, the first modern treatment of a variational problem is attributed to Fermat (see Goldstine [1; p.1]). Postulating that light follows a path of least possible time, in 1662 Fermat was able to derive the laws of refraction, thereby using methods which may already be termed analytic.

With the development of the Calculus by Newton and Leibniz, the basis was laid for a more systematic development of the calculus of variations. The brothers Johann and Jakob Bernoulli and Johann's student Leonhard Euler, all from the city of Basel in Switzerland, were to become the "founding fathers" (Hildebrandt-Tromba [1; p.21]) of this new discipline. In 1743 Euler [1] submitted "A method for finding curves enjoying certain maximum or minimum properties", published 1744, the first textbook on the calculus of variations. In an appendix to this book Euler [1; Appendix II, p. 298] expresses his belief that "every effect in nature follows a maximum or minimum rule" (see also Goldstine [1; p. 106]), a credo in the universality of the calculus of variations as a tool. The same conviction also shines through Maupertuis' [1] work on the famous "least action principle", also published in 1744. (In retrospect, however, it seems that Euler was the first to observe this important principle. See for instance Goldstine [1; p. 67 f. and p. 101 ff.] for a more detailed historical account.) Euler's book was a great source of inspiration for generations of mathematicians following.

Major contributions were made by Lagrange, Legendre, Jacobi, Clebsch, Mayer, and Hamilton to whom we owe what we now call "Euler-Lagrange equations", the "Jacobi differential equation" for a family of extremals, or "Hamilton-Jacobi theory".

The use of variational methods was not at all limited to 1-dimensional problems in the mechanics of mass-points. In the 19th century variational methods were also employed for instance to determine the distribution of an electrical charge on the surface of a conductor from the requirement that the energy of the associated electrical field be minimal ("Dirichlet's principle"; see Dirichlet [1] or Gauss [1]) or were used in the construction of analytic functions (Riemann [1]).

However, none of these applications was carried out with complete rigor. Often the model was confused with the phenomenon that it was supposed to describe and the fact (?) , for instance, that in nature there always exists a equilibrium distribution for an electrical charge on a conducting surface was taken as sufficient evidence for the corresponding mathematical problem to have a solution. A typical reasoning reads as follows:

"In any event therefore the integral will be non-negative and hence there must exist a distribution (of charge) for which this integral assumes its minimum value," (Gauss [1; p.232], translation by the author).

However, towards the end of the 19th century progress in abstraction and a better understanding of the foundations of the calculus opened such arguments to criticism. Soon enough, Weierstrass [1; pp. 52–54] found an example of a variational problem that did not admit a minimum solution. (Weierstrass challenged his colleagues to find a continuously differentiable function $u: [-1, 1] \to \mathbb{R}$ minimizing the integral

$$I(u) = \int_{-1}^{1} \left| x \frac{d}{dx} u \right|^2 dx$$

subject (for instance) to the boundary conditions $u(\pm 1) = \pm 1$. Choosing

$$u_\varepsilon(x) = \frac{\arctan(\frac{x}{\varepsilon})}{\arctan(\frac{1}{\varepsilon})}, \quad \varepsilon > 0,$$

as a family of comparison functions, Weierstrass was able to show that the infimum of I in the above class was 0; however, the value 0 is not attained; see also Goldstine [1; p. 371 f.].) Weierstrass' critique of Dirichlet's principle precipitated the calculus of variations into a Grundlagenkrise comparable to the crisis in set theory and logic after Russell's discovery of antinomies in Cantor's set theory or Gödel's incompleteness proof.

However, through the combined efforts of several mathematicians who did not want to give up the wonderful tool that Dirichlet's principle had been – including Weierstrass, Arzéia, Fréchet, Hilbert, and Lebesgue – the calculus of variations was revalidated and emerged from its crisis with new strength and vigor.

Hilbert's speech at the centennial assembly of the International Congress 1900 in Paris, where he proposed his famous 20 problems – two of which devoted to questions related to the calculus of variatons – marks this newly found confidence.

In fact, following Hilbert's [1] and Lebesgue's [1] solution of the Dirichlet problem, a development began which within a few decades brought tremendous success, highlighted by the 1929 theorem of Ljusternik and Schnirelman [1] on the existence of three distinct prime closed geodesics on any compact surface of genus zero, or the 1930/31 solution of Plateau's problem by Douglas [1], [2] and Radò [1].

The Ljusternik-Schnirelman result (and a previous result by Birkhoff [1], proving the existence of one closed geodesic on a surface of genus 0) also marks the beginning of global analyis. This goes beyond Dirichlet's principle as we no longer consider only minimizers (or maximizers) of variational integrals, but instead look at all their critical points. The work of Ljusternik and Schnirelman revealed that much of the complexity of a function space is invariably reflected in the set of critical points of any variational integral defined on it, an idea whose importance for the further development of mathematics can hardly be overestimated, whose implications even today may only be conjectured, and whose applications seem to be virtually unlimited. Later, Ljusternik and Schnirelman [2] laid down the foundations of their method in a general theory. In honor of their pioneering effort any method which seeks to draw information concerning the number of critical points of a functional from topological data today often is referred to as Ljusternik-Schnirelman theory.

Around the time of Ljusternik and Schnirelman's work, another – equally important – approach towards a global theory of critical points was pursued by Marston Morse [2]. Morse's work also reveals a deep relation between the topology of a space and the number and types of critical points of any function defined on it. In particular, this led to the discovery of unstable minimal surfaces through the work of Morse-Tompkins [1], [2] and Shiffman [1], [2]. Somewhat reshaped and clarified, in the 50's Morse theory was highly successful in topology (see Milnor [1] and Smale [2].) After Palais [1], [2] and Smale [1] in the 60's succeeded in generalizing Milnor's constructions to infinite-dimensional Hilbert manifolds – see also Rothe [1] for some early work in this regard – Morse theory finally was recognized as a useful (and usable) instrument also for dealing with partial differential equations.

However, applications of Morse theory seemed somewhat limited in view of prohibitive regularity and non-degeneracy conditions to be met in a variational problem, conditions which – by the way – were absent in Morse's original work. Today, inspired by the deep work of Conley [1], Morse theory seems to be turning back to its origins again. In fact, a Morse-Conley theory is emerging which one day may provide a tool as universal as Ljusternik-Schnirelman theory and still offer an even better resolution of the relation between the critical set of a functional and topological properties of its domain. However, in spite of encouraging results, for instance by Benci [4], Conley-Zehnder [1], Jost-Struwe [1], Rybakowski [1], [2], Rybakowski-Zehnder [1], Salamon [1], and – in particular – Floer [1], a general theory of this kind does not yet exist.

In these notes we want to give an overview of the state of the art in some areas of the calculus of variations. Chapter I deals with the classical direct methods and some of their recent extensions. In Chapters II and III we discuss minimax methods, that is Ljusternik-Schnirelman theory, with an emphasis on some limiting cases in the last chapter, leaving aside the issue of Morse theory which is currently changing all too rapidly.

Examples and applications are given to semilinear elliptic partial differential equations and systems, Hamiltonian systems, nonlinear wave equations, and problems related to harmonic maps of Riemannian manifolds or surfaces of prescribed mean curvature. Although our selection is of course biased by the interests of the author, an effort has been made to achieve a good balance between different areas of current research. Most of the results are known; some of the proofs have been reworked and simplified. Attributions are made to the best of the author's knowledge. No attempt has been made to give an exhaustive account of the field or a complete survey of the literature.

General references for related material are Berger-Berger [1], Berger [1], Chow-Hale [1], Eells [1], Nirenberg [1], Rabinowitz [11], Schwartz [2], Zeidler [1]; in particular, we recommend the recent books by Ekeland [2] and Mawhin-Willem [1] on variational methods with a focus on Hamiltonian systems and the forthcoming works of Chang [7] and Giaquinta-Hildebrandt. Moreover we mention the classical textbooks by Krasnoselskii [1] (see also Krasnoselskii-Zabraiko [1]), Ljusternik-Schnirelman [2], Morse [2], and Vainberg [1]. As for applications to Hamiltonian systems and nonlinear problems, the interested reader may also find additional references on a special topic in these fields in the short surveys by Ambrosetti [2], Rabinowitz [9], or Zehnder [1].

The material covered in these notes is designed for advanced graduate or Ph.D. students or anyone who wishes to acquaint himself with variational methods and possesses a working knowledge of linear functional analysis and linear partial differential equations. Being familiar with the definitions and basic properties of Sobolev spaces as provided for instance in the book by Gilbarg-Trudinger [1] is recommended. However, some of these prerequisites can also be found in the appendix.

In preparing this manuscript I have received help and encouragement from a number of friends and colleagues. In particular, I wish to thank Professors Herbert Amann and Hans-Wilhelm Alt for helpful comments concerning the first two sections of Chapter I. Likewise, I am indebted to Professor Jürgen Moser for useful suggestions concerning Section I.4 and to Professors Helmut Hofer and Eduard Zehnder for advice on Sections I.6, II.5, and II.8, concerning Hamiltonian systems.

Moreover, I am grateful to Gabi Hitz, Peter Bamert, Jochen Denzler, Martin Flucher, Frank Josellis, Thomas Kerler, Malte Schünemann, Miguel Sofer, Jean-Paul Theubet, and Thomas Wurms for going through a set of preliminary notes for this manuscript with me in a seminar at ETH Zürich during the winter term of 1988/89. The present text certainly has profited a great deal from their careful study and criticism.

My special thanks also go to Kai Jenni for the wonderful typesetting of this manuscript with the TEX text processing system.

I dedicate this book to my wife Anne.

Zürich, January 1990   Michael Struwe

Contents

Glossary of Notations

V, V^*	generic Banach space with dual V^*
$\lVert \cdot \rVert$	norm in V
$\lVert \cdot \rVert_*$	induced norm in V^*, often also denoted $\lVert \cdot \rVert$
$\langle \cdot, \cdot \rangle \colon V \times V^* \to \mathbb{R}$	dual pairing, occasionally also used to denote scalar product in $\mathbb{R}^n$
E	generic energy functional
DE	Fréchet derivative
$\mathrm{Dom}(E)$	domain of E
$\langle v, DE(u) \rangle = DE(u)v = D_v E(u)$	directional derivative of E at u in direction v
$L^p(\Omega; \mathbb{R}^n)$	space of Lebesgue-measurable functions $u \colon \Omega \to \mathbb{R}^n$ with finite L^p-norm

$$\lVert u \rVert_{L^p} = \Big(\int_\Omega |u|^p \, dx \Big)^{1/p}, \ 1 \le p < \infty$$

$L^\infty(\Omega; \mathbb{R}^n)$	space of Lebesgue-measurable and essentially bounded functions $u \colon \Omega \to \mathbb{R}^n$ with norm

$$\lVert u \rVert_{L^\infty} = \operatorname*{ess\,sup}_{x \in \Omega} |u(x)|$$

$H^{m,p}(\Omega; \mathbb{R}^n)$	Sobolev space of functions $u \in L^p(\Omega; \mathbb{R}^n)$ with $|\nabla^k u| \in L^p(\Omega)$ for all $k \in \mathbb{N}_0^n, |k| \le m$, with norm $\lVert u \rVert_{H^{m,p}} = \sum_{0 \le |k| \le m} \lVert \nabla^k u \rVert_{L^p}$
$H_0^{m,p}(\Omega; \mathbb{R}^n)$	completion of $C_0^\infty(\Omega; \mathbb{R}^n)$ in the norm $\lVert \cdot \rVert_{H^{m,p}}$; if Ω is bounded an equivalent norm is given by $\lVert u \rVert_{H_0^{m,p}} = \sum_{|k|=m} \lVert \nabla^k u \rVert_{L^p}$
$H^{-m,q}(\Omega; \mathbb{R}^n)$	dual of $H_0^{m,p}(\Omega; \mathbb{R}^n)$, where $\frac{1}{p} = \frac{1}{q} = 1$; q is omitted, if $p = q = 2$
$D^{m,p}(\Omega; \mathbb{R}^n)$	completion of $C_0^\infty(\Omega; \mathbb{R}^n)$ in the norm $\lVert u \rVert_{D^{m,p}} = \sum_{|k|=m} \lVert \nabla^k u \rVert_{L^p}$

$C^{m,\alpha}(\Omega; \mathbb{R}^n)$	space of m times continuously differentiable functions $u: \Omega \to \mathbb{R}^n$ whose m-th order derivatives are Hölder continuous with exponent $0 \le \alpha \le 1$
$C_0^\infty(\Omega; \mathbb{R}^n)$	space of smooth functions $u: \Omega \to \mathbb{R}^n$ with compact support in Ω
$\mathrm{supp}(u) = \overline{\{x \in \Omega \; ; \; u(x) \ne 0\}}$	support of a function $u: \Omega \to \mathbb{R}^n$
$\llcorner$	restriction of a measure
$\mathcal{L}^n$	Lebesgue measure on $\mathbb{R}^n$
$B_\rho(u; V) = \{v \in V \; ; \; \|u - v\| < \rho\}$	open ball of radius ρ around $u \in V$; in particular, if $V = \mathbb{R}^n$, then $B_\rho(x_0) = B_\rho(x_0; \mathbb{R}^n)$, $B_\rho = B_\rho(0)$
Re	real part
Im	imaginary part
c, C	generic constants
Cross-references	$(N.x.y)$ refers to formula (x, y) in Chapter N $(x.y)$ within Chapter N refers to formula $(N.x.y)$

Chapter I

The Direct Methods in the Calculus of Variations

A particular class of functional equations $F(u) = 0$, for u belonging to some Banach space V, is the class of Euler-Lagrange equations

$$DE(u) = 0$$

for a functional E on V which is Fréchet differentiable with derivative DE. We call such equations of *variational form.*

For equations of variational form an extensive theory has been developed and variational principles play an important role in mathematical physics and differential geometry, optimal control and numerical anlysis.

We briefly recall the basic definitions that will be needed in this and the following chapters, see Appendix C for details: Suppose E is a Fréchet differentiable functional on a Banach space V with normed dual V^* and duality pairing $\langle \cdot, \cdot \rangle : V \times V^* \to \mathbb{R}$, and let $DE : V \to V^*$ denote the Fréchet-derivative of E. Then the directional (Gateaux-) derivative of E at u in direction of v is given by

$$\frac{d}{d\varepsilon} E(u + \varepsilon v) \bigg|_{\varepsilon = 0} = \langle v, DE(u) \rangle = DE(u)\, v.$$

For such E, we call a point $u \in V$ *critical* if $DE(u) = 0$; otherwise, u is called *regular.* A number $\beta \in \mathbb{R}$ is a *critical value* of E if there exists a critical point u of E with $E(u) = \beta$. Otherwise, β is called *regular.* Of particular interest (also in the non-differentiable case) will be relative minima of E, possibly subject to constraints. Recall that for a set $M \subset V$ a point $u \in M$ is an *absolute minimizer* for E on M if for all $v \in M$ there holds $E(v) \geq E(u)$. A point $u \in M$ is a *relative minimizer* for E on M if for some neighborhood U of u in V it is absolutely E-minimizing in $M \cap U$. Moreover, in the differentiable case, we shall also be interested in the existence of *saddle points*, that is, critical points u of E such that any neighborhood U of u in V contains points v, w such that $E(v) < E(u) < E(w)$. In physical systems, saddle points appear as unstable equilibria or transient excited states.

In this chapter we review some basic methods for proving the existence of relative minimizers. Somewhat imprecisely we summarily refer to these methods as the *direct methods* in the calculus of variations. However, besides the classical lower semi-continuity and compactness method we also include the compensated compactness method of Murat and Tartar, and the concentration-compactness principle of P.L. Lions. Moreover, we recall Ekeland's variational principle and the duality method of Clarke and Ekeland.

Applications will be given to problems concerning minimal hypersurfaces, semilinear and quasi-linear elliptic boundary value problems, finite elasticity, Hamiltonian systems, and semilinear wave equations.

From the beginning it will be apparent that in order to achieve a satisfactory existence theory the notion of admissible function will have to be suitably relaxed. Hence, in general, the above methods will at first only yield generalized or "weak" solutions of our problems. A second step often will be necessary to show that these solutions are regular enough to be admitted as classical solutions. The regularity theory in many cases is very subtle and involves a delicate machinery. It would go beyond the scope of this book to cover this topic completely. However, for the problems that we will mostly be interested in the regularity question can be dealt with rather easily. The reader will find this material in Appendix B. References to more advanced texts on the regularity issue will be given where appropriate.

1. Lower Semi-Continuity

In this section we study the existence of relative minimizers of functionals E and we give sufficient conditions for a functional to be bounded from below and to attain its infimum.

The discussion can be made largely independent of any differentiability assumptions on E or structure assumptions on the underlying space of admissible functions M. In fact, we have the following classical result.

1.1 Theorem. *Let M be a topological Hausdorff space, and suppose $E : M \rightarrow \mathbb{R} \cup +\infty$ satisfies the condition of bounded compactness:*

For any $\alpha \in \mathbb{R}$ the set

(1.1) $$K_\alpha = \{u \in M \; ; \; E(u) \leq \alpha\}$$

is compact (Heine-Borel property).

Then E is uniformly bounded from below on M and attains its infimum. The conclusion remains valid if instead of (1.1) we suppose that any sub-level-set K_α is sequentially compact.

Remark. Necessity of condition (1.1) is illustrated by simple examples: The function $E(x) = x^2$, if $x \neq 0, = 1$ if $x = 0$ on $[-1, 1]$, or the exponential function $E(x) = \exp(x)$ on $\mathbb{R}$ are bounded from below but do not admit a minimizer. Note that the space M in the first example is compact while in the second example the function E is smooth – even analytic.

Proof of Theorem 1.1. Suppose (1.1) holds. We may assume $E \not\equiv +\infty$. Let

$$\alpha_0 = \inf_M E \geq -\infty,$$

and let (α_m) be the strictly decreasing sequence

$$\alpha_m \searrow \alpha_0 \qquad (m \to \infty) .$$

Let $K_m = K_{\alpha_m}$. By assumption, each K_m is compact and non-empty. More-over, $K_m \supset K_{m+1}$ for all m. By compactness of K_m there exists a point $u \in \bigcap_{m \in \mathbb{N}} K_m$, satisfying

$$E(u) \leq \alpha_m, \qquad \text{for all } m \geq m_0.$$

Passing to the limit $m \to \infty$ we obtain that

$$E(u) \leq \alpha_0 = \inf_M E,$$

and the claim follows.

If instead of (1.1) each K_α is sequentially compact, we choose a *minimizing sequence* (u_m) in M such that $E(u_m) \to \alpha_0$. Then for any $\alpha > \alpha_0$ the sequence (u_m) will eventually lie entirely within K_α. By sequential compactness of K_α therefore (u_m) will accumulate at a point $u \in \bigcap_{\alpha > \alpha_0} K_\alpha$ which is the desired minimizer. □

Remark that if $E : M \to \mathbb{R}$ satisfies (1.1), then for any $\alpha \in \mathbb{R}$ the set

$$\{u \in M ; \ E(u) > \alpha\} = M \setminus K_\alpha$$

is open, that is, E is *lower semi-continous*. (Respectively, if each K_α is sequentially compact, then E will be sequentially lower semi-continuous.) Conversely, if E is (sequentially) lower semi-continuous and for some $\overline{\alpha} \in \mathbb{R}$ the set $K_{\overline{\alpha}}$ is (sequentially) compact, then K_α will be (sequentially) compact for all $\alpha \leq \overline{\alpha}$ and again the conclusion of Theorem 1.1 will be valid.

Note that the lower semi-continuity condition can be more easily fulfilled the finer the topology on M. In contrast, the condition of compactness of the sub-level sets K_α , $\alpha \in \mathbb{R}$, calls for a coarse topology and both conditions are competing. In practice, there is often a natural weak Sobolev space topology where both conditions can be simultaneously satisfied. However, there are many interesting cases where condition (1.1) cannot hold in *any* reasonable topology (even though relative minimizers may exist). Later in this chapter we shall see some examples and some more delicate ways of handling the possible loss of compactness. See Section 4; see also Chapter III.

In applications, the conditions of the following special case of Theorem 1.1 can often be checked more easily.

1.2 Theorem. *Suppose V is a reflexive Banach space with norm $\|\cdot\|$, and let $M \subset V$ be a weakly closed subset of V. Suppose $E: M \to \mathbb{R} \cup +\infty$ is coercive on M with respect to V, that is*

($1°$) $E(u) \to \infty$ as $\|u\| \to \infty$, $u \in M$

and (sequentially) weakly lower semi-continous on M with respect to V, that is

($2°$) for any $u \in M$, any sequence (u_m) in M such that $u_m \to u$ weakly in V there holds:

$$E(u) \le \liminf_{m \to \infty} E(u_m) \ .$$

Then E is bounded from below on M and attains its infimum in M.

The concept of minimizing sequences offers a direct and (apparently) constructive proof.

Proof. Let $\alpha_0 = \inf_M E$ and let (u_m) be a minimizing sequence in M; that is, satisfying $E(u_m) \to \alpha_0$. By coerciveness, (u_m) is bounded in V. Since V is reflexive, by the Eberlein-Šmulian theorem (see Dunford-Schwartz [1; p. 430]) we may assume that $u_m \to u$ weakly for some $u \in V$. But M is weakly closed, therefore $u \in M$, and by weak lower semi-continuity

$$E(u) \le \liminf_{m \to \infty} E(u_m) = \alpha_0 \ . \hspace{2cm} \square$$

Examples. An important example of a sequentially weakly lower semi-continous functional is the norm in a Banach space V. Closed and convex subsets of Banach spaces are important examples of weakly closed sets. If V is the dual of a separable normed vector space, Theorem 1.2 and its proof remain valid if we replace weak by weak*-convergence.

     We present some simple applications.

Degenerate Elliptic Equations

1.3 Theorem. *Let Ω be a bounded domain in $\mathbb{R}^n$, $p \in [2, \infty[$ with conjugate exponent q satisfying $\frac{1}{p} + \frac{1}{q} = 1$, and let $f \in H^{-1,q}(\Omega)$, the dual of $H_0^{1,p}(\Omega)$, be given. Then there exists a weak solution $u \in H_0^{1,p}(\Omega)$ of the boundary value problem*

(1.2) $$-\nabla \cdot (|\nabla u|^{p-2} \nabla u) = f \hspace{1cm} in \ \Omega$$

(1.3) $$u = 0 \hspace{1cm} on \ \partial\Omega$$

in the sense that u satisfies the equation

(1.4) $$\int_\Omega (\nabla u |\nabla u|^{p-2} \nabla\varphi - f\varphi)dx = 0 \ , \hspace{1cm} \forall \varphi \in C_0^\infty(\Omega) \ .$$

Proof. Remark that the left part of (1.4) is the directional derivative of the C^1-functional

$$E(u) = \frac{1}{p} \int_{\Omega} |\nabla u|^p \, dx \; - \; \int_{\Omega} fu \, dx$$

on the Banach space $V = H_0^{1,p}(\Omega)$, in direction φ; that is, problem (1.2), (1.3) is of variational form.

Note that $H_0^{1,p}(\Omega)$ is reflexive. Moreover, E is coercive. In fact, we have

$$E(u) \geq \frac{1}{p}\|u\|_{H_0^{1,p}}^p \; - \; \|f\|_{H^{-1,q}} \, \|u\|_{H_0^{1,p}} \geq \frac{1}{p}\left(\|u\|_{H_0^{1,p}}^p - c\|u\|_{H_0^{1,p}}\right)$$

$$\geq c^{-1}\|u\|_{H_0^{1,p}}^p \; - \; C.$$

Finally, E is (sequentially) weakly lower semi-continuous: It suffices to show that for $u_m \rightharpoonup u$ weakly in $H_0^{1,p}(\Omega)$ we have

$$\int_{\Omega} f \, u_m \, dx \;\; \to \;\; \int_{\Omega} f \, u \, dx \; .$$

Since $f \in H^{-1,q}(\Omega)$, however, this follows from the very definition of weak convergence. Hence Theorem 1.2 is applicable and there exists a minimizer $u \in H_0^{1,p}(\Omega)$ of E, solving (1.4). □

Remark that the p-Laplacian is strongly monotone in the sense that

$$\int_{\Omega} \left(|\nabla u|^{p-2}\nabla u - |\nabla v|^{p-2}\nabla v\right) \cdot (\nabla u - \nabla v) \, dx \geq c\|u - v\|_{H_0^{1,p}}^p \; .$$

In particular, the solution u to (1.4) is unique.

If f is more regular, say $f \in C^{m,\alpha}(\overline{\Omega})$, we would expect the solution u of (1.4) to be more regular as well. This is true if $p = 2$, see Appendix B, but in the degenerate case $p > 2$, where the uniform ellipticity of the p-Laplace operator is lost at zeros of $|\nabla u|$, the best that one can hope for is $u \in C^{1,\alpha}(\overline{\Omega})$; see Uhlenbeck [1], Tolksdorf [2; p.128], Di Benedetto [1].

In Theorem 1.3 we have applied Theorem 1.2 to a functional on a reflexive space. An example in a non-reflexive setting is given next.

Minimal Partitioning Hypersurfaces

For a domain $\Omega \subset \mathbb{R}^n$ let $BV(\Omega)$ be the space of functions $u \in L^1(\Omega)$ such that

$$\int_{\Omega} |Du| = \sup\left\{ \int_{\Omega} \sum_{i=1}^{n} u D_i g_i \, dx \; ; \right.$$

$$\left. g = (g_1, \dots, g_n) \in C_0^1(\Omega; \mathbb{R}^n), \; |g| \leq 1 \right\} < \infty \; ,$$

endowed with the norm

$$\|u\|_{BV} = \|u\|_{L^1} + \int_\Omega |Du| \ .$$

$BV(\Omega)$ is a Banach space, embedded in $L^1(\Omega)$, and – provided Ω is bounded and sufficiently smooth – by Rellich's theorem the injection $BV(\Omega) \hookrightarrow L^1(\Omega)$ is compact; see for instance Giusti [1; Theorem 1.19, p.17]. Moreover, the function $u \mapsto \int_\Omega |Du|$ is lower semi-continuous with respect to L^1-convergence.

Let χ_G be the characteristic function of a set $G \subset \mathbb{R}^n$; that is, $\chi_G(x) = 1$ if $x \in G$, $\chi_G(x) = 0$ else. Also let $\mathcal{L}^n$ denote n-dimensional Lebesgue measure. We prove:

1.4 Theorem. *Let Ω be a smooth, bounded domain in $\mathbb{R}^n$. Then there exists a subset $G \subset \Omega$ such that*

(1°)
$$\mathcal{L}^n(G) = \mathcal{L}^n(\Omega \setminus G) = \frac{1}{2}\mathcal{L}^n(\Omega)$$

and such that its perimeter with respect to Ω,

(2°)
$$P(G, \Omega) = \int_\Omega |D\chi_G| \ ,$$

is minimal among all sets satisfying (1°).

Proof. Let $M = \{\chi_G \ ; \ G \subset \Omega$ is measurable and satisfies $(1°)\}$, endowed with the L^1-topology, and let $E : M \to \mathbb{R} \cup +\infty$ be given by

$$E(u) = \int_\Omega |Du| \ .$$

Since $\|\chi_G\|_{L^1} \leq \mathcal{L}^n(\Omega)$, the functional E is coercive on M with respect to the norm in $BV(\Omega)$. Since bounded sets in $BV(\Omega)$ are relatively compact in $L^1(\Omega)$ and since M is closed in $L^1(\Omega)$, by weak lower semi-continuity of E in $L^1(\Omega)$ the sub-level sets of E are compact. The conclusion now follows from Theorem 1.1.   □

The support of the distribution $D\chi_G$, where G has minimal perimeter (2°) with respect to Ω, can be interpreted as a minimal bisecting hypersurface, dividing Ω into two regions of equal volume. The regularity of the dividing hypersurface is intimately connected with the existence of minimal cones in $\mathbb{R}^n$. See Giusti [1] for further material on functions of bounded variation, sets of bounded perimeter, the area integrand, and applications.

A related setting for the study of minimal hypersurfaces and related objects is offered by geometric measure theory. Also in this field variational principles play an important role; see for instance Almgren [1], Morgan [1], or Simon [1] for introductory material and further references.

Our next example is concerned with a parametric approach.

Minimal Hypersurfaces in Riemannian Manifolds

Let Ω be a bounded domain in $\mathbb{R}^n$, and let S be a compact subset in $\mathbb{R}^N$. Also let $u_0 \in H^{1,2}(\Omega; \mathbb{R}^N)$ with $u_0(\Omega) \subset S$ be given. Define

$$H^{1,2}(\Omega; S) = \left\{ u \in H^{1,2}(\Omega; \mathbb{R}^N) \; ; \; u(\Omega) \subset S \text{ almost everywhere} \right\} .$$

and let

$$M = \left\{ u \in H^{1,2}(\Omega; S) \; ; \; u - u_0 \in H_0^{1,2}(\Omega; \mathbb{R}^n) \right\} .$$

Then, by Rellich's theorem, M is closed in the weak topology of $V = H^{1,2}(\Omega; \mathbb{R}^N)$. For $u = (u^1, \ldots, u^N) \in H^{1,2}(\Omega; S)$ let

$$E(u) = \sum_{i,j=1}^N \int_\Omega g_{ij}(u) \nabla u^i \nabla u^j \, dx ,$$

where $g = (g_{ij})_{1 \le i,j \le N}$ is a given positively definite symmetric matrix with coefficients $g_{ij}(u)$ depending continuously on $u \in S$. Note that since S is compact g is uniformly positive definite on S, and there exists $\lambda > 0$ such that $E(u) \ge \lambda \|\nabla u\|_{L^2}^2$ for $u \in H^{1,2}(\Omega; S)$. In addition, since S and Ω are bounded, we have that $\|u\|_{L^2} \le c$ uniformly, for $u \in H^{1,2}(\Omega; S)$. Hence E is coercive on $H^{1,2}(\Omega; S)$ with respect to the norm in $H^{1,2}(\Omega; \mathbb{R}^N)$.

Finally, E is lower semi-continuous in $H^{1,2}(\Omega; S)$ with respect to weak convergence in $H^{1,2}(\Omega; \mathbb{R}^N)$. Indeed, if $u_m \rightharpoonup u$ weakly in $H^{1,2}(\Omega; \mathbb{R}^N)$, by Rellich's theorem $u_m \to u$ strongly in L^2 and hence a subsequence (u_m) converges almost everywhere. By Egorov's theorem, given $\delta > 0$ there is an exceptional set Ω_δ of measure $\mathcal{L}^n(\Omega_\delta) < \delta$ such that $u_m \to u$ uniformly on $\Omega \setminus \Omega_\delta$. We may assume that $\Omega_\delta \subset \Omega_{\delta'}$ for $\delta \le \delta'$. By weak lower semi-continuity of the semi-norm on $H^{1,2}(\Omega; \mathbb{R}^N)$, defined by

$$|v|^2 = \int_{\Omega \setminus \Omega_\delta} g_{ij}(u) \nabla v^i \nabla v^j \, dx$$

then

$$\int_{\Omega \setminus \Omega_\delta} g_{ij}(u) \nabla u^i \nabla u^j \, dx$$

$$\le \liminf_{m \to \infty} \int_{\Omega \setminus \Omega_\delta} g_{ij}(u) \nabla u_m^i \nabla u_m^j \, dx$$

$$= \liminf_{m \to \infty} \int_{\Omega \setminus \Omega_\delta} g_{ij}(u_m) \nabla u_m^i \nabla u_m^j \, dx$$

$$\le \liminf_{m \to \infty} E(u_m) .$$

Passing to the limit $\delta \to 0$, from Beppo Levi's theorem we obtain

$$E(u) = \lim_{\delta \to 0} \int_{\Omega \setminus \Omega_\delta} g_{ij}(u) \nabla u^i \nabla u^j \, dx$$

$$\le \liminf_{m \to \infty} E(u_m) .$$

Applying Theorem 1.2 to E on M we obtain

1.5 Theorem.   *For any boundary data $u_0 \in H^{1,2}(\Omega; S)$ there exists an E-minimal extension $u \in M$.*

In differential geometry Example 1.5 arises in the study of harmonic maps $u : \Omega \to S$ from a domain Ω into an N-dimensional manifold S with metric g for prescribed boundary data $u = u_0$ on $\partial\Omega$. Like in the previous example, the regularity question is related to the existence of harmonic maps from special domains into the given manifold S; more precisely, singularities of harmonic maps are induced by harmonic mappings of spheres. For further references see Eells-Lemaire [1], [2], Hildebrandt [3], Jost [2]. For questions concerning regularity see Giaquinta-Giusti [1], Schoen-Uhlenbeck [1], [2].

A General Lower Semi-Continuity Result

We now conclude this short list of introductory examples and return to the development of the variational theory. Note that the property of E being lower semi-continuous with respect to some weak kind of convergence is at the core of the above existence result. In Theorem 1.6 below we establish a lower semi-continuity result for a very broad class of variational integrals, including and going beyond those encountered in Theorem 1.5, as Theorem 1.6 would also apply in the case of unbounded targets S and possibly degenerate or singular metrics g.
    We consider variational integrals

$$(1.5) \qquad\qquad E(u) = \int_{\Omega} F(x, u, \nabla u)\, dx$$

involving (vector-valued) functions $u : \Omega \subset \mathbb{R}^n \to \mathbb{R}^N$.

1.6 Theorem. *Let Ω be a domain in $\mathbb{R}^n$, and assume that $F : \Omega \times \mathbb{R}^N \times \mathbb{R}^{nN} \to \mathbb{R}$ is a Caratheodory function satisfying the conditions*
($1°$) $F(x, u, p) \geq \phi(x)$ for almost every x, u, p, where $\phi \in L^1(\Omega)$.
($2°$) $F(x, u, \cdot)$ is convex in p for almost every x, u.
Then, if u_m, $u \in H^{1,1}_{loc}(\Omega)$ and $u_m \to u$ in $L^1(\Omega')$, $\nabla u_m \rightharpoonup \nabla u$ weakly in $L^1(\Omega')$ for all bounded $\Omega' \subset\subset \Omega$, it follows that

$$E(u) \leq \liminf_{m \to \infty} E(u_m)\,,$$

where E is given by (1.5).

Notes. In the scalar case $N = 1$, weak lower semi-continuity results like Theorem 1.6 were first stated by L. Tonelli [1] and C.B. Morrey, Jr. [1]; these results were then extended and simplified by J. Serrin [1], [2] who showed that for non-negative, smooth functions $F(x, u, p) : \Omega \times \mathbb{R} \times \mathbb{R}^n \to \mathbb{R}$ which are convex in p, the functional E given by (1.5) is lower semi-continuous with respect to convergence in $L^1_{loc}(\Omega)$. A corresponding result in the vector-valued case $N > 1$

subsequently was derived by Morrey [4; Theorem 4.1.1]; however, Eisen [1] not only pointed out a gap in Morrey's proof but also gave an example showing that for $N > 1$ in general Theorem 1.6 ceases to be true without the assumption that the L^1-norms of ∇u_m are uniformly locally bounded. Theorem 1.6 is due to Berkowitz [1] and Eisen [2]. Related results can be found for instance in Morrey [4; Theorem 1.8.2], or Giaquinta [1]. Our proof is modelled on Eisen [2].

Proof. We may assume that $(E(u_m))$ is finite and convergent. Moreover, replacing F by $F - \phi$ we may assume that $F \geq 0$. Let $\Omega' \subset\subset \Omega$ be given. By weak local L^1-convergence $\nabla u_m \rightharpoonup \nabla u$, for any $m_0 \in \mathbb{N}$ there exists a sequence $(P^l)_{l \geq m_0}$ of convex linear combinations

$$P^l = \sum_{m=m_0}^{l} \alpha_m^l \nabla u_m \ , \ 0 \leq \alpha_m^l \leq 1 \ , \ \sum_{m=m_0}^{l} \alpha_m^l = 1 \ , \ l \geq m_0$$

such that $P^l \to \nabla u$ strongly in $L^1(\Omega')$ and pointwise almost everywhere; see for instance Rudin [1; Theorem 3.13]. By convexity, for any m_0, any $l \geq m_0$, and almost every $x \in \Omega'$:

$$F\left(x, u(x), P^l(x)\right) = F\left(x, u(x), \sum_{m=m_0}^{l} \alpha_m^l \nabla u_m(x)\right)$$

$$\leq \sum_{m=m_0}^{l} \alpha_m^l F\left(x, u(x), \nabla u_m(x)\right) \ .$$

Integrating over Ω' and passing to the limit $l \to \infty$, from Fatou's Lemma we obtain:

$$\int_{\Omega'} F\left(x, u(x), \nabla u(x)\right) \, dx \leq \liminf_{l \to \infty} \int_{\Omega'} F\left(x, u(x), P^l(x)\right) \, dx$$

$$\leq \sup_{m \geq m_0} \int_{\Omega'} F\left(x, u(x), \nabla u_m(x)\right) \, dx \ .$$

Since m_0 was arbitrary, this implies that

$$\int_{\Omega'} F\left(x, u(x), \nabla u(x)\right) \, dx \leq \limsup_{m \to \infty} \int_{\Omega'} F\left(x, u(x), \nabla u_m(x)\right) \, dx \ ,$$

for any bounded $\Omega' \subset\subset \Omega$.

Now we need the following result (Eisen [2; p.75]).

1.7 Lemma. *Under the hypotheses of Theorem 1.6 on F, u_m, and u there exists a subsequence (u_m) such that:*

$$F\left(x, u_m(x), \nabla u_m(x)\right) - F\left(x, u(x), \nabla u_m(x)\right) \to 0$$

in measure, locally in Ω.

Proof of Theorem 1.6 (completed). By Lemma 1.7 for any $\Omega' \subset\subset \Omega$, any $\varepsilon > 0$, and any $m_0 \in \mathbb{N}$ there exists $m \geq m_0$ and a set $\Omega'_{\varepsilon,m} \subset \Omega'$ with $\mathcal{L}^n\left(\Omega'_{\varepsilon,m}\right) < \varepsilon$ such that

(1.6) $\left|F\left(x, u_m(x), \nabla u_m(x)\right) - F\left(x, u(x), \nabla u_m(x)\right)\right| < \varepsilon$

for all $x \in \Omega' \setminus \Omega'_{\varepsilon,m}$. Replacing ε by $\varepsilon_m = 2^{-m}$ and passing to a subsequence, if necessary, we may assume that for each m there is a set $\Omega'_{\varepsilon_m,m} \subset \Omega'$ of measure $< \varepsilon_m$ such that (1.6) is satisfied (with ε_m) for all $x \in \Omega' \setminus \Omega'_{\varepsilon_m,m}$. Hence, for any given $\varepsilon > 0$, if we choose $m_0 = m_0(\varepsilon) > |\log_2 \varepsilon|$, $\Omega'_\varepsilon = \bigcup_{m \geq m_0} \Omega'_{\varepsilon_m,m}$, this set has measure $\mathcal{L}^n(\Omega'_\varepsilon) < \varepsilon$ and inequality (1.6) holds uniformly for all $x \in \Omega' \setminus \Omega'_\varepsilon$, and all $m \geq m_0(\varepsilon)$. Moreover, for $\varepsilon < \delta$ by construction $\Omega'_\varepsilon \subset \Omega'_\delta$.

Cover Ω by disjoint bounded sets $\Omega^{(k)} \subset\subset \Omega$, $k \in \mathbb{N}$. Let $\varepsilon > 0$ be given and choose a sequence $\varepsilon^{(k)} > 0$, such that $\sum_{k \in \mathbb{N}} \mathcal{L}^n\left(\Omega^{(k)}\right) \varepsilon^{(k)} \leq \varepsilon$. Passing to a subsequence, if necessary, for each $\Omega^{(k)}$ and $\varepsilon^{(k)}$ we may choose $m_0^{(k)}$ and $\Omega_\varepsilon^{(k)} \subset \Omega^{(k)}$ such that $\mathcal{L}^n\left(\Omega_\varepsilon^{(k)}\right) < \varepsilon^{(k)}$ and

$$\left|F\left(x, u_m(x), \nabla u_m(x)\right) - F\left(x, u(x), \nabla u_m(x)\right)\right| < \varepsilon^{(k)}$$

uniformly for $x \in \Omega^{(k)} \setminus \Omega_\varepsilon^{(k)}$, $m \geq m_0^{(k)}$. Moreover, we may assume that $\Omega_\varepsilon^{(k)} \subset \Omega_\delta^{(k)}$, if $\varepsilon < \delta$, for all k. Then for any $K \in \mathbb{N}$, letting $\Omega^K = \cup_{k=1}^K \Omega^{(k)}$, $\Omega_\varepsilon^K = \cup_{k=1}^K \Omega_\varepsilon^{(k)}$ we have

$$\int_{\Omega^K \setminus \Omega_\varepsilon^K} F(x, u, \nabla u)\, dx$$

$$\leq \limsup_{m \to \infty} \int_{\Omega^K \setminus \Omega_\varepsilon^K} F\left(x, u, \nabla u_m\right) dx$$

$$\leq \limsup_{m \to \infty} \int_{\Omega^K \setminus \Omega_\varepsilon^K} F\left(x, u_m, \nabla u_m\right) dx + \varepsilon$$

$$\leq \limsup_{m \to \infty} E(u_m) + \varepsilon\ .$$

Letting $\varepsilon \to 0$ and then $K \to \infty$, the claim follows from Beppo Levi's theorem, since $F \geq 0$ and $\Omega^K \setminus \Omega_\varepsilon^K$ is increasing as $\varepsilon \downarrow 0$, or $K \uparrow \infty$. $\square$

Proof of Lemma 1.7. We basically follow Eisen [2]. Suppose by contradiction that there exist $\Omega' \subset\subset \Omega$ and $\varepsilon > 0$ such that, letting

$$\Omega_m = \{x \in \Omega' \; ; \; |F(x, u_m, \nabla u_m) - F(x, u, \nabla u_m)| \geq \varepsilon\} \,,$$

there holds

$$\liminf_{m \to \infty} \mathcal{L}^n(\Omega_m) \geq 2\varepsilon \,.$$

The sequence (∇u_m), being weakly convergent, is uniformly bounded in $L^1(\Omega')$. In particular,

$$\mathcal{L}^n\{x \in \Omega' \; ; \; |\nabla u_m(x)| \geq l\} \leq l^{-1} \int_{\Omega'} |\nabla u_m| \, dx \leq \frac{C}{l} \leq \varepsilon \,,$$

if $l \geq l_0(\varepsilon)$ is large enough. Setting $\tilde{\Omega}_m := \{x \in \Omega_m \; ; \; |\nabla u_m(x)| \leq l_0(\varepsilon)\}$ therefore there holds

$$\liminf_{m \to \infty} \mathcal{L}^n\left(\tilde{\Omega}_m\right) \geq \varepsilon.$$

Hence also for $\Omega^M = \bigcup_{m \geq M} \tilde{\Omega}_m$ we have

$$\mathcal{L}^n(\Omega^M) \geq \varepsilon \,,$$

uniformly in $M \in \mathbb{N}$. Moreover, $\Omega' \supset \Omega^M \supset \Omega^{M+1}$ for all M and therefore $\Omega^\infty := \bigcap_{M \in \mathbb{N}} \Omega^M \subset \Omega'$ has $\mathcal{L}^n(\Omega^\infty) \geq \varepsilon$. Finally, neglecting a set of measure zero and passing to a subsequence, if necessary, we may assume that $F(x, z, p)$ is continuous in (z, p), that $u_m(x)$, $u(x)$, $\nabla u_m(x)$ are unambiguously defined and finite while $u_m(x) \to u(x)$ as $m \to \infty$ at every point $x \in \Omega^\infty$.

Remark that every point $x \in \Omega^\infty$ by construction belongs to infinitely many of the sets $\tilde{\Omega}_m$. Choose such a point x. Relabelling, we may assume $x \in \bigcap_{m \in \mathbb{N}} \tilde{\Omega}_m$. By uniform boundedness $|\nabla u_m(x)| \leq C$ there exists a subsequence $m \to \infty$ and a vector $p \in \mathbb{R}^{nN}$ such that $\nabla u_m(x) \to p$ $(m \to \infty)$. But then by continuity

$$F(x, u_m(x), \nabla u_m(x)) \to F(x, u(x), p)$$

while also

$$F(x, u(x), \nabla u_m(x)) \to F(x, u(x), p)$$

which contradicts the characterization of Ω_m given above.   □

1.8 Remarks. The following observations may be useful in applications.

(1°) Theorem 1.6 also applies to functionals involving higher (m^{th}-) order derivatives of a function u by letting $U = (u, \nabla u, \ldots, \nabla^{m-1}u)$ denote the (m-1)-jet of u. Note that convexity is only required in the highest-order derivatives $P = \nabla^m u$.

(2°) If (u_m) is bounded in $H^{1,1}(\Omega')$ for any $\Omega' \subset\subset \Omega$, by Rellich's Theorem and repeated selection of subsequences there exists a subsequence (u_m) which converges strongly in $L^1(\Omega')$ for any $\Omega' \subset\subset \Omega$.

Local boundedness in $H^{1,1}$ of a minimizing sequence (u_m) for E can be inferred from a coerciveness condition like

$$(1.7) \qquad F(x, z, p) \geq |p|^\mu - \phi(x), \ \mu \geq 1, \ \phi \in L^1 \ .$$

The delicate part in the hypotheses concerning (u_m) is the assumption that (∇u_m) converges weakly in L^1_{loc}. In case $\mu > 1$ in (1.7) this is clear, but in case $\mu = 1$ the weak limit of a minimizing sequence may lie in BV_{loc} instead of $H^{1,1}_{loc}$. See Theorem 1.4, for example; see also Section 3.

(3°) By convexity in p, continuity of F in (u, p) for almost every x is equivalent to the following condition which is easier to check in applications:

$F(x, \cdot, \cdot)$ is continuous, separately in $u \in \mathbb{R}^N$ and $p \in \mathbb{R}^{nN}$, for almost every $x \in \Omega$.

Indeed, for any fixed x, u, p and all $e \in \mathbb{R}^{nN}, |e| = 1, \alpha \in [0, 1]$, letting $q = p + \alpha e, \ p_+ = p + e, \ p_- = p - e$ and writing $F(x, u, p) = F(u, p)$ for brevity, by convexity we have

$$F(u, q) = F(u, \alpha p_+ + (1 - \alpha)p) \leq \alpha F(u, p_+) + (1 - \alpha)F(u, p) \ ,$$

$$F(u, p) = F(u, \frac{1}{1 + \alpha}q + \frac{\alpha}{1 + \alpha}p_-) \leq \frac{1}{1 + \alpha}F(u, q) + \frac{\alpha}{1 + \alpha}F(u, p_-) \ .$$

Hence

$$\alpha\left(F(u, p) - F(u, p_+)\right) \leq F(u, p) - F(u, q) \leq \alpha\left(F(u, p_-) - F(u, p)\right)$$

and it follows that

$$\sup_{|q-p|\leq 1} \frac{|F(u, q) - F(u, p)|}{|q - p|} \leq \sup_{|q-p|=1} |F(u, q) - F(u, p)| \ .$$

Since the sphere of radius 1 around p lies in the convex hull of finitely many vectors $q_0, q_1, \ldots, q_{nN}$, by continuity of F in u and convexity in p the right hand side of this inequality remains uniformly bounded in a neighborhood of (u, p). Hence $F(\cdot, \cdot)$ is locally Lipschitz continous in p, locally uniformly in $(u, p) \in \mathbb{R}^N \times \mathbb{R}^{nN}$. Therefore, if $u_m \to u$, $p_m \to p$ we have

$$|F\left(u_m, p_m\right) - F\left(u, p\right)| \le |F\left(u_m, p_m\right) - F\left(u_m, p\right)| + |F\left(u_m, p\right) - F\left(u, p\right)|$$
$$\le c|p_m - p| + o(1) \to 0 \qquad \text{as } m \to \infty,$$

where $o(1) \to 0$ as $m \to \infty$, as desired.

(4°) In the scalar case (N=1), if F is C^2 for example, the existence of a minimizer u for E implies that the *Legendre condition*

$$\sum_{\alpha,\beta=1}^{n} F_{p_\alpha p_\beta}\left(x, u, p\right) \xi_\alpha \xi_\beta \ge 0, \qquad \text{for all } \xi \in \mathbb{R}^n$$

holds at all points $(x, u = u(x), p = \nabla u(x))$, see for instance Giaquinta [1; p.11 f.]. This condition in turn implies the convexity of F in p.

The situation is quite different in the vector-valued case $N > 1$. In this case, in general only the *Legendre-Hadamard condition*

$$\sum_{i,j=1}^{N} \sum_{\alpha,\beta=1}^{n} F_{p_\alpha^i p_\beta^j}(x, u, p)\xi_\alpha \xi_\beta \eta^i \eta^j \ge 0, \qquad \text{for all } \xi \in \mathbb{R}^n, \ \eta \in \mathbb{R}^N$$

will hold at a minimizer, which is much weaker then convexity. (Giaquinta [1; p.12]).

In fact, below we shall see how, under certain additional structure conditions on F, the convexity assumption in Theorem 1.6 can be weakened in the vector-valued case.

2. Constraints

Applying the direct methods often involves a delicate interplay between the functional E, the space of admissible functions M, and the topology on M. In this section we will see how, by means of imposing constraints on admissible functions and/or by a suitable modification of the variational problem, the direct methods can be successfully employed also in situations where their use seems highly unlikely at first.

Note that we will not consider constraints that are dictated by the problems themselves, such as physical restrictions on the response of a mechanical system. Constraints of this type in general lead to variational inequalities, and we refer to Kinderlehrer-Stampacchia [1] for a comprehensive introduction to this field. Instead, we will show how certain variational problems can be solved by adding virtual – that is, purely technical – constraints to the conditions defining the admissible set, thus singling out distinguished solutions.

Semi-Linear Elliptic Boundary Value Problems

We start by deriving the existence of positive solutions to non-coercive, semili-near elliptic boundary value problems by a constrained minimization method. Such problems are motivated by studies of flame propagation (see for example Gel'fand [1; (15.5), p.357]) or arise in the context of the Yamabe problem (see Section 2.6 and Chapter III.3).

Let Ω be a smooth, bounded domain in $\mathbb{R}^n$, and let $p > 2$. If $n \geq 3$ we also assume that p satisfies the condition $p < 2^* = \frac{2n}{n-2}$. For $\lambda \in \mathbb{R}$ consider the problem

(2.1) $-\Delta u + \lambda u = u|u|^{p-2}$ *in* Ω ,

(2.2) $u > 0$ *in* Ω ,

(2.3) $u = 0$ *on* $\partial\Omega$.

Also let $0 < \lambda_1 < \lambda_2 \leq \lambda_3 \leq \ldots$ denote the eigenvalues of the operator $-\Delta$ on $H_0^{1,2}(\Omega)$. Then we have the following result:

2.1 Theorem. *For any $\lambda > -\lambda_1$ there exists a positive solution $u \in C^2(\Omega) \cap C^0(\overline{\Omega})$ to problem (2.1)–(2.3).*

Proof. Observe that Equation (2.1) is the Euler-Lagrange equation of the functional

$$\tilde{E}(u) = \frac{1}{2} \int_\Omega \left(|\nabla u|^2 + \lambda |u|^2\right) dx - \frac{1}{p} \int_\Omega |u|^p \, dx$$

on $H_0^{1,2}(\Omega)$ which is neither bounded from above nor from below on this space. However, using the homogeneity of (2.1) a solution of problem (2.1)–(2.3) can also be obtained by solving a constrained minimization problem for the functional

$$E(u) = \frac{1}{2} \int_\Omega \left(|\nabla u|^2 + \lambda |u|^2\right) dx$$

on the Hilbert space $H_0^{1,2}(\Omega)$, restricted to the set

$$M = \left\{ u \in H_0^{1,2}(\Omega) \; ; \; \int_\Omega |u|^p \, dx = 1 \right\} .$$

We verify that $E : M \to \mathbb{R}$ satisfies the hypotheses of Theorem 1.2. By the Rellich-Kondrakov theorem the injection $H_0^{1,2}(\Omega) \hookrightarrow L^p(\Omega)$ is completely continuous for $p < 2^*(n \geq 3)$, respectively any $p < \infty(n = 1, 2)$; see Theorem A.5 of the appendix. Hence M is weakly closed in $H_0^{1,2}(\Omega)$.

Recall the Rayleigh-Ritz characterization

(2.4) $\lambda_1 = \inf_{\substack{u \in H_0^{1,2}(\Omega) \\ u \neq 0}} \frac{\int_\Omega |\nabla u|^2 \, dx}{\int_\Omega |u|^2 \, dx}$

of the smallest Dirichlet eigenvalue. This gives the estimate

$$(2.5) \qquad E(u) \geq \frac{1}{2} \min \left\{ 1, \left(1 + \frac{\lambda}{\lambda_1} \right) \right\} \|u\|^2_{H_0^{1,2}} \ .$$

From this, coerciveness of E for $\lambda > -\lambda_1$ is immediate.

Weak lower semi-continuity of E follows from weak lower semi-continuity of the norm in $H_0^{1,2}(\Omega)$ and the Rellich-Kondrakov theorem. By Theorem 1.2 therefore E attains its infimum at a point $\underline{u}$ in M. Remark that since $E(u) = E(|u|)$ we may assume that $\underline{u} \geq 0$.

To derive the variational equation for E first note that E is Fréchet-differentiable in $H_0^{1,2}(\Omega)$ with

$$\langle v, DE(u) \rangle = \int_\Omega \left(\nabla u \nabla v + \lambda u v \right) dx \ .$$

Moreover, letting

$$G(u) = \int_\Omega |u|^p \, dx - 1 \ ,$$

$G : H_0^{1,2}(\Omega) \to \mathbb{R}$ also is Fréchet-differentiable with

$$\langle v, DG(u) \rangle = p \int_\Omega u |u|^{p-2} v \, dx \ .$$

In particular, at any point $u \in M$

$$\langle u, DG(u) \rangle = p \int_\Omega |u|^p \, dx = p \neq 0 \ ,$$

and by the implicit function theorem the set $M = G^{-1}(0)$ is a C^1-submanifold of $H_0^{1,2}(\Omega)$.

Now, by the Lagrange multiplier rule, there exists a parameter $\mu \in \mathbb{R}$ such that

$$\langle v, (DE(\underline{u}) - \mu DG(\underline{u})) \rangle = \int_\Omega \left(\nabla \underline{u} \nabla v + \lambda \underline{u} v - \mu \underline{u} |\underline{u}|^{p-2} v \right) dx$$
$$= 0, \qquad \text{for all } v \in H_0^{1,2}(\Omega) \ .$$

Inserting $v = \underline{u}$ into this equation yields that

$$2E(\underline{u}) = \int_\Omega \left(|\nabla \underline{u}|^2 + \lambda |\underline{u}|^2 \right) dx = \mu \int_\Omega |\underline{u}|^p \, dx = \mu \ .$$

Since $\underline{u} \in M$ cannot vanish identically, from (2.5) we infer that $\mu > 0$. Scaling with a suitable power of μ, we obtain a weak solution $u = \mu^{\frac{1}{p-2}} \cdot \underline{u} \in H_0^{1,2}(\Omega)$ of (2.1), (2.3) in the sense that

$$(2.6) \qquad \int_\Omega \left(\nabla u \nabla v + \lambda u v - u |u|^{p-2} v \right) dx = 0 \ , \qquad \text{for all } v \in H_0^{1,2}(\Omega) \ .$$

Moreover, (2.2) holds in the weak sense $u \geq 0$, $u \neq 0$. To finish the proof we use the regularity result Lemma B.3 of the appendix and the observations following it to obtain that $u \in C^2(\Omega)$. Finally, by the strong maximum principle $u > 0$ in Ω ; see Theorem B.4. $\square$

2.2 Symmetry. By a result of Gidas-Ni-Nirenberg [1; Theorem 2.1, p. 216, and Theorem 1, p.209], if Ω is symmetric with respect to a hyperplane, say $x_1 = 0$, any positive solution u of (2.1), (2.3) is even in x_1, that is, $u(x_1, x') = u(-x_1, x')$ for all $x = (x_1, x') \in \Omega$, and $\frac{\partial u}{\partial x_1} < 0$ at any point $x = (x_1, x') \in \Omega$ with $x_1 > 0$. In particular, if Ω is a ball, any positive solution u is radially symmetric. The proof of this result uses a variant of the Alexandrov reflexion principle.

Observe that, at least for the kind of nonlinear problems considered here, by Lemma B.3 of the appendix the regularity theory is taken care of and in the following we may concentrate on proving existence of (weak) solutions.

Perron's Method in a Variational Guise

In the previous example the constraint built into the definition of M had the effect of making the restricted functional $E = \tilde{E}|_M$ coercive. Moreover, this constraint only changed the Euler-Lagrange equations by a factor which could be scaled away using the homogeneity of the right hand side of (2.1).

In the next application we will see that sometimes also inequality constraints can be imposed without changing the Euler-Lagrange equations at a minimizer.

2.3 Weak sub- and super-solutions. Suppose Ω is a smooth, bounded domain in $\mathbb{R}^n$, and let $g : \Omega \times \mathbb{R} \to \mathbb{R}$ be a Carathéodory function. Let $u_0 \in H_0^{1,2}(\Omega)$ be given. Consider the equation

$$(2.7) \qquad\qquad -\Delta u = g(\cdot, u) \qquad \text{in } \Omega ,$$
$$(2.8) \qquad\qquad u = u_0 \qquad \text{on } \partial\Omega .$$

By definition $u \in H^{1,2}(\Omega)$ is a (weak) *sub-solution* to (2.7–8) if $u \leq u_0$ on $\partial\Omega$ and

$$\int_\Omega \nabla u \nabla \varphi \, dx - \int_\Omega g(\cdot, u)\varphi \, dx \leq 0 \qquad \text{for all } \varphi \in C_0^\infty(\Omega) , \ \varphi \geq 0 .$$

Similarly $u \in H^{1,2}(\Omega)$ is a (weak) *super-solution* to (2.7–8) if in the above the reverse inequalities hold.

2.4 Theorem. *Suppose $\underline{u} \in H^{1,2}(\Omega)$ is a sub-solution while $\overline{u} \in H^{1,2}(\Omega)$ is a super-solution to problem (2.7–8) and assume that with constants $\underline{c}, \overline{c} \in \mathbb{R}$ there holds $-\infty < \underline{c} \leq \underline{u} \leq \overline{u} \leq \overline{c} < \infty$, almost everywhere in Ω. Then there exists a weak solution $u \in H^{1,2}(\Omega)$ of (2.7–8), satisfying the condition $\underline{u} \leq u \leq \overline{u}$ almost everywhere in Ω.*

Proof. With no loss of generality we may assume $u_0 = 0$. Let $G(x, u) = \int_0^u g(x, v)\, dv$ denote a primitive of g. Note that (2.7–8) formally are the Euler-Lagrange equations of the functional

$$E(u) = \frac{1}{2} \int_\Omega |\nabla u|^2 \, dx - \int_\Omega G(x, u)\, dx \ .$$

However, our assumptions do not allow the conclusion that E is finite or even differentiable on $V := H_0^{1,2}(\Omega)$ – the smallest space where we have any chance of verifying coerciveness. Instead we restrict E to

$$M = \left\{ u \in H_0^{1,2}(\Omega) \ ; \ \underline{u} \leq u \leq \overline{u} \text{ almost everywhere} \right\} \ .$$

Since $\underline{u}, \overline{u} \in L^\infty$ by assumption, also $M \subset L^\infty$ and $G\big(x, u(x)\big) \leq c$ for all $u \in M$ and almost every $x \in \Omega$.

Now we can verify the hypotheses of Theorem 1.2: Clearly, $V = H_0^{1,2}(\Omega)$ is reflexive. Moreover, M is closed and convex, hence weakly closed. Since M is essentially bounded, our functional $E(u) \geq \frac{1}{2}\|u\|^2_{H_0^{1,2}(\Omega)} - c$ is coercive on M. Finally, to see that E is weakly lower semi-continuous on M, it suffices to show that

$$\int_\Omega G(x, u_m)\, dx \to \int_\Omega G(x, u)\, dx$$

if $u_m \rightharpoonup u$ weakly in $H_0^{1,2}(\Omega)$, where $u_m, u \in M$. But – passing to a subsequence, if necessary – we may assume that $u_m \to u$ pointwise almost everywhere and $|G\big(x, u_m(x)\big)| \leq c$ uniformly. Hence we may appeal to Lebesgue's theorem on dominated convergence.

From Theorem 1.2 we infer the existence of a relative minimizer $u \in M$. To see that u weakly solves (2.7), for $\varphi \in C_0^\infty(\Omega)$ and $\varepsilon > 0$ let $v_\varepsilon = \min\big\{\overline{u}, \max\{\underline{u}, u + \varepsilon\varphi\}\big\} = u + \varepsilon\varphi - \varphi^\varepsilon + \varphi_\varepsilon \in M$ with

$$\varphi^\varepsilon = \quad \max\big\{0, u + \varepsilon\varphi - \overline{u}\big\} \geq 0 \ ,$$
$$\varphi_\varepsilon = -\min\big\{0, u + \varepsilon\varphi - \underline{u}\big\} \geq 0 \ .$$

Note that $\varphi_\varepsilon, \varphi^\varepsilon \in H_0^{1,2} \cap L^\infty(\Omega)$.

E is differentiable in direction $v_\varepsilon - u$. Since u minimizes E in M we have

$$0 \leq \langle (v_\varepsilon - u), DE(u) \rangle = \varepsilon \langle \varphi, DE(u) \rangle - \langle \varphi^\varepsilon, DE(u) \rangle + \langle \varphi_\varepsilon, DE(u) \rangle \ ,$$

so that

$$\langle \varphi, DE(u) \rangle \geq \frac{1}{\varepsilon} \big[\langle \varphi^\varepsilon, DE(u) \rangle - \langle \varphi_\varepsilon, DE(u) \rangle \big] \ .$$

Now, since $\bar{u}$ is a supersolution to (2.7), we have

$$
\begin{aligned}
\langle \varphi^\varepsilon, DE(u) \rangle &= \langle \varphi^\varepsilon, DE(\bar{u}) \rangle + \langle \varphi^\varepsilon, DE(u) - DE(\bar{u}) \rangle \\
&\geq \langle \varphi^\varepsilon, DE(u) - DE(\bar{u}) \rangle \\
&= \int_{\Omega_\varepsilon} \big\{ \nabla(u - \bar{u}) \nabla(u + \varepsilon\varphi - \bar{u}) - \\
&\qquad - \big(g(x, u) - g(x, \bar{u})\big)(u + \varepsilon\varphi - \bar{u}) \big\}\, dx \\
&\geq \varepsilon \int_{\Omega_\varepsilon} \nabla(u - \bar{u}) \nabla\varphi\, dx - \varepsilon \int_{\Omega_\varepsilon} \big| g(x, u) - g(x, \bar{u}) \big|\, |\varphi|\, dx \ ,
\end{aligned}
$$

where $\Omega^\varepsilon = \big\{ x \in \Omega \ ; \ u(x) + \varepsilon\varphi(x) \geq \bar{u}(x) > u(x) \big\}$. Note that $\mathcal{L}^n(\Omega^\varepsilon) \to 0$ as $\varepsilon \to 0$. Hence by absolute continuity of the Lebesgue integral we obtain that

$$
\langle \varphi^\varepsilon, DE(u) \rangle \geq o(\varepsilon)
$$

where $o(\varepsilon)/\varepsilon \to 0$ as $\varepsilon \to 0$. Similarly, we conclude that

$$
\langle \varphi_\varepsilon, DE(u) \rangle \leq o(\varepsilon)
$$

whence

$$
\langle \varphi, DE(u) \rangle \geq 0
$$

for all $\varphi \in C_0^\infty(\Omega)$. Reversing the sign of φ and since $C_0^\infty(\Omega)$ is dense in $H_0^{1,2}(\Omega)$ we finally see that $DE(u) = 0$, as claimed. $\qquad\square$

2.5 A special case. Let Ω be a smooth bounded domain in $\mathbb{R}^n, n \geq 3$, and let

$$
(2.9) \qquad\qquad g(x, u) = k(x)u - u|u|^{p-2}
$$

where $p = \frac{2n}{n-2}$, and where k is a continuous function such that

$$
1 \leq k(x) \leq K < \infty
$$

uniformly in Ω. Suppose $u_0 \in C^1(\overline{\Omega})$ satisfies $u_0 \geq 1$ on $\partial\Omega$.
Then $\underline{u} \equiv 1$ is a sub-solution while $\bar{u} \equiv c$ for large $c > 1$ is a super-solution to equation (2.7–8). Consequently, (2.7–8) admits a solution $u \geq 1$.

2.6 Remark. The sub-super-solution method can also be applied to equations on manifolds. In the context of the Yamabe problem it has been used by Kazdan-Warner [1]; see Chapter III.3. The non-linear term in this case is precisely (2.9).

The Classical Plateau Problem

One of the great successes of the direct methods in the calculus of variations was the solution of Plateau's problem for minimal surfaces.

Let Γ be a smooth Jordan curve in $\mathbb{R}^3$. From his famous experiments with soap films Plateau became convinced that any such curve is spanned by a (not necessarily unique) surface of least area.

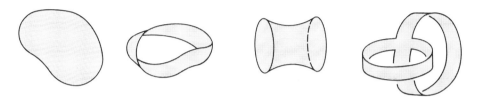

Fig. 2.1. Minimal surfaces of various topological types (disc, Möbius band, annulus, torus)

In the classical mathematical model the topological type of the surface is specified to be that of the disc

$$\Omega = \{z = (x, y) \; ; \; x^2 + y^2 < 1\} \; .$$

A naive approach to Plateau's conjecture would be to attempt to minimize the area

$$A(u) = \int_\Omega \sqrt{\det(\nabla u^t \nabla u)} \, dz = \int_\Omega \sqrt{|u_x|^2 |u_y|^2 - (u_x \cdot u_y)^2} \, dz$$

among "surfaces" $u \in H^{1,2} \cap C^0(\overline{\Omega}, \mathbb{R}^3)$ satisfying the Plateau boundary condition

(2.10) $u\big|_{\partial\Omega} : \partial\Omega \to \Gamma$ is a (weakly) monotone, orientation preserving parametrization.

However, A is invariant under arbitrary changes of parameter. Hence there is no chance of achieving bounded compactness in the original variational problem and some work was necessary in order to recast this problem in a way which is accessible by direct methods. Without entering into details let us briefly report the main ideas.

It had already beeen observed by Lagrange that if a (smooth) surface S is (locally) area-minimizing for fixed boundary Γ, necessarily the mean curvature of S vanishes. In isothermal coordinates $u(x, y)$ on S this amounts to the equations

(2.11) $\Delta u = 0 \; .$

(See Nitsche [2] or Osserman [1].) Moreover, our choice of parameter implies the conformality relations

$$(2.12) \qquad\qquad |u_x|^2 - |u_y|^2 = 0 = u_x \cdot u_y \text{ in } \Omega \,,$$

in addition to the Plateau boundary condition (2.10).

    We now take equations (2.10–12) as a definition for a minimal surface spanning Γ.

2.7 The variational problem. In their 1930 break-through papers, Douglas [1] and Radó [1] ingeniously proposed to solve (2.10–12) by minimizing Dirichlet's integral

$$E(u) = \frac{1}{2} \int_\Omega |\nabla u|^2 \, dz = \frac{1}{2} \int_\Omega (|u_x|^2 + |u_y|^2) \, dz$$

over the class

$$C(\Gamma) = \{u \in H^{1,2}(\Omega, \mathbb{R}^3) \ ; \ u\big|_{\partial\Omega} \in C^0(\partial\Omega, \mathbb{R}^3) \text{ satisfies } (2.10)\} \ .$$

It is easy to see that

$$E(u) \geq A(u) \,,$$

and equality holds if and only if u is conformal. Actually, we have

$$\inf_{C(\Gamma)} A(u) = \inf_{C(\Gamma)} E(u) \ .$$

This can be derived for instance from Morrey's "ε-conformality lemma" (Morrey [2; Theorem 1.2]). In Struwe [17; Appendix A] also a direct (variational) proof is given. Thus a minimizer of E automatically will satisfy (2.12) and also minimize A – hence will solve the original minimization problem.

    The solution of Plateau's problem is therefore reduced to the following theorem.

2.8 Theorem. *For any C^1-embedded curve Γ there exists a minimizer u of Dirichlet's integral E in $C(\Gamma)$.*

Note that $C(\Gamma) \neq \emptyset$ if $\Gamma \in C^1$. (Actually it suffices to assume that Γ is a rectifiable Jordan curve; see Douglas [1], Radó [1].)

    To show Theorem 2.8, observe that in replacing A by E we have succeeded in reducing the symmetries of the problem drastically. However, E is still conformally invariant, that is

$$E(u) = E(u \circ g)$$

for all $g \in \mathcal{G}$, where

$$\mathcal{G} = \left\{ g : z \mapsto g(z) = e^{i\phi} \frac{a+z}{1-\bar{a}z} \ ; \ a \in \mathbb{C}, \ |a| < 1, \ 0 < \phi < 2\pi \right\}$$

denotes the conformal group of Möbius transformations of the disc. $\mathcal{G}$ acts non-compact in the sense that for any $u \in C(\Gamma)$ the orbit $\{u \circ g \mid g \in \mathcal{G}\}$ weakly accumulates also at constant functions; see for instance Struwe [17; Lemma I.4.1] for a detailed proof. Hence $C(\Gamma)$ cannot be weakly closed in $H^{1,2}(\Omega, \mathbb{R}^3)$ and Theorem 1.2 cannot yet be applied.

Fortunately, we can also get rid of conformal invariance of D. Note that for any oriented triple $e^{i\phi_1}, e^{i\phi_2}, e^{i\phi_3} \in \partial\Omega, 0 \le \phi_1 < \phi_2 < \phi_3 < 2\pi$ there exists a unique $g \in \mathcal{G}$ such that $g\left(e^{\frac{2\pi ik}{3}}\right) = e^{i\phi_k}, k = 1, 2, 3$.

Fix a parametrization γ of Γ (we may assume that γ is a C^1-diffeomorphism $\gamma : \partial\Omega \to \Gamma$) and let

$$C^*(\Gamma) = \left\{ u \in C(\Gamma) ; \; u\left(e^{\frac{2\pi ik}{3}}\right) = \gamma\left(e^{\frac{2\pi ik}{3}}\right), \; k = 1, 2, 3\right\} ,$$

endowed with the $H^{1,2}$-topology. This is our space of admissible functions, normalized with respect to $\mathcal{G}$. Note that for any $u \in C(\Gamma)$ there is $g \in \mathcal{G}$ such that $u \circ g \in C^*(\Gamma)$.

The following classical result now is crucial for the proof of Theorem 2.8.

2.9 Courant-Lebesgue-Lemma. *The set $C^*(\Gamma)$ is weakly closed in $H^{1,2}$.*

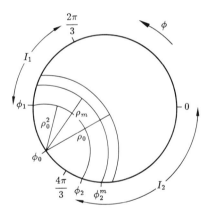

Fig. 2.2. The Courant-Lebesgue-Lemma

Proof. The proof in a subtle way uses a convexity argument like in the preceding example. To present this argument explicitly we use the fixed parametrization γ to associate with any $u \in C^*(\Gamma)$ a continuous map $\xi : \mathbb{R} \to \mathbb{R}$, such that

$$\gamma\left(e^{i\xi(\phi)}\right) = u\left(e^{i\phi}\right) , \; \xi(0) = 0 .$$

By (2.10) the functions ξ obtained in this manner are continuous, monotone and $\xi - id$ is 2π-periodic; moreover, $\xi\left(\frac{2\pi k}{3}\right) = \frac{2\pi k}{3}$, for all $k \in \mathbb{Z}$ by our 3-point normalization.

Now let

$$M = \left\{ \xi : \mathbb{R} \to \mathbb{R} \ ; \ \xi \text{ is continuous and monotone,} \right.$$

$$\left. \xi(\phi + 2\pi) = \xi(\phi) + 2\pi, \ \xi\left(\frac{2\pi k}{3}\right) = \frac{2\pi k}{3}, \text{ for all } \phi \in \mathbb{R}, k \in \mathbb{Z} \right\} .$$

Note that M is convex. Let (u_m) be a sequence in $C^*(\Gamma)$ with associated functions $\xi_m \in M$, and suppose $u_m \to u$ weakly in $H^{1,2}(\Omega)$. Since each ξ_m is monotone and satisfies the estimate $0 \leq \xi_m(\phi) \leq 2\pi$, for all $\phi \in [0, 2\pi]$, the family (ξ_m) is bounded in $H^{1,1}([0, 2\pi])$. Hence (a subsequence) $\xi_m \to \xi$ almost everywhere on $[0, 2\pi]$ and therefore – by periodicity – almost everywhere on $\mathbb{R}$, where ξ is monotone, $\xi - id$ is 2π-periodic, and ξ satisfies $\xi\left(\frac{2\pi k}{3}\right) = \frac{2\pi k}{3}$, for all $k \in \mathbb{Z}$.

Now if ξ is continuous, it follows from monotonicity that $\xi_m \to \xi$ uniformly. Thus, by continuity of γ, also u_m converges uniformly to u on $\partial\Omega$, and it follows that $u|_{\partial\Omega}$ is continuous and satisfies (2.10). That is, $u \in C^*(\Gamma)$, and the proof is complete in this case.

In order to exclude the remaining case, assume by contradiction that ξ is discontinuous at some point ϕ_0. We choose $k \in \mathbb{Z}$ such that $|\phi_0 - \frac{2\pi k}{3}| \leq \frac{\pi}{3}$ and let $\left]\frac{2\pi(k-1)}{3}, \frac{2\pi(k+1)}{3}\right[=: I_0$. By monotonicity, for almost every $\phi_1, \phi_2 \in I_0$ such that $\phi_1 < \phi_0 < \phi_2$ we have

$$\frac{2\pi(k-1)}{3} \leq \lim_{m\to\infty} \xi_m(\phi_1) = \xi(\phi_1) \leq \lim_{\phi\to\phi_0^-} \xi(\phi)$$

$$< \lim_{\phi\to\phi_0^+} \xi(\phi) \leq \xi(\phi_2) = \lim_{m\to\infty} \xi_m(\phi_2) \leq \frac{2\pi(k+1)}{3} .$$

For such ϕ_1, $\phi_2 \in I_0$ denote $I_1 = \{\phi \in I_0 \ ; \ \phi \leq \phi_1\}$, $I_2 = \{\phi \in I_0 \ ; \ \phi \geq \phi_2\}$. Then by monotonicity of ξ_m and using the fact that γ is a diffeomorphism we obtain

$$\limsup_{m\to\infty} \left(\inf_{\phi\in I_1, \ \psi\in I_2} \left| \gamma\left(e^{i\xi_m(\phi)}\right) - \gamma\left(e^{i\xi_m(\psi)}\right) \right| \right) \geq$$

$$\geq \inf_{\phi,\psi\in I_0, \phi<\phi_0<\psi} \left| \gamma\left(e^{i\xi(\phi)}\right) - \gamma\left(e^{i\xi(\psi)}\right) \right| > 0 .$$

In particular, there exists $\varepsilon > 0$ independent of ϕ_1, ϕ_2 such that

(2.13) $$|u_m(e^{i\phi}) - u_m(e^{i\psi})| \geq \varepsilon > 0$$

for all $\phi \in I_1$, $\psi \in I_2$ if $m \geq m_0(\phi_1, \phi_2)$ is sufficiently large.

Now let $z_0 = e^{i\phi_0}$ and for $\rho > 0$ denote

$$U_\rho = \{z \in \Omega \ ; \ |z - z_0| < \rho\}, \ C_\rho = \{z \in \overline{\Omega} \ ; \ |z - z_0| = \rho\} .$$

Note that for all $\rho < 1$ any point $z = e^{i\phi} \in C_\rho \cap \partial\Omega$ satisfies $\phi \in I_0$.

Following Courant [1; p. 103], we will use uniform boundedness of (u_m) in $H^{1,2}$ to show that for suitable numbers $\rho_0 \in]0, 1[$, $\rho_m \in [\rho_0^2, \rho_0]$ the oscillation of u_m on C_{ρ_m} can be made arbitrarily small, uniformly in $m \in \mathbb{N}$.

First note that by Fubini's Theorem, if we denote arc length on C_ρ by s, from the estimate

$$\infty > c \geq \int_\Omega |\nabla u_m|^2 \, dz \geq \int_0^1 \int_{C_\rho} |\frac{\partial}{\partial s} u_m|^2 \, ds \, d\rho$$

we obtain that

$$\int_{C_\rho} |\frac{\partial}{\partial s} u_m|^2 \, ds < \infty$$

for almost every $\rho < 1$ and all $m \in \mathbb{N}$.

Choosing $\rho_0 < 1$ we may refine this estimate as follows:

$$\int_\Omega |\nabla u_m|^2 \, dz \geq \int_{\rho_0^2}^{\rho_0} \left(\rho \int_{C_\rho} |\frac{\partial}{\partial s} u_m|^2 \, ds \right) \frac{d\rho}{\rho}$$

$$\geq |\log \rho_0| \operatorname*{ess\,inf}_{\rho_0^2 \leq \rho \leq \rho_0} \left(\rho \int_{C_\rho} |\frac{\partial}{\partial s} u_m|^2 \, ds \right).$$

Suppose $\rho_m \in [\rho_0^2, \rho_0]$ is such that

$$\rho_m \int_{C_{\rho_m}} |\frac{\partial}{\partial s} u_m|^2 \, ds \leq 2 \operatorname*{ess\,inf}_{\rho_0^2 \leq \rho \leq \rho_0} \left(\rho \int_{C_\rho} |\frac{\partial}{\partial s} u_m|^2 \, ds \right)$$

and denote

$$C = \sup_{m \in \mathbb{N}} \int_\Omega |\nabla u_m|^2 \, dz < \infty .$$

Fix $z_j = e^{i\phi_j}$, $j = 1, 2$, the points of intersection of $C_{\rho_0^2}$ with $\partial\Omega$, $\phi_1 < \phi_0 < \phi_2$. Also denote $z_j^m = e^{i\phi_j^m}$, $j = 1, 2$, with $\phi_1^m < \phi_0 < \phi_2^m$ the points of intersection of C_{ρ_m} with $\partial\Omega$.

Then $\phi_1^m \in I_1$, $\phi_2^m \in I_2$ while by Hölder's inequality

(2.14)
$$|u_m(z_1^m) - u_m(z_2^m)|^2 \leq \left(\int_{C_{\rho_m}} |\frac{\partial}{\partial s} u_m| \, ds \right)^2$$

$$\leq \pi \rho_m \int_{C_{\rho_m}} |\frac{\partial}{\partial s} u_m|^2 \, ds \leq \frac{2\pi C}{|\log \rho_0|} < \varepsilon$$

if $\rho_0 > 0$ is sufficiently small. This estimate being uniform in m, for large $m \geq m_0(\phi_1, \phi_2)$ we obtain a contradiction to (2.13) and the proof is complete.

$\square$

Remark. From (2.14), monotonicity (2.10), the assumption that Γ is a Jordan curve – that is, a homeomorphic image of the circle S^1 – and the three-point condition, also a direct proof of equi-continuity of the sub-level sets of E in $C^*(\Gamma)$ can be given; see Courant [1; Lemma 3.2, p.103] or Struwe [17; Lemma I.4.3]. Moreover, note that (2.14) implies a uniform estimate for the modulus of continuity of a function $u \in H_0^{1,2}(\Omega)$ on a sequence of concentric circular arcs around any fixed center $z_0 \in \overline{\Omega}$ in terms of its Dirichlet integral. This observation was used by Lebesgue [1] to obtain an equi-continuous minimizing sequence for Dirichlet's integral in his solution of the classical Dirichlet problem; see also Section 5.7.

Proof of Theorem 2.8. By (2.10) and the generalized Poincaré inequality Theorem A.9 of the appendix, for $u \in C^*(\Gamma)$ there holds

$$\int_\Omega |u|^2 \, dz \le c \left(\int_\Omega |\nabla u|^2 \, dz + \int_{\partial \Omega} |u|^2 \, do \right) \le cE(u) + c(\Gamma).$$

Thus E is coercive on $M = C^*(\Gamma)$ in $H^{1,2}(\Omega; \mathbb{R}^3)$. Moreover E is weakly lower semi-continuous on $H^{1,2}(\Omega; \mathbb{R}^3)$. By Theorem 1.2 and Lemma 2.9 therefore the functional E achieves its infimum in $C^*(\Gamma)$ which by conformal invariance equals that in $C(\Gamma)$. ☐

2.10 Regularity. As in the preceding examples we may ask what regularity properties the minimizer u and the parametrized surface possess. We note a few results:

(1°) $u|_{\partial \Omega}$ is strictly monotone (Douglas [1], Radó [1]).

(2°) If $\Gamma \in C^{m,\alpha}$, $m \ge 1$, $0 < \alpha < 1$, then also $u \in C^{m,\alpha}(\overline{\Omega}, \mathbb{R}^3)$; this result is due to Hildebrandt [1] (for $m \ge 4$) with later improvements by Nitsche [1] .

While Remarks (1°), (2°) apply to arbitrary solutions of the Plateau problem (2.10)–(2.12), minimality is crucial for the next observations concerning the geometric regularity of the parametrized solution surface.

(3°) A minimizer $u \in C^*(\Gamma)$ parametrizes an immersed minimal surface $S = u(\Omega) \subset \mathbb{R}^3$; see Osserman [2], Gulliver [1], Alt [1], Gulliver-Osserman-Royden[1]; if Γ is anlytic, S is immersed up to the boundary; see Gulliver-Lesley [1]. If Γ is extreme, that is $\Gamma \subset \partial K$ where $K \subset \mathbb{R}^3$ is convex, S is embedded; see Meeks-Yau [1]. Existence of embedded minimal surfaces bounded by extreme curves independently was obtained by Tomi-Tromba [1] and Almgren-Simon [1].

2.11 Note. In the history of the calculus of variations it seems that Plateau's problem has played a very prominent role. Important developments in the general stream of ideas often were prompted by insights gained from the study of minimal surfaces. As an example, consider the classical mountain pass lemma (see also Chapter II.1) which was used by Courant [1; Chapter VI.6–7] to establish the existence of unstable minimal surfaces, previously obtained by

Morse-Tompkins [1] and Shiffman [1] by a less direct, topological reasoning. However, since this material has been covered very extensively elsewhere (see Struwe [17]), here we will confine ourselves to the above remarks.

For an introduction to Plateau's problem and minimal surfaces, see for instance Osserman [1], or consult the encyclopedic book by Nitsche [2].

A truely remarkable – popular and profound – book on the subject is available by Hildebrandt-Tromba [1].

3. Compensated Compactness

As noted in Remark 1.8, in the vector-valued case lower semi-continuity results may hold true under a weaker convexity assumption than in Theorem 1.6, provided certain structure conditions are satisfied by the functional in variation. Weakening the convexity hypothesis is necessary in order to deal with problems arising for instance in 3-dimensional elasticity where we encounter energy functionals $\int_\Omega W(\nabla u)\, dx$ with a stored energy function W depending on the determinant, the minors and the eigenvalues of the deformation gradient ∇u. Since infinite volume distortion for elastic materials will afford an infinite amount of energy it is natural to suppose that $W \to \infty$ if $\det(\nabla u) \to \{0, \infty\}$, hence W cannot be convex in ∇u. However, there is a large class of materials which can be described by *polyconvex* stored energy functions which are of the form

$$W(\nabla u) = f(\text{subdeterminants of } \nabla u)$$

where f is convex in each of its variables. John Ball [1] was the first to see that lower semi-continuity results will hold for such functionals. The difficulty, of course, lies in proving for instance weak convergence $\det(\nabla u_m) \to \det(\nabla u)$ for a sequence $u_m \to u$ weakly in $H^{1,3}(\Omega, \mathbb{R}^3)$. Questions of this type had been investigated by Reshetnyak [1], [2]. A general frame for studying such problems is provided by the compensated compactness scheme of Murat and Tartar.

The basic principle of the compensated compactness method is given in the following lemma, see Tartar [2; p. 270 f.].

3.1 The compensated compactness lemma. *Let Ω be a domain in $\mathbb{R}^n$ and suppose that*

($1°$) $u^{(m)} = \left(u_1^{(m)}, \ldots, u_N^{(m)}\right) \to u$ weakly in $L^2(\Omega; \mathbb{R}^N)$.

($2°$) The set $\left\{\sum_{j,k} a_{jk} \dfrac{\partial u_j^{(m)}}{\partial x_k} \; ; \; m \in \mathbb{N}\right\}$ is relatively compact in $H_{loc}^{-1}(\Omega; \mathbb{R}^L)$

for a set of vectors $a_{jk} \in \mathbb{R}^L$; $1 \leq j \leq N$, $1 \leq k \leq n$. Let

$$\Lambda = \left\{\lambda \in \mathbb{R}^N \; ; \; \sum_{j,k} a_{jk} \lambda_j \xi_k = 0 \text{ for some } \xi \in \mathbb{R}^n \setminus \{0\}\right\}$$

and let Q be a (real) quadratic form such that $Q(\lambda) \geq 0$ for all $\lambda \in \Lambda$. Regarding $Q(u^{(m)}) \in L^1(\Omega)$ as Radon measures $Q(u^{(m)})dx \in \left(C^0(\Omega')\right)^*$, we may assume that $\left(Q(u^{(m)})\right)$ converges weak*, locally.

Then on any $\Omega' \subset\subset \Omega$ *we have*

$$\text{weak}^* - \lim_{m\to\infty} Q\big(u^{(m)}\big) \geq Q(u)$$

in the sense of measures. In particular, if $Q(\lambda) = 0$ *for all* $\lambda \in \Lambda$, *then*

$$\text{weak}^* - \lim_{m\to\infty} Q\big(u^{(m)}\big) = Q(u)$$

locally, in the sense of measures.

Proof. Choose $\varphi \in C^0(\Omega)$ with compact support and such that $0 \leq \varphi \leq 1$. We must show that

$$\liminf_{m\to\infty} \int_\Omega Q\big(u^{(m)}\big)\varphi \, dx \geq \int_\Omega Q(u)\varphi \, dx \ .$$

By translation we may assume that $u = 0$. Moreover, by additivity of the Lebesgue integral it suffices to assume that the supports of $u^{(m)}$ lie in a fixed cube $K \subset\subset \Omega$. In addition, since any function φ as above can be approximated in $L^\infty(\Omega)$ by step functions, we may assume that $\varphi \equiv 1$ on K. By translation and scaling, moreover, we can achieve that $K = [0, 2\pi]^n$. Let

$$u^{(m)}(x) = \sum_{\alpha\in\mathbb{Z}^n} \mu_\alpha^{(m)} e^{i\alpha\cdot x} \ , \qquad \mu_\alpha^{(m)} = \Big(\big(\mu_\alpha^{(m)}\big)_1, \ldots, \big(\mu_\alpha^{(m)}\big)_N\Big) \in \mathbb{C}^N \ ,$$

be the Fourier expansion for $u^{(m)}$. Since $u^{(m)}$ is real, we have $\mu_\alpha^{(m)} = \overline{\mu}_{-\alpha}^{(m)}$. The assertion then is equivalent to showing that

$$\liminf_{m\to\infty} \frac{1}{(2\pi)^n} \int_K Q\big(u^{(m)}\big) \, dx = \liminf_{m\to\infty} \sum_{\alpha\in\mathbb{Z}^n} \Big(Q\big(\text{Re }\mu_\alpha^{(m)}\big) + Q\big(\text{Im }\mu_\alpha^{(m)}\big)\Big) \geq 0 \ .$$

By weak convergence $u^{(m)} \to 0$ in L^2 we infer that $\sum_{\alpha\in\mathbb{Z}^n} |\mu_\alpha^{(m)}|^2 \leq c < \infty$ and $\mu_\alpha^{(m)} \to 0$ as $m \to \infty$, uniformly on bounded sets of indices α.

Moreover, by (2°) the set

$$\left\{ \sum_{\alpha\in\mathbb{Z}^n} \sum_{j,k} a_{jk}\big(\mu_\alpha^{(m)}\big)_j \alpha_k \, e^{i\alpha\cdot x} \ ; \ m \in \mathbb{N} \right\}$$

is relatively compact in H^{-1}, which implies that

$$\sum_{\substack{\alpha\in\mathbb{Z}^n \\ |\alpha|\geq\alpha_0}} \frac{\big|\sum_{j,k} a_{jk}\big(\mu_\alpha^{(m)}\big)_j \alpha_k\big|^2}{1 + |\alpha|^2} \to 0$$

as $\alpha_0 \to \infty$, uniformly in $m \in \mathbb{N}$. But this means that $\mu_\alpha^{(m)}$ can be decomposed $\mu_\alpha^{(m)} = \lambda_\alpha^{(m)} + \nu_\alpha^{(m)}$ with $\text{Re }\lambda_\alpha^{(m)}$, $\text{Im }\lambda_\alpha^{(m)} \in \Lambda$ and $\sum_{|\alpha|\geq\alpha_0} |\nu_\alpha^{(m)}|^2 \to 0$ as $\alpha_o \to \infty$, uniformly in $m \in \mathbb{N}$.

Indeed, for any $\alpha \in \mathbb{Z}^n$, $m \in \mathbb{N}$ decompose $\mu_\alpha^{(m)} = \lambda_\alpha^{(m)} + \nu_\alpha^{(m)}$, where Re $\lambda_\alpha^{(m)}$, Im $\lambda_\alpha^{(m)} \in \Lambda$ and $|\nu_\alpha^{(m)}|^2$ is minimal among all decompositions of this kind. We claim that there is a uniform constant C such that

$$(3.1) \qquad |\nu_\alpha^{(m)}|^2 \leq C \left| \sum_{j,k} a_{jk}(\mu_\alpha^{(m)})_j \frac{\alpha_k}{\sqrt{1+|\alpha|^2}} \right|^2 + o(1) \, |\mu_\alpha^{(m)}|^2$$

uniformly in $m \in \mathbb{N}$, $\alpha \in \mathbb{Z}^n$, where $o(1) \to 0$ as $|\alpha| \to \infty$.

Otherwise there exists a sequence $\alpha = \alpha(l)$, $l \in \mathbb{N}$, with $|\alpha(l)| \geq l$, and $m = m(l)$ such that

$$(3.2) \qquad \left| \mu_\alpha^{(m)} \right|^2 \geq \left| \nu_\alpha^{(m)} \right|^2 \geq l \left| \sum_{j,k} a_{jk}(\mu_\alpha^{(m)})_j \frac{\alpha_k}{\sqrt{1+|\alpha|^2}} \right|^2 .$$

(The first inequality follows from the choice of $\nu_\alpha^{(m)}$ above.) Let $\xi(l)$, $\eta(l)$ be the unit vectors

$$\xi(l) = \frac{\alpha(l)}{\sqrt{1+|\alpha(l)|^2}} \in S^{n-1}, \quad \eta(l) = \frac{\mu_{\alpha(l)}^{(m(l))}}{\left| \mu_{\alpha(l)}^{(m(l))} \right|} \in S^{N-1} ,$$

and denote by $A(l) \colon \mathbb{R}^N \to \mathbb{R}^L$ the linear map

$$\eta \mapsto \sum_{j,k} a_{jk} \eta_j \xi_k(l) .$$

We may assume that $\xi(l) \to \xi$ and $A(l) \to A$ as $l \to \infty$. Likewise, we may suppose that $\eta(l) \to \eta$. Passing to the limit in (3.2) it follows that $\eta \in \ker A$; that is, $\eta \in \Lambda$. Projecting $\mu_\alpha^{(m)}$ onto $\ker A$ for all $\alpha = \alpha(l)$, $m = m(l)$ we hence obtain a decomposition $\mu_\alpha^{(m)} = \tilde\lambda_\alpha^{(m)} + \tilde\nu_\alpha^{(m)}$ with Re $\tilde\lambda_\alpha^{(m)}$, Im $\tilde\lambda_\alpha^{(m)} \in \ker A \subset \Lambda$ and

$$\left| \tilde\nu_\alpha^{(m)} \right|^2 \leq C \left| A\mu_\alpha^{(m)} \right|^2 \leq C \left| \sum_{j,k} a_{jk}(\mu_\alpha^{(m)})_j \xi_k \right|^2$$

$$\leq C \left| \sum_{j,k} a_{jk}\mu_\alpha^{(m)} \frac{\alpha_k}{\sqrt{1+|\alpha|^2}} \right|^2 + o(1)|\mu_\alpha^{(m)}|^2 ,$$

where $o(1) \to 0$ $(l \to \infty)$. But by defintion $\left| \nu_\alpha^{(m)} \right|^2 \leq \left| \tilde\nu_\alpha^{(m)} \right|^2$, and we obtain (3.1). Hence for any $\alpha_0 \geq 0$:

$$\liminf_{m \to \infty} \sum_{\alpha \in \mathbb{Z}^n} \left(Q(\operatorname{Re} \mu_\alpha^{(m)}) + Q(\operatorname{Im} \mu_\alpha^{(m)}) \right) =$$

$$= \liminf_{m \to \infty} \sum_{|\alpha| \geq \alpha_0} \left(Q(\operatorname{Re} \mu_\alpha^{(m)}) + Q(\operatorname{Im} \mu_\alpha^{(m)}) \right)$$

$$= \liminf_{m \to \infty} \sum_{|\alpha| \geq \alpha_0} \left(Q(\operatorname{Re} \lambda_\alpha^{(m)}) + Q(\operatorname{Im} \lambda_\alpha^{(m)}) \right) + o(1) \geq o(1) ,$$

where $o(1) \to 0$ as $\alpha_0 \to \infty$. The conclusion follows. $\qquad\square$

As an application we mention the following well-known result.

3.2 The Div-Curl Lemma. *Suppose $u_m \rightharpoonup u$, $v_m \rightharpoonup v$ weakly in $L^2(\Omega; \mathbb{R}^3)$ on a domain $\Omega \subset \mathbb{R}^3$ while the sequences (div u_m) and (curl v_m) are bounded in $L^2(\Omega)$. Then for any $\varphi \in C_0^\infty(\Omega)$ we have*

$$\int_\Omega u_m \cdot v_m \varphi \, dx \to \int_\Omega u \cdot v \varphi \, dx$$

as $m \to \infty$.

Proof. Let $w^{(m)} = (u_m, v_m) \in L^2(\Omega; \mathbb{R}^6)$, and determine coefficients $a_{jk} \in \mathbb{R}^4$ such that $\sum_{jk} a_{jk} \frac{\partial w_j^{(m)}}{\partial x_k} = ($div u_m, curl $v_m)$. Let Q be the quadratic form $Q(u,v) = u \cdot v$, acting on vectors $w = (u,v) \in \mathbb{R}^6$. Note $a_{jk} = \left(\delta_{jk}, (\varepsilon_{ijk})_{1 \le i \le 3} \right)$ where $\delta_{jk} = 1$ if $j = k$, and $\delta_{jk} = 0$ else, $\varepsilon_{123} = 1$ and $\varepsilon_{ijk} = -\varepsilon_{jik} = \varepsilon_{jki}$. Hence

$$\Lambda = \left\{ \lambda = (\mu, \nu) \in \mathbb{R}^6 \; ; \; \exists \xi \in \mathbb{R}^3 \setminus \{0\} : \; (\xi \cdot \mu, \, \xi \wedge \nu) = 0 \right\}$$
$$= \left\{ \lambda = (\mu, \nu) \in \mathbb{R}^6 \; ; \; \mu \cdot \nu = 0 \right\},$$

and $Q \equiv 0$ on Λ. Thus the assertion follows from Lemma 3.1. ∎

The div-curl Lemma 3.2 shows how additional bounds on *some* derivatives allow to prove continuity of non-linear expressions (bi-linear in the above example) under weak convergence.

Applications in Elasticity

The most important applications of the compensated compactness method so far are in elasticity and hyperbolic systems, see Ball [1], [2], Di Perna [1]. DiPerna-Majda have applied compensated compactness methods to obtain the existence of weak solutions to the Euler equations for incompressible fluids, see for instance DiPerna-Majda [1]. Our interest lies with the extensions of the direct methods that compensated compactness implies. Thus we will concentrate on Ball's lower semi-continuity results for polyconvex materials in elasticity.

3.3 Theorem. *Suppose W is a function on (3×3)-matrices Φ, given by*

$$W(\Phi) = g(\Phi, adj \ \Phi, \det \ \Phi)$$

where g is a convex non-negative function in the sub-determinants of Φ. Let Ω be a domain in $\mathbb{R}^3$ and let u_m, $u \in H_{loc}^{1,3}(\Omega; \mathbb{R}^3)$. Suppose that $u_m \rightharpoonup u$ weakly in $H^{1,3}(\Omega'; \mathbb{R}^3)$ while $\det(\nabla u_m) \to \delta \in L_{loc}^1(\Omega)$ weakly in $L^1(\Omega')$, for all $\Omega' \subset\subset \Omega$. Then

$$\int_\Omega W(\nabla u) \, dx \le \liminf_{m \to \infty} \int_\Omega W(\nabla u_m) \, dx \ .$$

Proof. The proof of Theorem 1.6 can be carried over once we show that under the hypotheses made

$$adj(\nabla u_m) \rightharpoonup adj(\nabla u)$$
$$\det(\nabla u_m) \rightharpoonup \det(\nabla u)$$

weakly in $L^1(\Omega')$ for all $\Omega' \subset\subset \Omega$. The first assertion is a consequence of the divergence structure of the adjoint matrix $A_m = adj(\nabla u_m)$. Indeed, if indices i, j are counted *modulo* 3 we have

$$A_m^{ij} = \frac{\partial u_m^{i+1}}{\partial x^{j+1}} \frac{\partial u_m^{i+2}}{\partial x^{j+2}} - \frac{\partial u_m^{i+2}}{\partial x^{j+1}} \frac{\partial u_m^{i+1}}{\partial x^{j+2}}$$

$$= \frac{\partial}{\partial x^{j+1}} \left(u_m^{i+1} \frac{\partial u_m^{i+2}}{\partial x^{j+2}} \right) - \frac{\partial}{\partial x^{j+2}} \left(u_m^{i+1} \frac{\partial u_m^{i+2}}{\partial x^{j+1}} \right) .$$

Fix $\Omega' \subset\subset \Omega$. Note that $\left(A_m^{ij} \right)$ is bounded in $L^{3/2}(\Omega')$. Hence we may assume that $A_m^{ij} \rightharpoonup A^{ij}$ weakly in $L^{3/2}(\Omega')$. Moreover, by Rellich's theorem $u_m \to u$ in $L^3(\Omega')$, whence $u_m \nabla u_m \rightharpoonup u \nabla u$ weakly in $L^{3/2}(\Omega')$. By continuity of the distributional derivative with respect to weak convergence therefore $A_m^{ij} \rightharpoonup \left(adj(\nabla u) \right)^{ij}$ in the sense of distributions. Finally, by uniqueness of the distributional limit, $A^{ij} = \left(adj(\nabla u) \right)^{ij}$, and $adj(\nabla u_m) \rightharpoonup adj(\nabla u)$ weakly in $L^{3/2}(\Omega')$, in particular weakly in $L^1(\Omega')$, as claimed.

Similarly, expanding the determinant along the first row

$$\det(\nabla u_m) = \frac{\partial u_m^1}{\partial x^1} \left[\frac{\partial u_m^2}{\partial x^2} \frac{\partial u_m^3}{\partial x^3} - \frac{\partial u_m^2}{\partial x^3} \frac{\partial u_m^3}{\partial x^2} \right]$$

$$- \frac{\partial u_m^1}{\partial x^2} \left[\frac{\partial u_m^2}{\partial x^1} \frac{\partial u_m^3}{\partial x^3} - \frac{\partial u_m^2}{\partial x^3} \frac{\partial u_m^3}{\partial x^1} \right]$$

$$+ \frac{\partial u_m^1}{\partial x^3} \left[\frac{\partial u_m^2}{\partial x^1} \frac{\partial u_m^3}{\partial x^2} - \frac{\partial u_m^2}{\partial x^2} \frac{\partial u_m^3}{\partial x^1} \right]$$

$$= \frac{\partial}{\partial x^1} \left(u_m^1 \left[\frac{\partial u_m^2}{\partial x^2} \frac{\partial u_m^3}{\partial x^3} - \frac{\partial u_m^2}{\partial x^3} \frac{\partial u_m^3}{\partial x^2} \right] \right)$$

$$- \frac{\partial}{\partial x^2} \left(u_m^1 \left[\frac{\partial u_m^2}{\partial x^1} \frac{\partial u_m^3}{\partial x^3} - \frac{\partial u_m^2}{\partial x^3} \frac{\partial u_m^3}{\partial x^1} \right] \right)$$

$$+ \frac{\partial}{\partial x^3} \left(u_m^1 \left[\frac{\partial u_m^2}{\partial x^1} \frac{\partial u_m^3}{\partial x^2} - \frac{\partial u_m^2}{\partial x^2} \frac{\partial u_m^3}{\partial x^1} \right] \right) .$$

Thus convergence $\det(\nabla u_m) \to \det(\nabla u)$ in the sense of distributions follows by weak convergence in $L^{3/2}(\Omega')$ of the terms in brackets [- -], proved above, strong convergence $u_m \to u$ in $L^3(\Omega')$, and weak continuity of the distributional derivative. Finally, by uniqueness of the weak limit in the distribution sense, it follows that $\det(\nabla u) = \delta$ and $\det(\nabla u_m) \to \det(\nabla u)$ weakly in L^1_{loc}, as claimed.

□

The assumption $\det(\nabla u_m) \to \delta \in L^1_{loc}(\Omega)$ at first sight may appear rather awkward. However, examples by Ball-Murat [1] show that weak $H^{1,3}$-convergence in general does not imply weak L^1-convergence of the Jacobian. This difficulty does not arise if we assume weak convergence in $H^{1,3+\delta}$ for some $\delta > 0$.

Hence, adding appropriate growth conditions on W to ensure coerciveness of the functional $\int_\Omega W(\nabla u)\, dx$ on the space $H^{1,3+\delta}(\Omega; \mathbb{R}^3)$ for some $\delta > 0$, from Theorem 3.3 the reader can derive existence theorems for deformations of elastic materials involving polyconvex stored energy functions. As a further reference for such results, see Ciarlet [1] or Dacorogna [1], [2]. Recently, more general results on weak continuity of determinants and corresponding existence theorems in non-linear elasticity have been obtained by Giaquinta-Modica-Souček [1] and S. Müller [1], [2].

The regularity theory for problems in nonlinear elasticity is still evolving. Some material can be found in the references cited above. In particular, the question of cavitation of elastic materials has been studied. See for instance Giaquinta-Modica-Souček [1].

Convergence Results for Nonlinear Elliptic Equations

We close this section with another simple and useful example of how compensated compactness methods may be applied in a non-linear situation. The following result is essentially "Murat's lemma" from Tartar [2, p.278]:

3.4 Theorem. *Suppose $u_m \in H^{1,2}_0(\Omega)$ is a sequence of solutions to an elliptic equation*

$$-\Delta u_m = f_m \qquad in\ \Omega$$
$$u_m = 0 \qquad on\ \partial\Omega$$

in a smooth and bounded domain Ω in $\mathbb{R}^n$. Suppose $u_m \to u$ weakly in $H^{1,2}_0(\Omega)$ while (f_m) is bounded in $L^1(\Omega)$. Then for a subsequence $m \to \infty$ we have $\nabla u_m \to \nabla u$ in $L^q(\Omega)$ for any $q < 2$, and $\nabla u_m \to \nabla u$ pointwise almost everywhere.

Proof. Choose $p > n$ and let $\varphi_m \in H^{1,p}_0(\Omega)$ satisfy

$$\|\varphi_m\|_{H^{1,p}_0} \le 1$$

$$\int_\Omega (\nabla u_m - \nabla u)\nabla\varphi_m\, dx = \sup_{\varphi \in H^{1,p}_0(\Omega),\ \|\varphi\|_{H^{1,p}_0} \le 1} \int_\Omega (\nabla u_m - \nabla u)\nabla\varphi\, dx \quad .$$

By the Calderón-Zygmund inequality in L^p, see Simader[1], the latter

$$\sup_{\varphi \in H_0^{1,p}(\Omega),\ \|\varphi\|_{H_0^{1,p}} \leq 1} \int_\Omega (\nabla u_m - \nabla u)\nabla \varphi \, dx \geq c^{-1}\|u_m - u\|_{H_0^{1,q}}$$

where $\frac{1}{p} + \frac{1}{q} = 1$.

On the other hand, by Sobolev's embedding $H_0^{1,p}(\Omega) \hookrightarrow C^{1-\frac{n}{p}}(\overline{\Omega})$. Hence by the Arzéla-Ascoli theorem we may assume that $\varphi_m \rightharpoonup \varphi$ weakly in $H_0^{1,p}(\Omega)$ and uniformly in $\overline{\Omega}$. (See Theorem A.5.) Thus

$$\int_\Omega (\nabla u_m - \nabla u)\nabla \varphi_m \, dx = \int_\Omega (\nabla u_m - \nabla u)(\nabla \varphi_m - \nabla \varphi) \, dx + o(1)$$

$$= \lim_{l \to \infty} \int_\Omega (\nabla u_m - \nabla u_l)(\nabla \varphi_m - \nabla \varphi) \, dx + o(1)$$

$$= \lim_{l \to \infty} \int_\Omega (f_m - f_l)(\varphi_m - \varphi) \, dx + o(1)$$

$$\leq 2 \sup_{l \in \mathbb{N}} \|f_l\|_{L^1} \|\varphi_m - \varphi\|_{L^\infty} + o(1) = o(1),$$

where $o(1) \to 0$ as $m \to \infty$.

It follows that $\nabla u_m \to \nabla u$ in $L^{q_0}(\Omega)$ for some $q_0 \geq 1$. But then, by Hölder's inequality, for any $q < 2$ we have

$$\|\nabla u_m - \nabla u\|_{L^q} \leq \|\nabla u_m - \nabla u\|_{L^1}^\gamma \|\nabla u_m - \nabla u\|_{L^2}^{1-\gamma} \to 0,$$

where $\frac{1}{q} = \gamma + \frac{1-\gamma}{2}$. $\square$

Results like Theorem 3.4 are needed if one wants to solve non-linear partial differential equations

$$(3.3) \qquad\qquad -\Delta u = f(x, u, \nabla u)$$

with quadratic growth

$$|f(x, u, p)| \leq c(1 + |p|^2)$$

by approximation methods. Assuming some uniform control on approximate solutions u_m of (3.3) in $H^{1,2}$, Theorem 3.4 assures that

$$f(x, u_m, \nabla u_m) \to f(x, u, \nabla u) \quad \text{almost everywhere,}$$

where u is the weak limit of a suitable sequence (u_m). Given some further structure conditions on f, then there are various ways of passing to the limit $m \to \infty$ in equation (3.3); see for instance Frehse [2].

3.5 Example. As a model problem consider the equation

$$(3.4) \qquad A(u) = -\Delta u + u|\nabla u|^2 = h \quad \text{in } \Omega$$

$$(3.5) \qquad u = 0 \qquad \qquad \text{on } \partial\Omega$$

on a smooth and bounded domain $\Omega \subset \mathbb{R}^n$, with $h \in L^\infty(\Omega)$. This is a special case of a problem studied by Bensoussan-Boccardo-Murat [1; Theorem 1.1, p. 350]. Note that the non-linear term $g(u, p) = u|p|^2$ satisfies the condition

$$(3.6) \qquad g(u, p)u \geq 0 .$$

Approximate g by functions

$$g_\varepsilon(u, p) = \frac{g(u, p)}{1 + \varepsilon|g(u, p)|} , \varepsilon > 0 ,$$

satisfying $|g_\varepsilon| \leq \frac{1}{\varepsilon}$ and $g_\varepsilon(u, p) \cdot u \geq 0$ for all u, p.

Now, since g_ε is uniformly bounded, the map $H_0^{1,2}(\Omega) \ni u \mapsto g_\varepsilon(u, \nabla u) \in H^{-1}(\Omega)$ is compact and bounded for any $\varepsilon > 0$. Denote $A_\varepsilon(u) = -\Delta u + g_\varepsilon(u, \nabla u)$ the perturbed operator A. By Schauder's fixed point theorem, see for instance Deimling [1; Theorem 8.8, p.60], applied to the map $u \mapsto (-\Delta)^{-1}(h - g_\varepsilon(u, \nabla u))$ on a sufficiently large ball in $H_0^{1,2}(\Omega)$, there is a solution $u_\varepsilon \in H_0^{1,2}(\Omega)$ of the equation $A_\varepsilon u_\varepsilon = h$ for any $\varepsilon > 0$. In addition, since $g_\varepsilon(u, p) \cdot u \geq 0$ we have

$$\|u_\varepsilon\|_{H_0^{1,2}}^2 \leq \langle u_\varepsilon, A_\varepsilon u_\varepsilon \rangle = \langle u_\varepsilon, h \rangle \leq \|u_\varepsilon\|_{H_0^{1,2}}\|h\|_{H^{-1}} ,$$

and (u_ε) is uniformly bounded in $H_0^{1,2}(\Omega)$ for $\varepsilon > 0$. Moreover, since the non-linear term g_ε satisfies

$$g_\varepsilon(u, p) = \frac{u|p|^2}{1 + \varepsilon|u|\,|p|^2} \leq \frac{(1 + |u|^2)|p|^2}{1 + \varepsilon|u|\,|p|^2}$$
$$\leq |p|^2 + g_\varepsilon(u, p)u ,$$

we also deduce the uniform L^1-bound

$$\|g_\varepsilon(u_\varepsilon, \nabla u_\varepsilon)\|_{L^1} \leq \|u_\varepsilon\|_{H_0^{1,2}}^2 + \int_\Omega u_\varepsilon g_\varepsilon(u_\varepsilon, \nabla u_\varepsilon)\, dx$$
$$= \langle u_\varepsilon, A_\varepsilon u_\varepsilon \rangle \leq c .$$

We may assume that a sequence (u_{ε_m}) as $\varepsilon_m \to 0$ weakly converges in $H_0^{1,2}(\Omega)$ to a limit $u \in H_0^{1,2}(\Omega)$. By Theorem 3.4, moreover, we may assume that $u_m = u_{\varepsilon_m}$ converges strongly in $H_0^{1,q}(\Omega)$ and that u_m and ∇u_m converge pointwise almost everywhere. To show that u weakly solves (3.4), (3.5) we now use the "Fatou lemma technique" of Frehse [2]. As a preliminary step we establish a uniform L^∞-bound for the sequence (u_m).

Multiply the approximate equtions by u_m to obtain the differential inequality

$$-\Delta\left(\frac{|u_m|^2}{2}\right) \leq -\Delta\left(\frac{|u_m|^2}{2}\right) + |\nabla u_m|^2 + u_m g_{\varepsilon_m}(u_m, \nabla u_m)$$

$$= h u_m \leq C(\delta) + \delta\frac{|u_m|^2}{2},$$

for any $\delta > 0$. Choosing $\delta < \lambda_1$, the first eigenvalue of $-\Delta$ on $H_0^{1,2}(\Omega)$, the weak maximum principle implies that $u_m \in L^\infty$. (See Theorem B.7 and its application in the appendix.)

Next, testing the approximate equations $A_{\varepsilon_m}(u_m) = h$ with $\varphi = \xi \exp(\gamma u_m)$, where $\xi \in C_0^\infty(\Omega)$ is non-negative, upon integrating by parts we obtain

$$\int_\Omega \left(\gamma|\nabla u_m|^2 + g_{\varepsilon_m}(u_m, \nabla u_m)\right)\xi \exp(\gamma u_m)\, dx$$

$$+ \int_\Omega (\nabla u_m \nabla \xi - h\xi) \exp(\gamma u_m)\, dx = 0.$$

Note that on account of the growth condition

$$|g_{\varepsilon_m}(u, p)| \leq |u|\,|p|^2$$

and the uniform bound $\|u_m\|_{L^\infty} \leq C_0$ derived above, for $|\gamma| \geq C_0$ the term

$$\gamma|\nabla u_m|^2 + g_{\varepsilon_m}(u_m, \nabla u_m)$$

has the same sign as γ. Moreover, this term converges pointwise almost everywhere to the same expression involving u instead of u_m. Hence, by Fatou's lemma, upon passing to the limit $m \to \infty$ we obtain

$$\langle \xi \exp(\gamma u), A(u) - h \rangle \leq 0$$

if $\gamma \geq C_0$, respectively ≥ 0, if $\gamma \leq -C_0$. This holds for all non-negative $\xi \in C_0^\infty(\Omega)$ and hence also for $\xi \geq 0$ belonging to $H_0^{1,2} \cap L^\infty(\Omega)$. Setting $\xi = \xi_0 \exp(-\gamma u)$ we obtain

$$\langle \xi_0, A(u) - h \rangle \leq 0 \text{ and } \geq 0,$$

for any $\xi_0 \geq 0, \xi_0 \in C_0^\infty(\Omega)$. Hence u is a weak solution of (3.4), (3.5), as desired.

   More sophisticated variants and applications of Theorem 3.4 are given by Bensoussan-Boccardo-Murat [1], Boccardo-Murat-Puel [1], and Frehse [1], [2].

4. The Concentration-Compactness Principle

As we have seen in our analysis of the Plateau problem, Section 2.7, a very se-
rious complication for the direct methods to be applicable arises in the presence
of non-compact group actions.

   If, in a terminology borrowed from physics which we will try to make more
precise later, the action is a "manifest" symmetry – as in the case of the conformal group of the disc acting on Dirichlet's integral for minimal surfaces –
we may be able to eliminate the action by a suitable normalization. This was
the purpose for introducing the three-point-condition on admissible functions
for the Plateau problem in the proof of Theorem 2.8. However, if the action
is "hidden" such a normalization is not possible and there is no hope that
all minimizing sequences converge to a minimizer. Even worse, the variational
problem may not have a solution. For such problems, P.-L. Lions developed his
concentration-compactness principle. On the basis of this principle, for many
constrained minimization problems it is possible to state necessary and sufficient conditions for the convergence of all minimizing sequences satisfying the
given constraint. These conditions involve a delicate comparison of the given
functional in variation and a (family of) functionals "at infinity" (on which the
group action is manifest).

Rather than dwell on abstract notions we prefer to give an example – a variant
of problem (2.1), (2.3) – which will bring out the main ideas immediately.

4.1 Example. Let $a \colon \mathbb{R}^n \to \mathbb{R}$ be a continuous function $a > 0$ and suppose that

$$a(x) \to a_\infty > 0 \qquad (|x| \to \infty) \, .$$

We look for positive solutions u of the equation

$$(4.1) \qquad\qquad -\Delta u + a(x)u = u|u|^{p-2} \qquad \text{in } \mathbb{R}^n \, ,$$

decaying at infinity, that is

$$(4.2) \qquad\qquad u(x) \to 0 \qquad \text{as } |x| \to \infty \, .$$

Here $p > 2$ may be an arbitrary number, if $n = 1, 2$. If $n \geq 3$ we suppose that
$p < \frac{2n}{n-2}$. This guarantees that the imbedding

$$H^{1,2}(\Omega) \to L^p(\Omega)$$

is compact for any $\Omega \subset\subset \mathbb{R}^n$.

   Note that (4.1) is the Euler-Lagrange equation of the functional

$$E(u) = \frac{1}{2} \int_{\mathbb{R}^n} \left(|\nabla u|^2 + a(x)|u|^2 \right) dx$$

on $H^{1,2}(\mathbb{R}^n)$, restricted to the unit sphere

$$M = \{u \in H^{1,2}(\mathbb{R}^n) \; ; \; \int_{\mathbb{R}^n} |u|^p \, dx = 1\}$$

in $L^p(\mathbb{R}^n)$. Moreover, if $a(x) \equiv a_\infty$, E is invariant under translations

$$u \mapsto u_{x_0}(x) = u(x - x_0) .$$

In general, for any $u \in H^{1,2}(\mathbb{R}^n)$, after a substitution of variables

$$E(u_{x_0}) = \frac{1}{2} \int_{\mathbb{R}^n} \left(|\nabla u|^2 + a(\cdot + x_0)|u|^2 \right) dx \to \frac{1}{2} \int_{\mathbb{R}^n} \left(|\nabla u|^2 + a_\infty |u|^2 \right) dx$$

as $|x_o| \to \infty$, whence it may seem appropriate to call

$$E^\infty(u) := \frac{1}{2} \int_{\mathbb{R}^n} \left(|\nabla u|^2 + a_\infty |u|^2 \right) dx$$

the *functional at infinity* associated with E. The following result is due to Lions [2; Theorem I.2].

4.2 Theorem. *Suppose*

(4.3) $$I := \inf_M E < \inf_M E^\infty =: I^\infty ,$$

then there exists a positive solution $u \in H^{1,2}(\mathbb{R}^n)$ of equation (4.1). Moreover, condition (4.3) is necessary and sufficient for the relative compactness of all minimizing sequences for E in M.

Proof. Clearly, (4.3) is necessary for the convergence of all minimizing sequences in M. Indeed, suppose $I^\infty \leq I$ and let (u_m) be a minimizing sequence for E^∞. Then also $(\tilde{u}_m)$, given by $\tilde{u}_m = u_m(\cdot + x_m)$, is a minimizing sequence for E^∞, for any sequence (x_m) in $\mathbb{R}^n$. Choosing $|x_m|$ large enough such that

$$\left| E(\tilde{u}_m) - E^\infty(\tilde{u}_m) \right| \leq \frac{1}{m} ,$$

moreover, $(\tilde{u}_m)$ is a minimizing sequence for E. In addition, we can achieve that

$$\tilde{u}_m \to 0 \qquad \text{locally in } L^2 ,$$

whence $(\tilde{u}_m)$ cannot be relatively compact.

Note that this argument also proves that the weak inequality $I \leq I^\infty$ always holds true, regardless of the particular choice of the function a.

We now show that condition (4.3) is also sufficient. The existence of a positive solution to (4.1) then follows as in the proof of Theorem 2.1.

Let (u_m) be a minimizing sequence for E in M such that

$$E(u_m) \to I .$$

We may assume that $u_m \to u$ weakly in $L^p(\mathbb{R}^n)$. By continuity, a is uniformly positive on $\mathbb{R}^n$. Hence we also have

$$\|u_m\|_{H^{1,2}}^2 \le c\, E(u_m) \le C < \infty\ ,$$

and in addition we may assume that $u_m \rightharpoonup u$ weakly in $H^{1,2}(\mathbb{R}^n)$ and pointwise almost everywhere. Denote $u_m = v_m + u$. Observe that by Vitali's theorem

$$\int |u_m|^p\,dx - \int |u_m - u|^p\,dx = -\iint_0^1 \frac{d}{d\vartheta}|u_m - \vartheta u|^p\,d\vartheta\,dx$$

(4.4)
$$= p\iint_0^1 u(u_m - \vartheta u)|u_m - \vartheta u|^{p-2}\,d\vartheta\,dx$$

$$\rightarrow p\iint_0^1 u(u - \vartheta u)|u - \vartheta u|^{p-2}\,d\vartheta\,dx = \int |u|^p\,dx\ ,$$

where $\int \dots dx$ denotes integration over $\mathbb{R}^n$; that is

$$\int_{\mathbb{R}^n} |u|^p\,dx + \int_{\mathbb{R}^n} |v_m|^p\,dx \to 1\ .$$

Similarly

$$E(u_m) = E(v_m + u) =$$

(4.5)
$$= \frac{1}{2}\int_{\mathbb{R}^n} \left\{ \left(|\nabla u|^2 + 2\nabla u \nabla v_m + |\nabla v_m|^2 \right) \right.$$
$$\left. + a(x)\left(|u|^2 + 2uv_m + |v_m|^2 \right) \right\}\,dx$$

$$= E(u) + E(v_m) + \int_{\mathbb{R}^n} \left(\nabla u \nabla v_m + a(x)uv_m \right)\,dx$$

and the last term converges to zero by weak convergence $v_m = u_m - u \rightharpoonup 0$ in $H^{1,2}(\mathbb{R}^n)$.

Moreover, letting

$$\Omega_\varepsilon = \{x \in \mathbb{R}^n\ ;\ |a(x) - a_\infty| \ge \varepsilon\} \subset\subset \mathbb{R}^n,$$

since $v_m \to 0$ locally in L^2, the integral

$$\int_{\mathbb{R}^n} (a(x) - a_\infty)|v_m|^2\,dx \le$$

$$\le \varepsilon \int_{\mathbb{R}^n} |v_m|^2 + \sup_{\mathbb{R}^n}|a(x)|\int_{\Omega_\varepsilon} |v_m|^2\,dx$$

$$\le c\varepsilon + o(1)\ .$$

Here and in the following, $o(1)$ denotes error terms such that $o(1) \to 0$ as $m \to \infty$. Hence this integral can be made arbitrarily small if we first choose $\varepsilon > 0$ and then let $m \ge m_0(\varepsilon)$ be sufficiently large. That is, we have

$$E(u_m) = E(u) + E^\infty(v_m) + o(1)\ .$$

By homogeneity, if we denote $\lambda = \int_{\mathbb{R}^n} |u|^p \, dx$,

$$E(u) = \lambda^{2/p} E\left(\lambda^{-1/p} u\right) \geq \lambda^{2/p} I, \quad \text{if } \lambda > 0,$$

$$E^\infty(v_m) = (1-\lambda)^{2/p} E^\infty\left((1-\lambda)^{-1/p} v_m\right) \geq (1-\lambda)^{2/p} I^\infty + o(1), \quad \text{if } \lambda < 1.$$

Hence we obtain the estimate for $\lambda \in [0,1]$

$$\begin{aligned} I = E(u_m) + o(1) &= E(u) + E^\infty(v_m) + o(1) \\ &\geq \lambda^{2/p} I + (1-\lambda)^{2/p} I^\infty + o(1) \\ &\geq \left(\lambda^{2/p} + (1-\lambda)^{2/p}\right) I + o(1). \end{aligned}$$

Since $p > 2$ this implies that $\lambda \in \{0,1\}$. But if $\lambda = 0$, we obtain that

$$I \geq I^\infty + o(1) > I$$

for large m; a contradiction.

Therefore $\lambda = 1$; that is, $u_m \to u$ in L^p, and $u \in M$. By convexity of E, moreover,

$$E(u) \leq \liminf_{m \to \infty} E(u_m) = I,$$

and u minimizes E in M. Hence also $E(u_m) \to E(u)$. Finally, by (4.5)

$$\|u_m - u\|_{H^{1,2}}^2 \leq c E(u_m - u)$$

$$= c\left(E(u_m) - E(u)\right) + o(1) \to 0,$$

and $u_m \to u$ strongly in $H^{1,2}(\mathbb{R}^n)$. The proof is complete. □

Regarding $|u_m|^p \, dx$ as a measure on $\mathbb{R}^n$, a systematic approach to such problems is possible via the following lemma (P.L. Lions [1; p. 115 ff.]).

4.3 Concentration-Compactness Lemma I . *Suppose μ_m is a sequence of probability measures on $\mathbb{R}^n$: $\mu_m \geq 0$, $\int_{\mathbb{R}^n} d\mu_m = 1$. There is a subsequence (μ_m) such that one of the following three conditions holds:*

(1°) (Compactness) There exists a sequence $x_m \subset \mathbb{R}^n$ such that for any $\varepsilon > 0$ there is a radius $R > 0$ with the property that

$$\int_{B_R(x_m)} d\mu_m \geq 1 - \varepsilon$$

for all m.
(2°) (Vanishing) For all $R > 0$ there holds

$$\lim_{m \to \infty} \left(\sup_{x \in \mathbb{R}^n} \int_{B_R(x)} d\mu_m \right) = 0.$$

(3°) *(Dichotomy) There exists a number* λ, $0 < \lambda < 1$, *such that for any* $\varepsilon > 0$ *there is a number* $R > 0$ *and a sequence* (x_m) *with the following property: Given* $R' > R$ *there are non-negative measures* μ_m^1, μ_m^2 *such that*

$$0 \leq \mu_m^1 + \mu_m^2 \leq \mu_m \ ,$$

$$\operatorname{supp}(\mu_m^1) \subset B_R(x_m), \ \operatorname{supp}(\mu_m^2) \subset \mathbb{R}^n \setminus B_{R'}(x_m) \ ,$$

$$\limsup_{m \to \infty} \left(\left| \lambda - \int_{\mathbb{R}^n} d\mu_m^1 \right| + \left| (1 - \lambda) - \int_{\mathbb{R}^n} d\mu_m^2 \right| \right) \leq \varepsilon \ .$$

Proof. The proof is based on the notion of concentration function

$$Q(r) = \sup_{x \in \mathbb{R}^n} \left(\int_{B_r(x)} d\mu \right)$$

of a non-negative measure, introduced by P. Lévy [1].

Let Q_m be the concentration functions associated with μ_m. Note that (Q_m) is a sequence of non-decreasing, non-negative bounded functions on $[0, \infty[$ with $\lim_{R \to \infty} Q_m(R) = 1$. Hence, (Q_m) is locally bounded in BV on $[0, \infty[$ and there exists a subsequence (μ_m) and a bounded, non-negative, non-decreasing function Q such that

$$Q_m(R) \to Q(R) \qquad (m \to \infty) \ ,$$

for almost every $R > 0$. Let

$$\lambda = \lim_{R \to \infty} Q(R) \ .$$

Clearly $0 \leq \lambda \leq 1$. If $\lambda = 0$, we have "vanishing", case (2°). Suppose $\lambda = 1$. Then for some $R_0 > 0$ we have $Q(R_0) > \frac{1}{2}$. For any $m \in \mathbb{N}$ let x_m satisfy

$$Q_m(R_0) \leq \int_{B_{R_0}(x_m)} d\mu_m + \frac{1}{m} \ .$$

Now for $0 < \varepsilon < \frac{1}{2}$ fix R such that $Q(R) > 1 - \varepsilon > \frac{1}{2}$ and let y_m satisfy

$$Q_m(R) \leq \int_{B_R(y_m)} d\mu_m + \frac{1}{m} \ .$$

Then for large m we have

$$\int_{B_R(y_m)} d\mu_m + \int_{B_{R_0}(x_m)} d\mu_m > 1 = \int_{\mathbb{R}^n} d\mu_m \ .$$

It follows that for such m

$$B_R(y_m) \cap B_{R_0}(x_m) \neq \emptyset \ .$$

That is, $B_R(y_m) \subset B_{2R+R_0}(x_m)$ and hence

$$1 - \varepsilon \leq \int_{B_{2R+R_0}(x_m)} d\mu_m$$

for large m. Choosing R even larger, if necessary, we can achieve that $(1°)$ holds for all m.

If $0 < \lambda < 1$, given $\varepsilon > 0$ choose R and a sequence (x_m) – depending on ε and R – such that

$$\int_{B_R(x_m)} d\mu_m > \lambda - \varepsilon \, ,$$

if $m \geq m_0(\varepsilon)$. Enlarging $m_0(\varepsilon)$, if necessary, we can also find a sequence $R_m \to \infty$ such that

$$Q_m(R) \leq Q_m(R_m) < \lambda + \varepsilon \, ,$$

if $m \geq m_0(\varepsilon)$. Now let $\mu_m^1 = \mu_m \llcorner B_R(x_m)$, the restriction of μ_m to $B_R(x_m)$. Similarly, define $\mu_m^2 = \mu_m \llcorner (\mathbb{R}^n \setminus B_{R_m}(x_m))$. Obviously

$$0 \leq \mu_m^1 + \mu_m^2 \leq \mu_m \, ,$$

and, given $R' > R$, for large m we also have

$$\text{supp}(\mu_m^1) \subset B_R(x_m), \ \text{supp}(\mu_m^2) \subset \mathbb{R}^n \setminus B_{R_m}(x_m) \subset \mathbb{R}^n \setminus B_{R'}(x_m) \, .$$

Finally, for $m \geq m_0(\varepsilon)$ we can achieve

$$\left| \lambda - \int_{\mathbb{R}^n} d\mu_m^1 \right| + \left| 1 - \lambda - \int_{\mathbb{R}^n} d\mu_m^2 \right| =$$

$$= \left| \lambda - \int_{B_R(x_m)} d\mu_m \right| + \left| \int_{B_{R_m}(x_m)} d\mu_m - \lambda \right| < 2\varepsilon \, ,$$

which concludes the proof.   □

In the context of Theorem 4.2 Lemma 4.3 may be applied to $\mu_m = |u_m|^p \, dx$, $m \in \mathbb{N}$. Dichotomy in this case is made explicit in (4.4). In view of the compactness of the embedding $H^{1,2}(\Omega) \hookrightarrow L^p(\Omega)$ on bounded domains Ω for all $p < \frac{2n}{n-2}$ the situation dealt with in Example 4.1 is referred to as the *locally compact case*.

If a problem is conformally invariant, in particular invariant under the non-compact group of dilatations of $\mathbb{R}^n$ acting via

$$u \to u_R(x) = u(x/R), \ R > 0 \, ,$$

not even local compactness can hold.

Existence of Extremal Functions for Sobolev Embeddings

A typical example is the case of Sobolev's embedding on a (possibly unbounded) domain $\Omega \subset \mathbb{R}^n$.

4.4 Sobolev embeddings. For $u \in C_0^\infty(\Omega)$, $k \geq 1$, $p \geq 1$, let

$$\|u\|_{D^{k,p}}^p = \sum_{|\alpha|=k} \int_\Omega |D^\alpha u|^p \, dx \ ,$$

and let $D^{k,p}(\Omega)$ denote the completion of $C_0^\infty(\Omega)$ in this norm. Suppose $kp < n$.

By Sobolev's embedding, $D^{k,p}(\Omega) \hookrightarrow L^q(\Omega)$ where $\frac{1}{q} = \frac{1}{p} - \frac{k}{n}$, and there exists a (maximal) constant $S = S(k,n,p)$ such that

(4.6) $S\|u\|_{L^q}^p \leq \|u\|_{D^{k,p}}^p$, for all $u \in D^{k,p}(\Omega)$.

For $k = 1$ best constants and extremal functions (on $\Omega = \mathbb{R}^n$) can be computed classically, using Schwarz-symmetrization. See for instance Polya-Szegö [1; Note A.5, p. 189 ff.], or Talenti [1]; the earliest result in this regard seems to be due to Rodemich[1]. But for $k > 1$ this method can no longer be applied. Using the concentration-compactness principle, however, the existence of extremal functions for Sobolev's embedding can be established in general; see Theorem 4.9 below.

First we note an important property of the embedding (4.6).

4.5 Scale invariance. By invariance of the norms in $D^{k,p}(\mathbb{R}^n)$, respectively $L^q(\mathbb{R}^n)$, under scaling

(4.7) $u \mapsto u_{R,x_0}(x) = R^{-n/q} u\left(\dfrac{x - x_0}{R}\right)$

the Sobolev constant S is independent of Ω. Indeed, for any domain Ω, extending a function $u \in C_0^\infty(\Omega)$ by 0 outside Ω, we may regard $C_0^\infty(\Omega)$ as a subset of $C_0^\infty(\mathbb{R}^n)$. Similarly, we may regard $D^{k,p}(\Omega)$ as a subset of $D^{k,p}(\mathbb{R}^n)$. Hence we have

$$S(\Omega) = \inf\left\{\|u\|_{D^{k,p}}^p \ ; \ u \in D^{k,p}(\Omega), \ \|u\|_{L^q} = 1\right\}$$
$$\geq S(\mathbb{R}^n) \ .$$

Conversely, if $u_m \in D^{k,p}(\mathbb{R}^n)$ is a minimizing sequence for $S(\mathbb{R}^n)$, by density of $C_0^\infty(\mathbb{R}^n)$ in $D^{k,p}(\mathbb{R}^n)$ we may assume that $u_m \in C_0^\infty(\mathbb{R}^n)$. After translation, moreover, we have $0 \in \Omega$. Scaling with (4.7), for sufficiently small R_m we can achieve that $v_m = (u_m)_{R_m} \in C_0^\infty(\Omega)$. But by invariance of $\|\cdot\|_{D^{k,p}}, \|\cdot\|_{L^q}$ under (4.7) there now results

$$S(\Omega) \leq \liminf_{m\to\infty} \|v_m\|_{D^{k,p}}^p = S(\mathbb{R}^n) \ ,$$

and $S(\Omega) = S(\mathbb{R}^n) = S$, as claimed.

4.6 The case $k = 1$. For $k = 1$ the Sobolev inequality has an underlying geometric meaning which allows to analyze this case completely. Consider first the case $p = 1$. We claim: For $u \in D^{1,1}(\mathbb{R}^n)$, $q_1 = \frac{n}{n-1}$, there holds

$$(4.8) \qquad \left(\int_{\mathbb{R}^n} |u|^{q_1} \, dx \right)^{1/q_1} \leq \frac{1}{n^{1/q_1} \omega_{n-1}^{1/n}} \int_{\mathbb{R}^n} |\nabla u| \, dx \ ,$$

where ω_{n-1} denotes the $(n-1)$-dimensional measure of the unit sphere in $\mathbb{R}^n$. Observe that equality holds if (and only if) u is a scalar multiple of the characteristic function of a ball in $\mathbb{R}^n$. This reflects the fact that

$$K(n) = \frac{1}{n^{1/q_1} \omega_{n-1}^{1/n}} = \sup_{\Omega \subset\subset \mathbb{R}^n} \frac{\left(\mathcal{L}^n(\Omega) \right)^{1/q_1}}{P(\Omega; \mathbb{R}^n)}$$

equals the isoperimetric constant in $\mathbb{R}^n$, and $K(n)$ is achieved if and only if Ω is a ball in $\mathbb{R}^n$. (The perimeter $P(\Omega; \mathbb{R}^n)$ was defined in Theorem 1.4.)

The following proof of (4.8), based on Talenti [2; p. 404], reveals the deep relation between isoperimetric inequalities and best constants for Sobolev embeddings more clearly. (See also Cianchi [1; Lemma 1].)

Proof of (4.8). For $u \in C_0^\infty(\mathbb{R}^n)$, and $t \geq 0$ let

$$\Omega(t) = \{ x \in \mathbb{R}^n \ ; \ |u(x)| > t \} \ .$$

Then

$$|u(x)| = \int_0^\infty \chi_{\Omega(t)}(x) \, dt$$

for almost every $x \in \Omega$, and hence by Minkowsky's inequality

$$\|u\|_{L^{q_1}} \leq \int_0^\infty \|\chi_{\Omega(t)}\|_{L^{q_1}} \, dt$$

$$= \int_0^\infty \mathcal{L}^n \left(\Omega(t) \right)^{1/q_1} \, dt$$

$$\leq K(n) \int_0^\infty P\left(\Omega(t); \mathbb{R}^n \right) \, dt \ .$$

Finally, by the co-area formula (see for instance Federer [1; Theorem 3.2.11] or Giusti [1; Theorem 1.23])

$$\int_{\mathbb{R}^n} |\nabla u| \, dx = \int_{\mathbb{R}^n} |\nabla |u|| \, dx = \int_0^\infty P\left(\Omega(t); \mathbb{R}^n \right) \, dt \ ,$$

and (4.8) follows.   □

From (4.8) the general case $p \geq 1$ can be derived by applying Hölder's inequality. Denote $q = \frac{np}{n-p} = sq_1$, where $s = \frac{np-p}{n-p} \geq 1$. Then for $u \in C_0^\infty(\mathbb{R}^n)$ we can estimate

$$\|u\|_{L^q} = \left\||u|^s\right\|_{L^{q_1}}^{1/s} \le \left(K(n)\right)^{1/s} \left(\int_{\mathbb{R}^n} \left|\nabla|u|^s\right| dx\right)^{1/s}$$

$$\le \left(sK(n)\right)^{1/s} \left(\int_{\mathbb{R}^n} |\nabla u|\, |u|^{s-1}\, dx\right)^{1/s}$$

$$\le \left(sK(n)\right)^{1/s} \|u\|_{D^{1,p}}^{1/s} \|u\|_{L^q}^{1-1/s}$$

and

(4.9)
$$\|u\|_{L^q} \le sK(n)\|u\|_{D^{1,p}} .$$

(We do not claim that this constant is sharp.)

4.7 Bounded domains. Note that, in contrast to the case $p = 1$, for $p > 1$ the best constant in inequality (4.9) is never achieved on a (smooth) domain Ω different from $\mathbb{R}^n$, in particular is never achieved on a bounded domain. Indeed, if $u \in D^{1,p}(\Omega)$ achieves S, a multiple of u weakly solves the equation

(4.10)
$$-\nabla\left(|\nabla u|^{p-2}\nabla u\right) = u|u|^{q-2} \quad \text{in } \Omega$$

with vanishing boundary data on $\partial\Omega$. But since $S(\Omega) = S(\mathbb{R}^n)$, u then also solves this equation on $\mathbb{R}^n$, which contradicts the strong maximum principle for equation (4.10); see for instance Tolksdorf [1].

Of course, we suspect invariance under scaling (4.7) to hold responsible for this defect. Note that for any $u \in D^{k,p}(\mathbb{R}^n)$ there holds

$$u_{R,x_0} \rightharpoonup 0 \text{ weakly in } D^{k,p}(\mathbb{R}^n) \text{ as } R \to \infty , \text{ or } |x_0| \to \infty ,$$

while S is invariant under dilatations and translations; hence compactness of minimizing sequences that are not normalized with respect to these actions cannot be expected. In particular, invariance under dilatations may lead to a new type of loss of compactness, as compared with Example 4.1 or Lemma 4.3.

To handle this difficulty we employ the second concentration-compactness lemma from P.L. Lions [3; Lemma I.1]. Denote

$$\sum_{|\alpha|=k} |D^\alpha u|^p = |D^k u|^p ,$$

for convenience.

4.8 Concentration-Compactness Lemma II. *Let* $k \in \mathbb{N}$, $p \ge 1$, $kp < n$, $\frac{1}{q} = \frac{1}{p} - \frac{k}{n}$. *Suppose* $u_m \rightharpoonup u$ *weakly in* $D^{k,p}(\mathbb{R}^n)$ *and* $\mu_m = |\nabla^k u_m|^p\, dx \rightharpoonup \mu$, $\nu_m = |u_m|^q\, dx \rightharpoonup \nu$ *weakly in the sense of measures where* μ *and* ν *are bounded non-negative measures on* $\mathbb{R}^n$.
Then we have:
(1°) There exists some at most countable set J, *a family* $\{x^{(j)} ; j \in J\}$ *of distinct points in* $\mathbb{R}^n$, *and a family* $\{\nu^{(j)} ; j \in J\}$ *of positive numbers such that*

$$\nu = |u|^q \, dx + \sum_{j \in J} \nu^{(j)} \delta_{x^{(j)}} \, ,$$

where δ_x is the Dirac-mass of mass 1 concentrated at $x \in \mathbb{R}^n$.
(2°) In addition we have

$$\mu \geq |\nabla^k u|^p \, dx + \sum_{j \in J} \mu^{(j)} \delta_{x^{(j)}}$$

for some family $\{\mu^{(j)} \; ; \; j \in J\}$, $\mu^{(j)} > 0$ satisfying

$$S \left(\nu^{(j)} \right)^{p/q} \leq \mu^{(j)} \, , \qquad \text{for all } j \in J \, .$$

In particular, $\sum_{j \in J} \left(\nu^{(j)} \right)^{p/q} < \infty$.

Proof. Let $v_m = u_m - u \in D^{k,p}(\mathbb{R}^n)$. Then $v_m \rightharpoonup 0$ weakly in $D^{k,p}$. By (4.4), thus we have

$$\omega_m := \nu_m - |u|^q \, dx = \left(|u_m|^q - |u|^q \right) dx$$
$$= |u_m - u|^q \, dx + o(1) = |v_m|^q \, dx + o(1) \, ,$$

where $o(1) \to 0$ as $m \to \infty$. Also let $\lambda_m := |\nabla^k v_m|^p \, dx$. We may assume that $\lambda_m \to \lambda$, while $\omega_m \to \omega = \nu - |u|^q \, dx$ weakly in the sense of measures, where $\lambda, \omega \geq 0$.

Choose $\xi \in C_0^\infty(\mathbb{R}^n)$. Then

$$\int_{\mathbb{R}^n} |\xi|^q \, d\omega = \lim_{m \to \infty} \int_{\mathbb{R}^n} |\xi|^q \, d\omega_m = \lim_{m \to \infty} \int_{\mathbb{R}^n} |v_m \, \xi|^q \, dx$$
$$\leq S^{-q/p} \liminf_{m \to \infty} \left(\int_{\mathbb{R}^n} |\nabla^k(v_m \xi)|^p \, dx \right)^{q/p}$$
$$= S^{-q/p} \liminf_{m \to \infty} \left(\int_{\mathbb{R}^n} |\xi|^p |\nabla^k v_m|^p \, dx \right)^{q/p}$$
$$= S^{-q/p} \left(\int_{\mathbb{R}^n} |\xi|^p \, d\lambda \right)^{q/p} \, .$$

Observe that by Rellich's theorem any lower order terms like $|\nabla^l \xi||\nabla^{k-l} v_m| \to 0$ in L^p, as $m \to \infty$. That is, there holds

(4.11) $$S \left(\int_{\mathbb{R}^n} |\xi|^q \, d\omega \right)^{p/q} \leq \int_{\mathbb{R}^n} |\xi|^p \, d\lambda$$

for all $\xi \in C_0^\infty(\mathbb{R}^n)$. Now let $\{x^{(j)} \; ; \; j \in J\}$ be the atoms of the measure ω and decompose $\omega = \omega_0 + \sum_{j \in J} \nu^{(j)} \delta_{x^{(j)}}$, with ω_0 free of atoms. Since $\int_{\mathbb{R}^n} d\omega < \infty$, J is an at most countable set. Moreover, $\omega_0 \geq 0$. Choosing ξ such that $0 \leq \xi \leq 1$, $\xi(x^{(j)}) = 1$, from (4.11) we see that

$$\lambda \geq S\big(\nu^{(j)}\big)^{p/q} \delta_{x^{(j)}}, \qquad \text{for all } j \in J .$$

Since $|\nabla^k u_m|^p - |\nabla^k v_m|^p$ is of lower order than $|\nabla^k v_m|^p$ at points of concentration, the latter estimate also holds for μ.

On the other hand, by weak lower semi–continuity we have

$$\mu \geq |\nabla^k u|^p \, dx .$$

The latter measure and the measures $\delta_{x^{(j)}}$ being relatively singular, $(2°)$ follows.

Now, for any open set $\Omega \subset \mathbb{R}^n \setminus \{x^{(j)} \,;\, j \in J\}$ such that $\int_\Omega d\lambda \leq S$, by (4.11) with $\xi = \xi_k$ converging to the characteristic function of Ω as $k \to \infty$, we have

$$(4.12) \qquad \int_\Omega d\omega \leq \left(\int_\Omega d\omega \right)^{p/q} \leq S^{-1} \int_\Omega d\lambda \leq 1 .$$

That is, ω_0 is absolutely continuous with respect to λ and by the Radon–Nikodym theorem there exists $f \in L^1(\mathbb{R}^n; \lambda)$ such that $\omega_0 = f \lambda$, λ-almost everywhere. Moreover, for λ-almost every $x \in \mathbb{R}^n$ we have

$$f(x) = \lim_{\rho \to 0} \left(\frac{\int_{B_\rho(x)} d\omega_0}{\int_{B_\rho(x)} d\lambda} \right) .$$

But then by (4.11), if x is not an atom of λ,

$$S \, f(x)^{p/q} = \lim_{\rho \to 0} \left(\frac{S \left(\int_{B_\rho(x)} d\omega_0 \right)^{p/q}}{\left(\int_{B_\rho(x)} d\lambda \right)^{p/q}} \right) \leq \lim_{\rho \to 0} \left(\int_{B_\rho(x)} d\lambda \right)^{\frac{q-p}{q}} = 0 ,$$

λ-almost everywhere. Since λ has only countably many atoms and ω_0 has no atoms this implies that $\omega_0 = 0$, that is, $(1°)$. $\qquad\square$

Finally, we can state the following result; see P.L. Lions [3; Theorem I.1]:

4.9 Theorem. *Let $k \in \mathbb{N}$, $p > 1$, $kp < n$, $\frac{1}{q} = \frac{1}{p} - \frac{k}{n}$. Suppose (u_m) is a minimizing sequence for S in $D^{k,p} = D^{k,p}(\mathbb{R}^n)$ with $\|u_m\|_{L^q} = 1$. Then (u_m) up to translation and dilatation is relatively compact in $D^{k,p}$.*

Proof. Choose $x_m \in \mathbb{R}^n$, $R_m > 0$ such that for the rescaled sequence

$$v_m(x) = R_m^{-n/q} u_m \left(\frac{x - x_m}{R_m} \right)$$

there holds

$$(4.13) \qquad Q_m(1) = \sup_{x \in \mathbb{R}^n} \int_{B_1(x)} |v_m|^q \, dx = \int_{B_1(0)} |v_m|^q \, dx = \frac{1}{2} .$$

Since $p > 1$ we may assume that $v_m \rightharpoonup v$ weakly in $L^q(\mathbb{R}^n)$ and weakly in $D^{k,p}(\mathbb{R}^n)$. Consider the families of measures

$$\mu_m = |\nabla^k v_m|^p \, dx$$
$$\nu_m = |v_m|^q \, dx$$

and apply Lemma 4.3 to the sequence (ν_m). Vanishing is ruled out by our above normalization. If we have dichotomy, let $\lambda \in]0,1[$ be as in Lemma 4.3.(3°) and for $\varepsilon > 0$ determine $R > 0$, a sequence (x_m), and measures ν_m^1, ν_m^2 as in that lemma such that

$$0 \le \nu_m^1 + \nu_m^2 \le \nu_m \ ,$$
$$\operatorname{supp}(\nu_m^1) \subset B_R(x_m), \ \operatorname{supp}(\nu_m^2) \subset \mathbb{R}^n \setminus B_{2R}(x_m) \ ,$$
$$\limsup_{m \to \infty} \left\{ \left| \int_{\mathbb{R}^n} d\nu_m^1 - \lambda \right| + \left| \int_{\mathbb{R}^n} d\nu_m^2 - (1-\lambda) \right| \right\} \le \varepsilon \ .$$

Choosing a sequence $\varepsilon_m \to 0$, corresponding $R_m > 0$, and passing to a subsequence (ν_m) if necessary, we can achieve that

$$\operatorname{supp}(\nu_m^1) \subset B_{R_m}(x_m), \ \operatorname{supp}(\nu_m^2) \subset \mathbb{R}^n \setminus B_{2R_m}(x_m)$$

and

$$\limsup_{m \to \infty} \left\{ \left| \int_{\mathbb{R}^n} d\nu_m^1 - \lambda \right| + \left| \int_{\mathbb{R}^n} d\nu_m^2 - (1-\lambda) \right| \right\} = 0 \ .$$

Moreover, in view of Lemma 4.3, we may suppose that $R_m \to \infty$ $(m \to \infty)$.
    Choose $\varphi \in C_0^\infty(B_2(0))$ such that $\varphi \equiv 1$ in $B_1(0)$ and let $\varphi_m(x) = \varphi\left(\frac{x - x_m}{R_m}\right)$. Decompose

$$v_m = v_m \varphi_m + v_m(1 - \varphi_m) \ .$$

Then

$$\int_{\mathbb{R}^n} |\nabla^k v_m|^p \, dx = \int_{\mathbb{R}^n} |\nabla^k(v_m \varphi_m)|^p \, dx + \int_{\mathbb{R}^n} \left|\nabla^k\left(v_m(1 - \varphi_m)\right)\right|^p \, dx + \delta_m$$

where the error terms δ_m can be estimated from below

$$\delta_m \ge -C \sum_{l < k} \int_{B_{2R_m}(x_m) \setminus B_{R_m}(x_m)} |\nabla^l v_m|^p |\nabla^{k-l} \varphi_m|^p \, dx \ .$$

Here we also use the fact that $0 \le \varphi \le 1$ and $p \ge 1$. Let A_m denote the annulus $A_m = B_{2R_m}(x_m) \setminus B_{R_m}(x_m)$. Estimating $|\nabla^{k-l} \varphi_m| \le C\, R_m^{l-k}$, by interpolation (see for instance Adams [1; Theorem 4.14]) we can bound any term

$$\left\| |\nabla^l v_m|\, |\nabla^{k-l} \varphi_m| \right\|_{L^p(A_m)} \le C\, R_m^{l-k} \|\nabla^l v_m\|_{L^p(A_m)}$$

$$\le C\, K\, \gamma \|\nabla^k v_m\|_{L^p(A_m)}$$

$$+ C\, K\, R_m^{-k} \gamma^{-\frac{l}{k-l}} \|v_m\|_{L^p(A_m)} \ .$$

(4.14)

Here γ can be chosen arbitrarily in $]0, 1]$, while the constant K depends only on k and n. (Note that estimate (4.14) is invariant under dilatations.) Moreover, by Hölder's inequality

$$R_m^{-k} \|v_m\|_{L^p(A_m)} \leq R_m^{-k} \left(\mathcal{L}^n(A_m) \right)^{\frac{1}{p} - \frac{1}{q}} \|v_m\|_{L^q(A_m)}$$

$$= C \|v_m\|_{L^q(A_m)}$$

$$\leq C \left[\int_{\mathbb{R}^n} d\nu_m - \left(\int_{\mathbb{R}^n} d\nu_m^1 + \int_{\mathbb{R}^n} d\nu_m^2 \right) \right]^{\frac{1}{q}} .$$

Hence this term tends to 0 as $m \to \infty$, while $\|\nabla^k v_m\|_{L^p(A_m)}^p \leq \|v_m\|_{D^{k,p}}^p$ remains uniformly bounded. Choosing a suitable sequence $\gamma_m \to 0$, from (4.14) we thus obtain that $\delta_m \geq o(1)$, where $o(1) \to 0$ $(m \to \infty)$. Now by Sobolev's inequality

$$\|v_m\|_{D^{k,p}}^p = \|v_m \varphi_m\|_{D^{k,p}}^p + \|v_m(1 - \varphi_m)\|_{D^{k,p}}^p + \delta_m$$

$$\geq S \left(\|v_m \varphi_m\|_{L^q}^p + \|v_m(1 - \varphi_m)\|_{L^q}^p \right) + \delta_m$$

$$\geq S \left[\left(\int_{B_{R_m}(x_m)} d\nu_m \right)^{p/q} + \left(\int_{\mathbb{R}^n \setminus B_{2R_m}(x_m)} d\nu_m \right)^{p/q} \right] + \delta_m$$

$$\geq S \left[\left(\int_{\mathbb{R}^n} d\nu_m^1 \right)^{p/q} + \left(\int_{\mathbb{R}^n} d\nu_m^2 \right)^{p/q} \right] + \delta_m$$

$$\geq S \left(\lambda^{p/q} + (1 - \lambda)^{p/q} \right) - o(1) ,$$

where $o(1) \to 0$ $(m \to \infty)$. But for $0 < \lambda < 1$ and $p < q$ we have $\lambda^{p/q} + (1 - \lambda)^{p/q} > 1$, contradicting the initial assumption that $\|v_m\|_{D^{k,p}}^p = \|u_m\|_{D^{k,p}}^p \to S$. It remains the case $\lambda = 1$, that is, case (1°) of Lemma 4.3.

Let x_m be as in that lemma and for $\varepsilon > 0$ choose $R = R(\varepsilon)$ such that

$$\int_{B_R(x_m)} \geq 1 - \varepsilon .$$

If $\varepsilon < \frac{1}{2}$ our normalization condition (4.13) implies $B_R(x_m) \cap B_1(0) \neq \emptyset$. Hence the conclusion of Lemma 4.3.(1°) also holds with $x_m = 0$, replacing $R(\varepsilon)$ by $2R(\varepsilon) + 1$ if necessary. Thus, if $\nu_m \rightharpoonup \nu$ weakly, it follows that

$$\int_{\mathbb{R}^n} d\nu = 1 .$$

By Lemma 4.8 we may assume that

$$\mu_m \rightharpoonup \mu \geq |\nabla^k v|^p \, dx + \sum_{j \in J} \mu^{(j)} \delta_{x^{(j)}}$$

$$\nu_m \rightharpoonup \nu = |v|^q \, dx + \sum_{j \in J} \nu^{(j)} \delta_{x^{(j)}}$$

for certain points $x^{(j)} \in \mathbb{R}^n$, $j \in \mathcal{J}$, and positive numbers $\mu^{(j)}, \nu^{(j)}$ satisfying

$$S\left(\nu^{(j)}\right)^{p/q} \leq \mu^{(j)} , \qquad \text{for all } j \in \mathcal{J} .$$

By Sobolev's inequality then

$$S + o(1) = \|v_m\|_{D^{k,p}}^p = \int_{\mathbb{R}^n} d\mu_m \geq \|v\|_{D^{k,p}}^p + \sum_{j \in \mathcal{J}} \mu^{(j)} + o(1)$$

$$\geq S \left(\|v\|_{L^q}^{p/q} + \sum_{j \in \mathcal{J}} \left(\nu^{(j)}\right)^{p/q} \right) + o(1)$$

where $o(1) \to 1$ $(m \to \infty)$. By strict concavity of the map $\lambda \to \lambda^{p/q}$ now the latter will be

$$\geq S \left(\|v\|_{L^q}^q + \sum_{j \in \mathcal{J}} \nu^{(j)} \right)^{p/q} + o(1)$$

(4.15)

$$= S \left(\int_{\mathbb{R}^n} d\nu \right)^{p/q} + o(1) = S + o(1)$$

and equality holds if and only if at most one of the terms $\|v\|_{L^q}$, $\nu^{(j)}$, $j \in \mathcal{J}$, is different from 0.

Note that our normalization (4.13) assures that

$$\nu^{(j)} \leq \frac{1}{2} \qquad \text{for all } j \in \mathcal{J} .$$

Hence all $\nu^{(j)}$ must vanish, $\|v\|_{L^q} = 1$, and $v_m \to v$ strongly in $L^q(\mathbb{R}^n)$. But then by Sobolev's inequality $\|v\|_{D^{k,p}}^p \geq S$ and $\|v_m\|_{D^{k,p}} \to \|v\|_{D^{k,p}}$ as $m \to \infty$. It follows that $v_m \to v$ in $D^{k,p}(\mathbb{R}^n)$, as desired. The proof is complete.   □

Remark. As Lemma 4.8 requires weak convergence $u_m \rightharpoonup u$ in $D^{k,p}(\mathbb{R}^n)$ the above proof of Theorem 4.9 cannot be extended to the case $p = 1$. In fact, we have seen that the best constant for Sobolev's embedding $D^{1,1}(\mathbb{R}^n) \hookrightarrow L^{\frac{n}{n-1}}(\mathbb{R}^n)$ is attained on characteristic functions of balls; that is, in $BV_{loc}(\mathbb{R}^n)$.

5. Ekeland's Variational Principle

In general it is not clear that a bounded and lower semi-continuous functional E actually attains its infimum. The analytic function $f(x) = \arctan x$, for example, neither attains its infimum nor its supremum on the real line.

A variant due to Ekeland [1] of Dirichlet's principle, however, permits to construct minimizing sequences for such functionals E whose elements u_m each minimize a functional E_m, for a sequence of functionals E_m converging locally uniformly to E.

5.1 Theorem. *Let M be a complete metric space with metric d, and let $E\colon M \to \mathbb{R} \cup +\infty$ be lower semi-continuous, bounded from below, and $\not\equiv \infty$. Then for any $\varepsilon, \delta > 0$, any $u \in M$ with*

$$E(u) \leq \inf_M E + \varepsilon,$$

there is an element $v \in M$ strictly minimizing the functional

$$E_v(w) \equiv E(w) + \frac{\varepsilon}{\delta} d(v, w) \ .$$

Moreover, we have

$$E(v) \leq E(u), \quad d(u, v) \leq \delta \ .$$

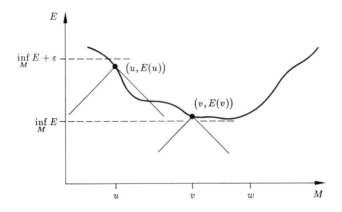

Fig. 5.1. Comparing E with E_v. v is a strict absolute minimizer of E_v if and only if the downward cone of slope ε/δ with vertex at $\big(v, E(v)\big)$ lies entirely below the graph of E.

Proof. Denote $\alpha = \frac{\varepsilon}{\delta}$ and define a partial ordering on $M \times \mathbb{R}$ by letting

(5.1) $$(v, \beta) \leq (v', \beta') \Leftrightarrow (\beta' - \beta) + \alpha \, d(v, v') \leq 0 \ .$$

This relation is easily seen to be reflexive, identitive, and transitive:

$$(v, \beta) \leq (v, \beta) \ ,$$
$$(v, \beta) \leq (v', \beta') \wedge (v', \beta') \leq (v, \beta) \Leftrightarrow v = v', \ \beta = \beta',$$
$$(v, \beta) \leq (v', \beta') \wedge (v', \beta') \leq (v'', \beta'') \Rightarrow (v, \beta) \leq (v'', \beta'') \ .$$

Moreover, if we denote

$$S = \{(v, \beta) \in M \times \mathbb{R} \ ; \ E(v) \leq \beta\} \ ,$$

by lower semi-continuity of E, S is closed in $M \times \mathbb{R}$. To complete the proof we need a lemma.

5.2 Lemma. *S contains a maximal element (v, β) with respect to the partial ordering $\leq$ on $M \times \mathbb{R}$ such that $(u, E(u)) \leq (v, \beta)$.*

Proof. Let $(v_1, \beta_1) = (u, E(u))$ and define a sequence (v_m, β_m) inductively as follows: Given (v_m, β_m) define

$$S_m = \{(v, \beta) \in S \; ; \; (v_m, \beta_m) \leq (v, \beta)\}$$
$$\mu_m = \inf\{\beta \; ; \; (v, \beta) \in S_m\} \geq$$
$$\geq \inf\{E(v) \; ; \; (v, \beta) \in S_m\} \geq \inf_M E =: \mu_0 \; .$$

Note that $\mu_m \leq \beta_m$. Now let $(v_{m+1}, \beta_{m+1}) \in S_m$ be chosen such that

(5.2)
$$\beta_m - \beta_{m+1} \geq \frac{1}{2}(\beta_m - \mu_m) \; .$$

Note that by transitivity of $\leq$ the sequence S_m is nested: $S_1 \supset S_2 \supset \dots \supset S_m \supset S_{m+1} \supset \dots$. Hence also $\dots \leq \mu_m \leq \mu_{m+1} \leq \dots \leq \beta_{m+1} \leq \beta_m \leq \dots$. By induction, from (5.2) we obtain

$$\beta_{m+1} - \mu_{m+1} \leq \beta_{m+1} - \mu_m$$
$$\leq \frac{1}{2}(\beta_m - \mu_m) \leq \dots \leq \left(\frac{1}{2}\right)^m (\beta_1 - \mu_1) \; .$$

Therefore, by definition of S_m, for any $m \in \mathbb{N}$ and any $(v, \beta) \in S_m$ we have

(5.3)
$$|\beta_m - \beta| = \beta_m - \beta \leq \beta_m - \mu_m \leq C \left(\frac{1}{2}\right)^m \; ,$$
$$d(v_m, v) \leq \alpha^{-1}(\beta_m - \beta) \leq C\alpha^{-1} \left(\frac{1}{2}\right)^m \; .$$

In particular, $\big((v_m, \beta_m)\big)_{m \in \mathbb{N}}$ is a Cauchy sequence in $M \times \mathbb{R}$. Thus, by completeness of M, $\big((v_m, \beta_m)\big)$ converges to some limit $(\overline{v}, \overline{\beta}) \in \bigcap_{m \in \mathbb{N}} S_m$. By transitivity, clearly $(u, E(u)) = (v_1, \beta_1) \leq (\overline{v}, \overline{\beta})$. Moreover, $(\overline{v}, \overline{\beta})$ is maximal. In fact, if $(\overline{v}, \overline{\beta}) \leq (\tilde{v}, \tilde{\beta})$ for some $(\tilde{v}, \tilde{\beta}) \in M \times \mathbb{R}$, then also $(v_m, \beta_m) \leq (\tilde{v}, \tilde{\beta})$ for all m, and $(\tilde{v}, \tilde{\beta}) \in S_m$ for all m. Letting $(v, \beta) = (\tilde{v}, \tilde{\beta})$ in (5.3) we infer that $(v_m, \beta_m) \to (\tilde{v}, \tilde{\beta})$ $(m \to \infty)$, whence $(\tilde{v}, \tilde{\beta}) = (\overline{v}, \overline{\beta})$, as desired. $\qquad\square$

Proof of Theorem 5.1 (Completed). Let (v, β) be maximal in S with $(u, E(u)) \leq (v, \beta)$. Comparing with $(v, E(v)) \in S$ at once yields $\beta = E(v)$. By definition (5.1) the statement $(u, E(u)) \leq (v, E(v))$ translates into the estimate

$$E(v) - E(u) + \alpha \, d(u, v) \leq 0 \; ;$$

in particular this implies

$$E(v) \leq E(u)$$

and

$$d(u, v) \le \alpha^{-1} \left(E(u) - E(v) \right) \le \frac{\delta}{\varepsilon} \left(\inf_M E + \varepsilon - \inf_M E \right) = \delta .$$

Finally, if $w \in M$ satisfies

$$E_v(w) = E(w) + \alpha \, d(v, w) \le E(v) = E_v(v) ,$$

by definition (5.1) we have $(v, E(v)) \le (w, E(w))$. Hence $w = v$ by maximality of $(v, E(v))$; that is, v is a strict minimizer of E_v, as claimed.   □

5.3 Corollary. *If V is a Banach space and $E \in C^1(V)$ is bounded from below, there exists a minimizing sequence (v_m) for E in B such that*

$$E(v_m) \to \inf_V E, \quad DE(v_m) \to 0 \qquad \text{in } V^* .$$

Proof. Choose a sequence (ε_m) of numbers $\varepsilon_m > 0$, $\varepsilon_m \to 0$ $(m \to \infty)$. For $m \in \mathbb{N}$ choose $u_m \in V$ such that

$$E(u_m) \le \inf_V E + \varepsilon_m^2 .$$

For $\varepsilon = \varepsilon_m^2$, $\delta = \varepsilon_m$, $u = u_m$ determine an element $v_m = v$ according to Theorem 5.1, satisfying

$$E(v_m) \le E(v_m + w) + \varepsilon_m \, \|w\|_V$$

for all $w \in B$. Hence

$$\|DE(v_m)\|_{V^*} = \lim_{\delta \to 0} \sup_{\|w\|_V \le \delta, \ w \neq 0} \frac{E(v_m) - E(v_m + w)}{\|w\|_V} \le \varepsilon_m \to 0 ,$$

as claimed.   □

In Chapter II we will re-encounter the special minimizing sequences of Corollary 5.3 as "Palais-Smale sequences". Compactness of such sequences by Corollary 5.3 turns out to be a sufficient condition for the existence of a minimizer for a differentiable functional E which is bounded from below on a Banach space V. Moreover, we shall see that the compactness of Palais-Smale sequences (under suitable assumptions on the topology of the level sets of E) will also guarantee the existence of critical points of saddle-type.

However, before turning our attention to critical points of general type we sketch another application of Ekeland's variational principle.

Existence of Minimizers for Quasi-Convex Functionals

Theorem 5.1 may be used to construct minimizing sequences for variational integrals enjoying better smoothness properties than can a priori be expected. We present an example due to Marcellini-Sbordone [1].

5.4 Example. Let Ω be a bounded domain in $\mathbb{R}^n$ and let $f \colon \Omega \times \mathbb{R}^n \times \mathbb{R}^{nN} \to \mathbb{R}$ be a Carathéodory function satisfying the coercivity and growth condition

$$(5.4) \qquad |p|^s \leq f(x, u, p) \leq C\big(1 + |u|^s + |p|^s\big) \text{ for some } s > 1 \,, C \in \mathbb{R}.$$

Moreover, suppose f is quasi-convex in the sense of Morrey [3]; that is, for almost every $x_0 \in \Omega$, and any $u_0 \in \mathbb{R}^N, p_o \in \mathbb{R}^{nN}$ there holds

$$(5.5) \qquad \frac{1}{\mathcal{L}^n(\Omega)} \int_\Omega f\big(x_0, u_0, p_0 + D\varphi(x)\big) \, dx \geq$$
$$\geq f(x_0, u_0, p_0), \qquad \text{for all } \varphi \in H_0^{1,s}(\Omega) \,.$$

For $u \in H^{1,s}(\Omega; \mathbb{R}^N)$ now set

$$E(u) = E(u; \Omega) = \int_\Omega f\big(x, u(x), \nabla u(x)\big) \, dx \,.$$

Remark that by Jensen's inequality condition (5.5) is weaker than convexity of f in p. Moreover, an example by Tartar and Murat shows that – in contrast to the case $f = f(p)$; see Acerbi-Fusco [1] – conditions (5.4), (5.5) are in general not sufficient to guarantee weak lower semi-continuity of E in $H^{1,s}(\Omega; \mathbb{R}^N)$; see for instance Murat [2; Section 4] or Marcellini-Sbordone [1; p. 3]. However, by a result of Fusco [1], see also Marcellini-Sbordone [1; Section 2], the functional E is weakly lower semi-continuous in $H^{1,q}(\Omega; \mathbb{R}^N)$ for any $q > s$.

The following result therefore guarantees that E assumes its infimum on suitable subsets of $H^{1,s}(\Omega; \mathbb{R}^N)$. Its proof is based on Ekeland's variational principle. Moreover, it employs a profound measure-theoretical result, Lemma 5.6, due to Giaquinta-Modica [1], following an idea of Gehring [1].

5.5 Theorem. *Under the above hypotheses (5.4), (5.5) on f, for any $u_0 \in H^{1,s}(\Omega; \mathbb{R}^N)$ there is a minimizing sequence (u_m) for E on $\{u_0\} + H_0^{1,s}(\Omega; \mathbb{R}^N)$ which is locally bounded in $H^{1,q}$ for sone $q > s$.*

Proof. Choose
$$M = \{u_0\} + H_0^{1,1}(\Omega; \mathbb{R}^N)$$
with metric d derived from the $H_0^{1,1}$-norm
$$d(u, v) = \int_\Omega |\nabla u - \nabla v| \, dx \,.$$

Note that by Fatou's lemma $E: M \to \mathbb{R} \cup +\infty$ is lower semi-continuous with respect to d. Let $u_m \in \{u_0\} + H_0^{1,s}(\Omega; \mathbb{R}^N) \subset M$ be a minimizing sequence. By Theorem 5.1, if we let

$$\varepsilon_m^2 = E(u_m) - \inf\{E(u) \; ; \; u \in M\} \, ,$$

$\delta_m = \varepsilon_m$, we can choose a new minimizing sequence (v_m) in M such that each v_m minimizes the functional

$$E_m(w) = E(w) + \varepsilon_m \, d(v_m, w) \, .$$

In particular, for each $\Omega' \subset\subset \Omega$ and any $w \in \{v_m\} + H_0^{1,s}(\Omega'; \mathbb{R}^N)$ there holds

(5.6) $$E(v_m, \Omega') \leq E(w; \Omega') + \varepsilon_m \int_{\Omega'} |\nabla w - \nabla v_m| \, dx \, .$$

Choose $x_0 \in \Omega$ and for $R < \frac{1}{2} \operatorname{dist}(x_0, \partial\Omega)$ choose $\varphi \in C_0^\infty(B_{2R}(x_0))$ satisfying $0 \leq \varphi \leq 1$, $\varphi \equiv 1$ on $B_R(x_0)$, $|\nabla\varphi| \leq \frac{c}{R}$ with c independent of R. Let

$$\bar{v}_m = \frac{1}{\mathcal{L}^n(B_{2R} \setminus B_R(x_0))} \int_{B_{2R} \setminus B_R(x_0)} v_m \, dx$$

denote the mean value of u over the annulus $B_{2R} \setminus B_R(x_0)$. Define

$$w = (1 - \varphi)v_m + \varphi \bar{v}_m \, .$$

Then by $(2°)$ and (5.6) we have

$$\int_{B_{2R}(x_0)} |\nabla v_m|^s \, dx \leq E(v_m; B_{2R}(x_0)) \leq$$

$$\leq E\big(w; B_{2R}(x_0)\big) + \varepsilon_m \bigg(\int_{B_{2R}(x_0)} \varphi |\nabla v_m| \, dx$$

$$+ \int_{B_{2R} \setminus B_R(x_0)} |\nabla\varphi| \, |v_m - \bar{v}_m| \, dx \bigg) \, .$$

By Poincaré's inequality

$$\int_{B_{2R} \setminus B_R(x_0)} |v_m - \bar{v}_m|^q \, dx \leq cR^q \int_{B_{2R} \setminus B_R(x_0)} |\nabla v_m|^q \, dx, \quad 1 \leq q < \infty \, ;$$

see Theorem A.10 of the appendix. Hence, in particular, the last term is $\leq$ $\varepsilon_m \int_{B_{2R}(x_0)} |\nabla v_m| \, dx$.

By choice of w and condition $(2°)$, moreover, we may estimate

$$E\big(w; B_{2R}(x_0)\big) \le C \int_{B_{2R}(x_0)} \big(1 + |w|^s + |\nabla w|^s\big)\, dx$$

$$\le C \int_{B_{2R}(x_0)} \big(1 + |v_m|^s + \varphi^s |v_m - \bar{v}_m|^s\big)\, dx$$

$$+ C \int_{B_{2R} \setminus B_R(x_0)} \big(|\nabla v_m|^s + |\nabla \varphi|^s |v_m - \bar{v}_m|^s\big)\, dx$$

$$\le C \int_{B_{2R}(x_0)} \big(1 + |v_m|^s\big)\, dx + C \int_{B_{2R} \setminus B_R(x_0)} |\nabla v_m|^s\, dx\ .$$

Note that in the last line we have again used Poincaré's inequality. Moreover, we have estimated

$$\int_{B_{2R}(x_0)} |\bar{v}_m|^s\, dx \le C R^{n(1-s)} \left(\int_{B_{2R}(x_0)} v_m\, dx \right)^s$$

$$\le C \int_{B_{2R}(x_0)} |v_m|^s\, dx$$

by Hölder's inequality. Hence with a uniform constant C_0 there holds

$$\int_{B_R(x_0)} |\nabla v_m|^s\, dx \le \int_{B_{2R}(x_0)} |\nabla v_m|^s\, dx \le$$

$$\le C \int_{B_{2R}(x_0)} \big(1 + |v_m|^s + |\nabla v_m|\big)\, dx$$

$$+ C_0 \int_{B_{2R} \setminus B_R(x_0)} |\nabla v_m|^s\, dx\ .$$

Now add C_0 times the left hand side to this inequality to "fill in" the annulus on the right. (This idea in a related context first appears in Widman [1], Hildebrandt-Widman[1].) Dividing by $C_0 + 1$ we then obtain that

$$\int_{B_R(x_0)} |\nabla v_m|^s\, dx \le C \int_{B_{2R}(x_0)} \big(1 + |v_m|^s + |\nabla v_m|\big)\, dx$$

$$+ \theta \int_{B_{2R}(x_0)} |\nabla v_m|^s\, dx$$

with constants C and $\theta = \frac{C_0}{C_0+1} < 1$ independent of x_0, R, and m.

Note that by Sobolev's embedding theorem $|v_m|^s \in L^{\frac{n}{n-s}}(\Omega)$. Uniform local boundedness of (v_m) in $H^{1,q}$ for some $q > s$ hence follows from the next lemma, due to Giaquinta-Modica[1], if we let $g = |\nabla v_m|^s$, $h = 1 + |v_m|^s + |\nabla v_m|$, $p = \min\{\frac{n}{n-s}, s\}$.

5.6 Lemma. *Suppose Ω is a domain in $\mathbb{R}^n$ and $g \in L^1(\Omega)$, $h \in L^p(\Omega)$ for some $p > 1$. Assume that for any $x_0 \in \Omega$ and $0 < R < \frac{1}{2}\operatorname{dist}(x_0, \partial\Omega)$ there holds*

$$\fint_{B_R(x_0)} g \, dx \le b \left\{ \fint_{B_{2R}(x_0)} h \, dx \right\} + \theta \fint_{B_{2R}(x_0)} g \, dx$$

with uniform constants $\theta < 1$, b independent of x_0 and R. Then there exists $\varepsilon > 0$, depending only on b, θ, p, Ω, and n, such that $g \in L^t_{loc}(\Omega)$ for $1 < t < 1 + \varepsilon$, and for any $x_0 \in \Omega$, $0 < R < \frac{1}{2}\operatorname{dist}(x, \partial\Omega)$, and any such t there holds

$$\fint_{B_R(x_0)} g^t \, dx \le C \left\{ \fint_{B_R(x_0)} h^t \, dx + \left(\fint_{B_{2R}} g \, dx \right)^t \right\}$$

with C possibly depending also on t. Here, for $\Omega' \subset \Omega$ and $u \in \mathcal{L}^1_{loc}(\Omega)$, we let

$$\fint_{\Omega'} u \, dx = \frac{1}{\mathcal{L}^n(\Omega')} \int_{\Omega'} u \, dx$$

denote the mean value of u over Ω'.

The proof of Lemma 5.6 goes beyond the scope of this book; a reference is Giaquinta [1; Proposition V.1.1, p.122]. Regularity results for minimizers of (strictly) quasi-convex functionals have been obtained by Evans [1], Evans-Gariepy [1], and Giaquinta-Modica [2].

5.7 Note. Besides various other applications, Ekeland's variational principle has given rise to new interpretations of known results; see for instance De Figueiredo [1] or Mawhin-Willem [1; Chapter 4.1] for expositions of these developments.

The idea of choosing special minimizing sequences to ensure convergence towards a minimizer already appears in the work of Hilbert [1] and Lebesgue [1]. In their solution of Dirichlet's problem they use barriers, respectively a variant of the Courant-Lebesgue Lemma 2.8 mentioned earlier, to ensure the equicontinuity and hence compactness of a suitably constructed minimizing sequence for Dirichlet's integral. (The compactness criterion for families of continuous functions on a compact domain was known from an earlier – though unsuccessful – attempt at solving Dirichlet's problem by Arzéla [1] in 1897.)

6. Duality

Let V be a Banach space and suppose $G: V \to \mathbb{R} \cup +\infty$ is lower semi-continuous and convex. Geometrically, convexity of G is equivalent to convexity of the epigraph of G

$$\operatorname{epi}(G) = \left\{ (v, \beta) \in V \times \mathbb{R} \; ; \; \beta \ge G(v) \right\} ,$$

while lower semi-continuity is equivalent to the closedness of $\operatorname{epi}(G)$. By the Hahn-Banach separation theorem any closed convex set can be represented as the intersection of the closed half-spaces which contain it, bounded by support hyperplanes. Hence, for any lower semi-continuous, convex $G\colon V \to \mathbb{R} \cup +\infty$ there exists a set L_G of affine maps such that

$$G(v) = \sup \{l(v) \; ; \; l \in L_G\} \,,$$

see for instance Ekeland-Temam [1; Proposition I.3.1], and at any $v \in V$ where G is locally bounded there is a "support function" $l_v \in L_G$ such that $l_v(v) = G(v)$. The set of slopes of support functions

$$\partial G(v) = \{Dl \; ; \; l \in L_G \, , \; l(v) = G(v)\}$$

is called the subdifferential of G at v.

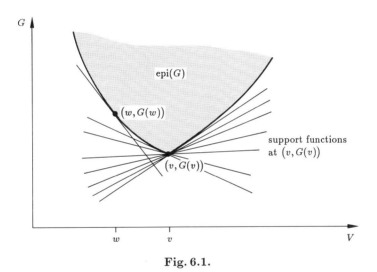

Fig. 6.1.

6.1 Lemma. *Suppose $G\colon V \to \mathbb{R} \cup +\infty$ is lower semi-continuous and convex. If G is (Gâteaux) differentiable at v, then $\partial G(v) = \{DG(v)\}$. Conversely, if G is locally bounded near v and if $\partial G(v) = \{v^*\}$ is single-valued, then G is Gâteaux differentiable at v with $D_w G(v) = \langle w, v^* \rangle$ for all $w \in V$.*

(See for instance Ekeland-Temam [1; Proposition I.5.3] for a proof.)

6.2 The Legendre-Fenchel transform. For a function $G\colon V \to \mathbb{R} \cup +\infty$, $G \not\equiv +\infty$, not necessarily convex, the function $G^*\colon V^* \to \mathbb{R} \cup +\infty$ given by

$$G^*(v^*) = \sup \{\langle v, v^* \rangle - G(v) \; ; \; v \in V\}$$

defines the Legendre-Fenchel transform of G.

Note that G^* is the pointwise supremum of affine maps, hence G^* is lower semi-continous and convex; moreover, the affine function

$$l^*(v^*) = \langle v, v^* \rangle - G(v)$$

is a support function of G^* at v^* if and only if

(6.1) $G^*(v^*) + G(v) = \langle v, v^* \rangle$.

The situation becomes symmetric in G and G^* if we also introduce $G^{**} = (G^*)^*|_V$. Note that the Legendre-Fenchel transform reverses order, that is, $G \leq \tilde{G}$ implies $G^* \geq \tilde{G}^*$. Hence, applying the transform twice preserves order: $G \leq \tilde{G}$ implies $G^{**} \leq \tilde{G}^{**}$. Moreover, for affine functions we have $l^{**} = l$, while in general

$$G^{**}(v) = \sup_{v^* \in V^*} \left\{ \langle v, v^* \rangle - \sup_{w \in V} \left\{ \langle w, v^* \rangle - G(w) \right\} \right\} \leq G(v) .$$

(Choose $w = v$ inside $\{\ldots\}$.)

Hence G^{**} is the largest lower semi-continuous convex function $\leq G$, and $G^{**} = G$ if and only if G is lower semi-continuous and convex. Moreover, for lower semi-continuous and convex functions G, relation (6.1) becomes $G^*(v^*) + G^{**}(v) = \langle v, v^* \rangle$, and our previous discussion implies:

6.3 Lemma. *Suppose $G: V \to \mathbb{R} \cup +\infty$, $G \not\equiv +\infty$, is lower semi-continuous and convex, and let G^* be its Legendre-Fenchel transform. Then (6.1) is equivalent to either one of the relations $v \in \partial G^*(v^*)$ or $v^* \in \partial G(v)$.*

In particular, if $G \in C^1(V)$ is strictly convex, which implies that

$$\langle v - w, DG(v) - DG(w) \rangle > 0 \qquad \text{if } v \neq w ,$$

then DG is injective, G^* is finite on the range of DG, and $\partial G^*\big(DG(v)\big) = \{v\}$ for any $v \in V$.

If in addition DG is strongly monotone and coercive in the sense that for all $v, w \in V$ there holds

$$\langle v - w, DG(v) - DG(w) \rangle \geq \alpha\big(\|v - w\|\big)\|v - w\|$$

with a non-decreasing function $\alpha: [0, \infty[\to [0, \infty[$ vanishing only at 0 and such that $\alpha(r) \to \infty$ as $r \to \infty$, then $DG: V \to V^*$ is also surjective; see for instance Brezis [1; Corollary 2.4, p.31]. Moreover, ∂G^* – and hence G^* – are locally bounded near any $v^* \in V^*$. Thus, by Lemma 6.1, G^* is Gâteaux differentiable with $DG^*(v^*) = v$ for any $v^* = DG(v)$. Finally, from the estimate

$$\|(v^*) - (w^*)\| \geq \frac{\langle DG^*(v^*) - DG^*(w^*), v^* - w^* \rangle}{\|DG^*(v^*) - DG^*(w^*)\|} \geq \alpha\big(\|DG^*(v^*) - DG^*(w^*)\|\big)$$

for all $v^*, w^* \in V^*$ it follows that DG^* is continuous; that is, $G^* \in C^1(V^*)$. We conclude that for any strictly convex function $G \in C^1(V)$ such that DG is strongly monotone, the differential DG is a homeomorphism of V onto its dual V^*. (In the following we apply these results only in a finite-dimensional setting. A convenient reference in this case is Rockafellar [1; Theorem 26.5, p.258].)

Hamiltonian Systems

We now apply these concepts to the solution of Hamiltonian systems: For a Hamiltonian $H \in C^2(\mathbb{R}^{2n})$ and the standard symplectic form $\mathcal{J}$ on $\mathbb{R}^{2n} = \mathbb{R}^n \times \mathbb{R}^n$,

$$\mathcal{J} = \begin{pmatrix} 0 & -id \\ id & 0 \end{pmatrix} ,$$

where id is the identity map on $\mathbb{R}^n$, consider the ordinary differential equation

(6.2) $$\dot{x} = \mathcal{J} \, \nabla H(x) .$$

Note that by anti-symmetry of $\mathcal{J}$ we have

$$\frac{d}{dt} H\big(x(t)\big) = \nabla H(x) \cdot \dot{x}(t) = 0 ,$$

that is, $H\big(x(t)\big) \equiv const.$ along any solution of (6.2) and any "energy-surface" $H = const.$ is invariant under the flow (6.2).

6.4 Periodic solutions. One would like to understand the global structure of the set of trajectories of (6.2) and their asymptotic behavior. This is motivated of course by celestial mechanics, where questions of "stable and ramdon motion" (see Moser [3]) also seem to be of practical importance. However, with exception of the – very particular – "completely integrable" case this program is far too complex to be dealt with as a whole. Therefore, one is interested in sub-systems of the flow (6.2) such as stationary points, periodic orbits, invariant tori, or quasi-periodic solutions. While stationary points in general do not reveal too much about the system, it turns out that – C^1-generically at least – periodic orbits of (6.2) are dense on a compact energy surface $H = const.$; see Pugh-Robinson [1]. Such a result seems to have already been envisioned by Poincaré [1; Tome 1, Article 36]. For particular systems, however, such results are much harder to obtain. In fact, the question whether any *given* energy surface carries a periodic solution of (6.2) is one of the most challenging problems in the field today.

However, for Hamiltonians with particular structures certain existence results are known. The following result is due to Rabinowitz [5] and Weinstein [2], extending earlier work of Seifert [1].

6.5 Theorem.   *Suppose* $H \in C^1(\mathbb{R}^{2n})$ *is strictly convex, non-negative and coercive with* $H(0) = 0$. *Then for any* $\alpha > 0$ *there is a periodic solution* $x \in C^1(\mathbb{R} ; \mathbb{R}^{2n})$ *of (6.2) with* $H(x(t)) = \alpha$ *for all* t. *The period* T *is not specified.*

Remarks.   Seifert and Weinstein essentially approached problem (6.2) using differential geometric methods; that is, by interpreting solutions of (6.2) as geodesics in a suitable Riemannian or Finsler metric (the so-called Jacobi metric). Rabinowitz' proof of Theorem 6.5 revolutionized the study of Hamiltonian systems in the large as it introduced variational methods to this field.

Note that (6.2) is the Euler-Lagrange equation of a functional

$$(6.3) \qquad\qquad E(x) = \frac{1}{2} \int_0^1 \langle x, \mathcal{J}\dot{x} \rangle \, dt$$

on the class

$$C_\alpha = \{x \in C^1(\mathbb{R}; \mathbb{R}^{2n}) \; ; \; x(t+1) = x(t), \; \int_0^1 H(x(t)) \, dt = \alpha\}.$$

Indeed, at a critical point $x \in C_\alpha$ of E, by the Lagrange multiplier rule there exists $T \neq 0$ such that

$$\dot{x} = T\mathcal{J}\nabla H(x) \,,$$

and scaling time by a factor T we obtain a T-periodic solution of (6.2) on the energy surface $H = \alpha$.

However, the integral (6.3) is not bounded from above or below. In fact, E is a quadratic form given by an operator $x \mapsto \mathcal{J}\dot{x}$ with *infinitely* many positive *and* negative eigenvalues. Due to this complication, actually, for a long time it was considered hopeless to approach the existence problem for periodic solutions of (6.2) via the functional (6.3). Surprisingly, by methods as will be presented in Chapter II below, and by using a delicate approximation procedure, Rabinowitz was able to overcome these difficulties. His result even is somewhat more general than stated above as it applies to compact, strictly star-shaped energy hypersurfaces.

On a compact, convex energy hypersurface – as was first observed by Clarke [2], [4] – by duality methods his original proof can be considerably simplified and the problem of finding a periodic solution of (6.2) can be recast in a way such that a solution again may be sought as a minimizer of a suitable "dual" variational problem. See also Clarke-Ekeland [1]. This is the proof we now present.

Later we shall study the existence of periodic solutions of Hamiltonian systems under much more general hypotheses. In fact, Theorem 6.5 can be re-obtained as a corollary to the existence result by Hofer and Zehnder that we state in Chapter II.8. Moreover, we shall study the existence of multiple periodic orbits; see Chapter II.5.

Proof of Theorem 6.5. In a first step we reformulate the problem in a way that duality methods can be applied.

Note that by strict convexity and coerciveness of H the level surface $S_\alpha = H^{-1}(\{\alpha\})$ bounds a strictly convex neighborhood of 0 in $\mathbb{R}^{2n}$. Thus, for any ξ in the unit sphere $S^{2n-1} \subset \mathbb{R}^{2n}$ there exists a unique number $r(\xi) > 0$ such that $x = r(\xi)\xi \in S_\alpha$. By the implicit function theorem $r \in C^1(S^{2n-1})$. Replace H by the function

$$\tilde{H}(\rho\xi) = \begin{cases} \alpha \left(\frac{\rho}{r(\xi)} \right)^q, & \text{if } \rho > 0, \; \xi \in S^{2n-1} \\ 0, & \text{if } \rho = 0 \end{cases}$$

where q is a fixed number $1 < q < 2$.

Note that $\tilde{H} \in C^1(\mathbb{R}^{2n})$ and is homogeneous of degree q on half-rays from the origin. Moreover, if we let $\tilde{S}_\alpha = \{x \in \mathbb{R}^{2n} \; ; \; \tilde{H}(x) = \alpha\}$ we have $\tilde{S}_\alpha = S_\alpha$ and hence $\nabla\tilde{H}(x)$ is proportional to $\nabla H(x)$, say

$$\nabla H(x) = \lambda(x)\nabla\tilde{H}(x) \;, \quad \text{at any } x \in S_\alpha \;.$$

A periodic solution $\tilde{x}$ on $\tilde{S}_\alpha$ to (6.2) for $\tilde{H}$ after a change of parameter thus will yield a periodic solution

$$x(t) = \tilde{x}\big(s(t)\big)$$

to (6.2) on S_α for the original function H, where s solves

$$\dot{s} = \lambda\big(\tilde{x}(s)\big) \;.$$

Incidentally, this short computation shows that whether or not a level surface $H = const.$ carries a periodic solution of (6.2) is a question concerning the *surface* and the symplectic structure $\mathcal{J}$ – not the particular Hamiltonian H.

Finally, $\tilde{H}$ is strictly convex. Indeed, at any point $x = \rho\xi \in \mathbb{R}^{2n}$, letting $\beta = \tilde{H}(x) = \left(\frac{\rho}{r(\xi)} \right)^q \alpha$, we have

$$\tilde{S}_\beta = \{x \in \mathbb{R}^{2n} \; ; \; \tilde{H}(x) = \beta\} = \left(\frac{\beta}{\alpha} \right)^{1/q} S_\alpha \;.$$

Thus, the hyperplane through $(x, \beta) \in \mathbb{R}^{2n+1}$, parallel to the hyperplane spanned by

$$T_x\tilde{S}_\beta \cong T_{r(\xi)\xi}S_\alpha \subset \mathbb{R}^{2n} \times \{0\} \subset \mathbb{R}^{2n+1}$$

and the vector

$$(x, x \cdot \nabla\tilde{H}(x)) = (x, q\tilde{H}(x)) = (x, q\beta) \in \mathbb{R}^{2n+1} \;,$$

is a support hyperplane which touches the graph of $\tilde{H}$ precisely at (x, β).

Hence in the following we may assume that $H = \tilde{H}$.

Let H^* be the Legendre-Fenchel transform of H. Note that, since H is homogeneous on rays of degree $q > 1$, the function H^* is everywhere finite.

Moreover, $H^*(0) = 0$, $H^* \geq 0$. Also note that for a function H on $\mathbb{R}^{2n}$ which is homogeneous of degree $q > 1$, strict convexity implies strong monotonicity of the gradient. Hence, by the discussion following Lemma 6.3, $H^* \in C^1(\mathbb{R}^{2n})$. Finally, letting $p = \frac{q}{q-1} > 2$ be the conjugate exponent of q, we have

$$
\begin{aligned}
\frac{H^*(y)}{|y|^p} &= \sup \left\{ \left\langle \frac{x}{|y|^{p-1}}, \frac{y}{|y|} \right\rangle - \frac{H(x)}{|y|^p} \; ; \; x \in \mathbb{R}^{2n} \right\} \\
(6.4) \qquad &= \sup \left\{ \left\langle \frac{x}{|y|^{p-1}}, \frac{y}{|y|} \right\rangle - H\left(\frac{x}{|y|^{p-1}} \right) \; ; \; x \in \mathbb{R}^{2n} \right\} \\
&= H^* \left(\frac{y}{|y|} \right) \; ;
\end{aligned}
$$

that is, H^* is homogeneous on rays of degree $p > 2$.

(At this point we should remark that the components of the variable x above include both position and momentum variables. Thus, although these are certainly related, the conjugate H^* of H differs from the usual Legendre transform of H which customarily only involves the momentum variables.)

Introduce the space

$$
L_0^p = \left\{ y \in L^p\left([0,1]; \mathbb{R}^{2n}\right) \; ; \; \int_0^1 y \, dt = 0 \right\} .
$$

Now, if $x \in C^1\left([0,1]; \mathbb{R}^{2n}\right)$ is a 1-periodic solution of (6.2), the function $y = -\mathcal{J}\dot{x} \in L_0^p$ solves the system of equations

$$
\begin{aligned}
(6.5) \qquad & y = -\mathcal{J}\dot{x} \, , \\
(6.6) \qquad & y = \nabla H(x) \, .
\end{aligned}
$$

Equation (6.5) can be inverted (up to an integration constant $x_0 \in \mathbb{R}^n$) by introducing the integral operator

$$
K \colon L_0^p \to H^{1,p}\left([0,1]; \mathbb{R}^{2n}\right) , \qquad (Ky)(t) = \int_0^t \mathcal{J}y \, dt \, .
$$

By Lemma 6.3 relation (6.6) is equivalent to the relation $x = \nabla H^*(y)$. That is, system (6.5), (6.6) is equivalent to the system

$$
\begin{aligned}
(6.5') \qquad & x = Ky + x_0 \\
(6.6') \qquad & x = \nabla H^*(y)
\end{aligned}
$$

for some $x_0 \in \mathbb{R}^{2n}$. The latter can be summarized in the single equation

$$
(6.7) \qquad \int_0^1 \left(\nabla H^*(y) - Ky \right) \cdot \eta \, dt = 0 \, , \qquad \forall \eta \in L_0^p \, .
$$

Indeed, if $y \in L_0^p$ satisfies (6.7), it follows that

$$\nabla H^*(y) - Ky = const. = x_0 \in \mathbb{R}^n .$$

Hence y solves (6.5'), (6.6') for some $x \in H^{1,p}([0,1];\mathbb{R}^{2n})$. Transforming back to (6.5), (6.6), from (6.6) we see that $y \in H^{1,p} \hookrightarrow C^0$, and therefore $x \in C^1([0,1];\mathbb{R}^{2n})$ is a 1-periodic solution of (6.2). Thus, (6.2) and its weak "dual" form (6.7) are in fact equivalent.

Now we can conclude the proof of Theorem 6.5: We recognize (6.7) as the Euler-Lagrange equation of the functional E^* on L_0^p, given by

$$E^*(y) = \int_0^1 \left(H^*(y) - \frac{1}{2}\langle y, Ky \rangle \right) dt .$$

Note that by (6.4) the functional E^* is Fréchet differentiable and coercive on L_0^p; see also Theorem C.1 of the appendix. Moreover, E^* is the sum of a continuous convex and a compact quadratic term, hence weakly lower semi-continuous. Thus, by Theorem 1.2, a minimizer $y^* \in L_0^p$ of E^* exists, solving (6.7). By (6.4) the quadratic term $-\int_0^1 \langle y, Ky \rangle \, dt$ in E^* dominates near $y = 0$. Since K also possesses positive eigenvalues, $\inf E^* < 0$, and $y^* \neq 0$. By the above discussion there is a constant x_0 such that the function $x = Ky^* + x_0$ solves (6.2). Since $y^* \neq 0$, also x is non-constant; hence $H(x(t)) = \beta > 0$. But $H = \tilde{H}$ is homogeneous on rays. Thus a suitable multiple $\tilde{x}$ of x will satisfy (6.2) with $H(\tilde{x}) = \alpha$, as desired.   □

Periodic Solutions of Nonlinear Wave-Equations

As a second example we consider the problem of finding a non-constant, time-periodic solution $u = u(x,t)$, $0 \le x \le \pi$, $t \in \mathbb{R}$, of the problem

(6.8) $Au = u_{tt} - u_{xx} = -u|u|^{p-2}$ in $]0,\pi[\times\mathbb{R}$

(6.9) $u(0,\cdot) = u(\pi,\cdot) = 0$

(6.10) $u(\cdot, t+T) = u(\cdot, t)$ for all $t \in \mathbb{R}$,

where $p > 2$ and the period T are given.

6.6 Theorem. *Suppose $\frac{T}{\pi} \in \mathbb{Q}$; then there exists a non-constant T-periodic weak solution $u \in L^p([0,\pi] \times \mathbb{R})$ of problem (6.8–10).*

Remark. For simplicity, we consider only the case $T = 2\pi$; the general case $T/\pi \in \mathbb{Q}$ can be handled in a similar way. The situation, however, changes completely if T is not a rational multiple of π and whether or not Theorem 6.7 holds true in this case is an open problem which seems to call for techniques totally different from those we are going to describe.

Proof. Problem (6.8–10) can be interpreted as the Euler-Lagrange equations associated with a constrained minimization problem for the functional

$$E(u) = \frac{1}{2} \int_0^{2\pi} \int_0^{\pi} \left(|u_x|^2 - |u_t|^2 \right) \, dx \, dt$$

on the space

$$H = \left\{ u \in H_{loc}^{1,2}([0,\pi] \times \mathbb{R}) \; ; \; u \text{ satisfies } (6.9), (6.10) \right\} ,$$

endowed with the $H^{1,2}$-norm on $\Omega = [0,\pi] \times [0,2\pi]$, subject to the constraint

$$\|u\|_{L^p(\Omega)} = 1 .$$

However, E is unbounded on this set. Moreover, the operator $A = \partial_t^2 - \partial_x^2$ related to the second variation of E has infinitely many positive and negative eigenvalues and also possesses an infinite-dimensional kernel. Therefore – as in the case of Hamiltonian systems considered above – the direct methods do not immediately apply.

In order to convert (6.8–10) into a problem that we can handle, we write (6.8) as a system

(6.11) $$v = Au$$

(6.12) $$-v = u|u|^{p-2} = \nabla G(u) ,$$

where $G(u) = \frac{1}{p}|u|^p$. Since G is strictly convex, (6.12) may be inverted using the Legendre-Fenchel transform of G,

$$G^*(v) = \sup \left\{ uv - \frac{1}{p}|u|^p \; ; \; u \in \mathbb{R} \right\} = \frac{1}{q}|v|^q ,$$

where $1 < q < 2$ is the exponent conjugate to p, satisfying $\frac{1}{p} + \frac{1}{q} = 1$. By Lemma 6.3 then, (6.12) is equivalent to the equation

(6.13) $$u = \nabla G^*(v) = v|v|^{q-2} .$$

In order to invert (6.11) we need to collect some facts about the wave operator A. In our exposition we basically follow Brezis-Coron-Nirenberg [1]. The representation formula (6.14)–(6.16) is due to Lovicarová [1].

6.7 Estimates for the wave operator A.

For $T = 2\pi$ the spectrum $\sigma(A)$ and kernel N of A, acting on functions in $L^1(\Omega)$ satisfying (6.9), (6.10), can be characterized as follows:

$$\sigma(A) = \{ j^2 - k^2 \; ; \; j \in \mathbb{N}, \; k \in \mathbb{N}_0 \} ,$$

$$N = \left\{ p(t+x) - p(t-x) \; ; \; p \in L_{loc}^1(\mathbb{R}), \; p(s+2\pi) = p(s) \text{ for almost all } s, \right.$$

$$\left. \int_0^{2\pi} p \, dx = 0 \right\} .$$

The last condition appearing in the definition of N is a normalization condition. N is closed in $L^1(\Omega)$; moreover, given $f \in L^1(\Omega)$ such that $\int_\Omega f \varphi \, dx \, dt = 0$

for all $\varphi \in N \cap L^\infty(\Omega)$, there exists a unique function $u \in C(\overline{\Omega})$, satisfying (6.9), (6.10), such that $Au = f$ and $\int_\Omega u\varphi\, dx\, dt = 0$ for all $\varphi \in N$. In fact, u is given explicitly as follows:

$$u(x,t) = \psi(x,t) + \big(p(t+x) - p(t-x)\big),$$

where ψ is constructed from a 2π-periodic extension of f to $[0, \pi] \times \mathbb{R}$ using the fundamental solution of the wave operator; that is,

$$(6.14) \qquad \psi(x,t) = -\frac{1}{2} \int_x^\pi \left(\int_{t-(\xi-x)}^{t+(\xi-x)} f(\xi,\tau)\, d\tau \right) d\xi + c\frac{(\pi - x)}{\pi},$$

with

$$(6.15) \qquad c = \frac{1}{2} \int_0^\pi \left(\int_{t-\xi}^{t+\xi} f(\xi,\tau)\, d\tau \right) d\xi.$$

Note that c is constant; here, the fact that f is orthogonal to N is used. The choice of c now guarantees that u satisfies the boundary condition (6.9); moreover, periodicity of f implies (6.10). Finally, choosing

$$(6.16) \qquad p(s) = \frac{1}{2\pi} \int_0^\pi \big[\psi(\xi, s - \xi) - \psi(\xi, s + \xi) \big]\, d\xi$$

ensures that u is L^2-orthogonal to N, as desired.

Formulas (6.14–16) determine an operator $K = A^{-1}$ from the weak orthogonal complement of N

$$N^\perp = \left\{ f \in L^1(\Omega) \,;\, \int_\Omega f\varphi\, dx\, dt = 0 \qquad \text{for all } \varphi \in N \cap L^\infty(\Omega) \right\}$$

into $C(\overline{\Omega})$ satisfying the condition

$$(6.17) \qquad \|Kf\|_{L^\infty} \le c\,\|f\|_{L^1}.$$

Moreover, for $f \in N^\perp \cap L^q(\Omega)$, $q > 1$, we have $Kf \in C^\alpha(\overline{\Omega})$, with $\alpha = 1 - 1/q > 0$ and

$$(6.18) \qquad \|Kf\|_{C^\alpha} \le c\,\|f\|_{L^q};$$

in particular, K is a compact, selfadjoint linear operator of $N^\perp \cap L^2(\Omega)$ into itself, with eigenvalues $1/(j^2 - k^2)$, $j \in \mathbb{N}$, $k \in \mathbb{N}_0$, $j \ne k$. (6.17) and (6.18) are easy consequences of (6.14)–(6.16) and Hölder's inequality.

Proof of Theorem 6.6. Fix $q = \frac{p}{p-1}$ and let $V = N^\perp \cap L^q(\Omega)$, endowed with the L^q-norm. By (6.18) the operator $K: V \to L^p(\Omega)$ is compact. Define

$$E^*(v) = \frac{1}{2} \int_\Omega (Kv)\, v\, dx\, dt;$$

clearly $E^* \in C^1(V)$. Moreover, since K is compact, it follows that E^* is weakly lower semi-continuous. Restrict E^* to the unit sphere

$$M = \{v \in V \; ; \; \|v\|_{L^q} = 1\}$$

in L^q and consider a minimizing sequence (v_m) for E^* in M. We may assume that $v_m \rightharpoonup v^*$ weakly in L^q, whence by weak lower semi-continuity

$$(6.19) \qquad E^*(v^*) \leq \liminf_{m \to \infty} E^*(v_m) = \inf \{E^*(v) \; ; \; v \in M\} < 0 \; .$$

(To verify the last inequality recall that K possesses also negative eigenvalues.)

In particular, $v^* \neq 0$ and $v^*/\|v^*\|_{L^q} \in M$. But then, since $E^*(\rho v) = \rho^2 E^*(v)$ for all v, by (6.19) we must have $\|v^*\|_{L^q} = 1$, and $v^* \in M$ minimizes E^* on M.

By the Lagrange multiplier rule, v^* satisfies the equation

$$(6.20) \qquad \int_\Omega \left(Kv^* + \mu v^* |v^*|^{q-2}\right)\varphi \, dx \, dt = 0 \; , \qquad \text{for all } \varphi \in V \; ,$$

with a Lagrange parameter $\mu \in \mathbb{R}$. Choosing $\varphi = v^*$ in (6.20) we realize that $\mu = -2E^*(v^*) > 0$. Scaling v^* suitably, we obtain a non-constant function $v \in V$, satisfying (6.20) with $\mu = 1$. But then v satisfies

$$Kv + v|v|^{q-2} \in N \cap L^p \; .$$

Letting $u = -v|v|^{q-2} \in L^p$, thus there exists $\psi \in N \cap L^p$ such that

$$u = Kv + \psi \; ,$$
$$u = -v|v|^{q-2} \; .$$

But then u is a non-constant solution of

$$Au + u|u|^{p-2} = 0$$

satisfying the boundary and periodicity conditions (6.9), (6.10), as desired. □

6.8 Notes. Actually, the solution obtained above is of class L^∞, see Brezis-Coron-Nirenberg [1; p. 672 f.].

Theorem 6.6 remains valid for a large class of semilinear equations

$$(6.21) \qquad\qquad u_{tt} - x_{xx} + g(u) = 0$$

involving functions g sharing the qualitative behavior of a superlinear monomial $g(u) = u|u|^{p-2}$; see Rabinowitz [6], Brezis-Coron-Nirenberg [1]. It is not even in general necessary that g is monotone (Coron [1]) – although clearly the proof given above cannot be extended to such a case.

In case g is smooth (C^∞) and strictly increasing, it was shown by Rabino-witz [6] and Brezis-Nirenberg [1] that any bounded solution u to (6.21), with boundary conditions (6.9), (6.10), is of class C^∞.

Further aspects of problem (6.8)–(6.10) have been studied by Salvatore [1] and Tanaka [1]. See Chapter II, Remark 7.3 and Notes 9.6 for references.

Other applications of duality methods to semilinear wave equations and related problems have been given by Willem [1], [2].

Chapter II

Minimax Methods

In the preceding chapter we have seen that (weak sequential) lower semi-continuity and (weak sequential) compactness of the sub-level sets of a functional E on a Banach space V suffice to guarantee the existence of a minimizer of E.

To prove the existence of saddle points we will now strengthen the regularity hypothesis on E and in general require E to be of class $C^1(V)$, that is continuously Fréchet differentiable. In this case, the notion of critical point is defined and it makes sense to classify such points as relative minima or saddle points as we did in the introduction to Chapter I.

Moreover, we will impose a certain compactness assumption on E, to be stated in Section 2. First, however, we recall a classical result in finite dimensions.

1. The Finite-Dimensional Case

In the finite dimensional case, the existence of saddle points can be obtained for instance as follows (see for instance Courant [1; p.223 ff.]):

1.1 Theorem. *Suppose $E \in C^{1,1}(\mathbb{R}^n)$ is coercive and suppose that E possesses two distinct strict relative minima x_1 and x_2. Then E possesses a third critical point x_3 which is not a relative minimizer of E and hence distinct from x_1, x_2, characterized by the minimax principle*

$$E(x_3) = \inf_{p \in P} \max_{x \in p} E(x) =: \beta \ ,$$

where

$$P = \{p \subset \mathbb{R}^n \ ; \ x_1, x_2 \in p, \ p \text{ is compact and connected}\}$$

is the class of "paths" connecting x_1 and x_2.

Proof. Let (p_m) be a minimizing sequence in P

$$\max_{x \in p_m} E(x) \to \beta \qquad (m \to \infty) \ .$$

Since E is coercive, the sets p_m are uniformly bounded. Hence

$$p = \bigcap_{m \in \mathbb{N}} \overline{\bigcup_{l \geq m} p_l} \ ,$$

the set of accumulation points of (p_m), is the intersection of a decreasing sequence of compact and connected sets, hence compact and connected. Moreover, by construction $x_1, x_2 \in p_m$ for every m. Hence $x_{1,2} \in p$, and $p \in P$. Thus

$$\max_{x \in p} E(x) \geq \inf_{p' \in P} \max_{x \in p'} E(x) = \beta \ .$$

By continuity, on the other hand,

$$\max_{x \in p} E(x) \leq \limsup_{m \to \infty} (\max_{x \in p_m} E(x)) = \beta \ ,$$

and $\max_{x \in p} E(x) = \beta$.

Remark that in particular, since $x_{1,2}$ are strict relative minima joined by p, it now also follows that $\beta > \max\{E(x_1), E(x_2)\}$.

To see that there is a critical point $x_3 \in p$ such that $E(x_3) = \beta$, we argue indirectly: Note that by continuity of E and compactness of p the set $K = \{x \in p \ ; \ E(x) = \beta\}$ is compact. Suppose $DE(x) \neq 0$ for any $x \in K$. Then there is a uniform number $\delta > 0$ such that $|DE(x)| \geq \delta$ for all $x \in K$.

By continuity, there exists a neighborhood

$$U_\varepsilon = \{x \in \mathbb{R}^n \ ; \ \exists y \in K : |x - y| < \varepsilon\}$$

of K such that $|DE(x)| \geq \delta/2$ in U_ε. We may assume $x_1, x_2 \notin U_\varepsilon$. Let η be a continuous cut-off function with support in U_ε such that $0 \leq \eta \leq 1$ and $\eta \equiv 1$ in a neighborhood of K. Let $\nabla E(x)$ denote the gradient of E at x, characterized by the condition

$$\nabla E(x) \cdot v = DE(x)v \ , \text{ for all } v \in \mathbb{R}^n \ .$$

Define a continuous map $\Phi \colon \mathbb{R}^n \times \mathbb{R} \to \mathbb{R}^n$ by letting

$$\Phi(x, t) = x - t\eta(x) \, \nabla E(x) \ .$$

Note that Φ is continuously differentiable in t and

$$\frac{d}{dt} E\left(\Phi(x, t)\right)\Big|_{t=0} = - <\eta(x)\nabla E(x), DE(x)> = -\eta(x)|\nabla E(x)|^2.$$

Moreover, $|\nabla E(x)|^2 \geq \frac{\delta^2}{4} > 0$ on $\text{supp}(\eta) \subset U_\varepsilon$. By continuity then, there exists $T > 0$ such that

$$\frac{d}{dt} E\left(\Phi(x, t)\right) \leq -\frac{\eta(x)}{2}|\nabla E(x)|^2$$

for all $t \in [0, T]$. Thus if we choose

$$p_T = \{\Phi(x, T) \ ; \ x \in p\} \ ,$$

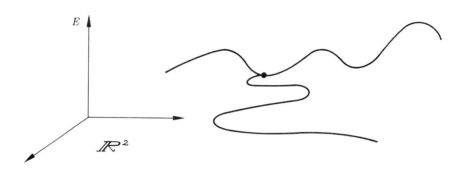

Fig. 1.1. A mountain pass in the E-landscape

for any point $\Phi(x, T) \in p_T$ we compute that

$$E\left(\Phi(x, T)\right) = E(x) + \int_0^T \frac{d}{dt} E\left(\Phi(x, t)\right) dt$$

$$\leq E(x) - \frac{T}{2}\eta(x) \left|\nabla E(x)\right|^2 ,$$

and the latter is either $\leq E(x) < \beta$, if $x \notin K$, or $\leq \beta - \frac{T}{2}\delta^2 < \beta$, if $x \in K$. Hence

$$\max_{x \in p_T} E(x) < \beta .$$

But by continuity of Φ it follows that p_T is compact and connected, while by choice of U_ε and η also $x_i = \Phi(x_i, T) \in p_T$, $i = 1, 2$. Hence $p_T \in P$, contradicting the definition of β.

Finally, if all critical points u of E in p with $E(u) = \beta$ were relative minima, the set $\tilde{K}$ of such points would be open in p and (by continuity of E and DE) also closed . Moreover, by the preceding argument $\tilde{K} \neq \emptyset$. But p is connected. Thus $p = \tilde{K}$, contradicting the fact that $E(x_0), E(x_1) < \beta$. This concludes the proof. ☐

1.2 Interpretation. It is useful to think of $E(x)$ as measuring the elevation at a point x in a landscape. Our two minima x_1, x_2 then correspond to two villages at the deepest points of two valleys, separated from each other by a mountain ridge. If now we walk along a path p from x_1 to x_2 with the property that the maximal elevation $E(x)$ at points x on p is minimal among all such paths we will cross the ridge at a mountain pass x_3 which is a saddle point of E. Because of this geometric interpretation Theorem 1.2 is sometimes called the finite-dimensional "mountain pass theorem".

2. The Palais-Smale Condition

From the experience in the preceding section we expect a functional to possess critical points of saddle type whenever the set of points with energy less than a certain value is disconnected or has a non-trivial topology. However, even in the finite-dimensional setting of Theorem 1.1 and with suitable assumptions about the topology of the sub-level sets of E, saddle points in general need not exist unless a certain compactness property holds. This is illustrated by the following simple example.

Example. Let $E \in C^1(\mathbb{R}^2)$, $E(x,y) = exp(-x) - y^2$. $E_0 := \{(x,y)|E(x,y) < 0\}$ is disconnected, while there is no "mountain pass" of minimal height 0.

Note, however, that there is a sequence of paths p_m, $p_m(t) = (m,t)$, $-1 \le t \le 1$, connecting the two components of E_0, such that E achieves its maximum on p_m at points $z_m = (m,0)$, satisfying $E(z_m) \to 0$, $DE(z_m) \to 0$ as $m \to \infty$. The latter observation will turn out to hold true in general, see for instance Remark 3.5.($3°$) or Mawhin-Willem [1; Theorem 4.3]. □

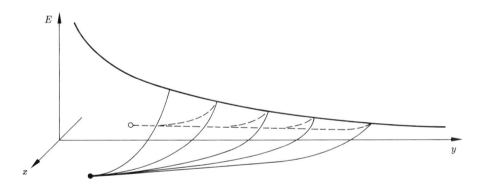

Fig. 2.1. Searching in vain for an optimal mountain pass

As we have seen, one way of inducing the necessary compactness in the finite-dimensional case is by requiring E to be coercive, which generalizes to the condition of bounded compactness in the infinite-dimensional case.

However, as remarked earlier, in infinite dimensions the requirements of bounded compactness and our regularity assumption $E \in C^1(V)$ are incompatible. Moreover, we would like to apply minimax methods to functionals which in general are neither bounded from above nor below. Thus, any of the former conditions on E is too restrictive. Instead, we will require the so-called *Palais-Smale* condition to be satisfied by E. Originally, in the work of Palais and Smale this assumption is stated as follows:

(C)   If S is a subset of V on which $|E|$ is bounded but on which $\|DE\|$ is not bounded away from zero, then there is a critical point in the closure of S.

(See Palais [1], [2], Smale [2], Palais-Smale [1].)

We will replace condition (C) by a slightly stronger condition which is more easy to work with. It is convenient to introduce the following concept.

Definition. *A sequence* (u_m) *in* V *is a Palais-Smale sequence for* E *if* $|E(u_m)| \leq c$, *uniformly in* m, *while* $\|DE(u_m)\| \to 0$ *strongly, as* $m \to \infty$.

In terms of this definition our compactness condition may be phrased as follows.

(P.-S.)   Any Palais-Smale sequence has a (strongly) convergent subsequence.

Condition (P.-S.) implies condition (C): From any set S as in condition (C), we may extract a Palais-Smale sequence. However, the functional $E \equiv 0$ satisfies (C) but in general will not satisfy (P.-S.).

Note that (P.-S.) implies that any set of critical points of uniformly bounded energy is relatively compact, see Lemma 2.3.($1°$). In fact, if we were to strengthen condition (C) by this requirement, this new condition would be equivalent to (P.-S.).

In finite dimensisons, a large class of functionals satisfying (P.-S.) can be characterized as follows:

2.1 Proposition. *Suppose* $E \in C^1(\mathbb{R}^n)$ *and assume the function* $\|DE\| + |E|: \mathbb{R}^n \to \mathbb{R}$ *is coercive. Then (P.-S.) holds for* E.

Proof. If $\|DE\| + |E|$ is coercive, clearly a Palais-Smale sequence will be bounded, hence will contain a convergent subsequence by the Bolzano-Weierstrass theorem.   □

Example. Suppose $E: \mathbb{R}^n \to \mathbb{R}$ is a quadratic polynomial

$$E(x) = \sum_{i,j=1}^{n} a_{ij}\, x_i\, x_j \; + \; \sum_{i=1}^{n} b_i\, x_i \; + \; c$$

in $x = (x_1, \ldots, x_n) \in \mathbb{R}^n$, which is non-degenerate in the sense that $D^2 E(x) = (a_{ij})_{1 \leq i,j \leq n}$ induces an invertible linear map $\mathbb{R}^n \to \mathbb{R}^n$. Then E satisfies (P.-S.).

We may ask whether a similar non-degeneracy condition in general will guarantee that a polynomial map satisfies (P.-S.). Suppose E is a polynomial of degree m in $x = (x_1, \ldots, x_n) \in \mathbb{R}^n$:

$$E(x) = \sum_{|\alpha| \le m} a_\alpha x^\alpha \ ,$$

where $\alpha = (\alpha_1, \ldots, \alpha_n)$, $x^\alpha = x_1^{\alpha_1} \cdots x_n^{\alpha_n}$, $|\alpha| = \alpha_1 + \ldots + \alpha_n$. Suppose $D^2 E(x) \colon \mathbb{R}^n \times \mathbb{R}^n \to \mathbb{R}$ is non-degenerate for any $x \in \mathbb{R}^n$. Does E satisfy (P.-S.)? (The answer seems to be unknown. In fact, this question seems related to another puzzling problem in algebraic geometry, see for instance Bass-Connell-Wright [1].)

In general, we can say the following:

2.2 Proposition. *Suppose $E \in C^1(V)$ on a Banach space V. Suppose that E has the property that*
(1°) Any Palais-Smale sequence for E is bounded in V.
Further assume that
(2°) for any $u \in V$ we can decompose

$$DE(u) = L + K(u) \ ,$$

where $L \colon V \to V^*$ is a fixed boundedly invertible linear map and the operator K maps bounded sets in V to relatively compact sets in V^*. Then E satisfies (P.-S.).

Proof. Any (P.-S.)-sequence (u_m) is bounded by assumption. Moreover

$$DE(u_m) = Lu_m + K(u_m) \to 0$$

implies that

$$u_m = o(1) - L^{-1} K(u_m) \ ,$$

where $o(1) \to 0$ in B as $m \to \infty$. By boundedness of (u_m) and compactness of K the sequences $(L^{-1} K(u_m))$ – and hence (u_m) – are relatively compact.   □

The Palais-Smale condition permits to distinguish a certain family of neighborhoods of critical points of a functional E and thus offers a useful characterization of regular values of E.

For $\beta \in \mathbb{R}$, $\delta > 0$, $\rho > 0$ let

$$\begin{aligned}
E_\beta &= \{u \in V \ ; \ E(u) < \beta\}, \\
K_\beta &= \{u \in V \ ; \ E(u) = \beta, \ DE(u) = 0\} \ , \\
N_{\beta,\delta} &= \{u \in V \ ; \ |E(u) - \beta| < \delta, \ \|DE(u)\| < \delta\} \ , \\
U_{\beta,\rho} &= \bigcup_{u \in K_\beta} \{v \in V \ ; \ \|v - u\| < \rho\} \ .
\end{aligned}$$

That is, K_β is the set of critical points of E having "energy" β, $\{U_{\beta,\rho}\}_{\rho>0}$ is the family of norm-neighborhoods of K_β.

2.3 Lemma. *Suppose E satisfies (P.-S.). Then for any $\beta \in \mathbb{R}$ the following holds:*

(1°) K_β is compact.

(2°) The family $\{U_{\beta,\delta}\}_{\rho>0}$ is a fundamental system of neighborhoods of K_β.

(3°) The family $\{N_{\beta,\delta}\}_{\delta>0}$ is a fundamental system of neighborhoods of K_β.

Proof. (1°) Any sequence (u_m) in K_β by (P.-S.) has a convergent subsequence. By continuity of E and DE any accumulation point of such a sequence also lies in K_β, and K_β is compact.

(2°) Any $U_{\beta,\rho}$, $\rho > 0$, is a neighborhood of K_β. Conversely, let N be any open neighborhood of K_β. Suppose by contradiction that for $\rho_m \to 0$ there is a sequence of points $u_m \in U_{\beta,\rho_m} \setminus N$. Let $v_m \in K_\beta$ be such that $\|u_m - v_m\| \leq \rho_m$. Since K_β is compact by (1°), we may assume that $v_m \to v \in K_\beta$. But then also $u_m \to v$, and $u_m \in N$ for large m, contrary to assumption.

(3°) Similarly, each $N_{\beta,\delta}$, for $\delta > 0$, is a neighborhood of K_β. Conversely, suppose that for some neighborhood N of K_β and $\delta_m \to 0$ there is a sequence $u_m \in N_{\beta,\delta_m} \setminus N$. By (P.-S.) the sequence (u_m) accumulates at a critical point $u \in K_\beta \subset N$. The contradiction proves the lemma. $\qquad\square$

2.4 Remarks. (1°) In particular, if $K_\beta = \emptyset$ for some $\beta \in \mathbb{R}$ there exists $\delta > 0$ such that $N_{\beta,\delta} = \emptyset$; that is, the differential $DE(u)$ is uniformly bounded in norm away from 0 for all $u \in V$ with $E(u)$ close to β.

(2°) The conclusion of Lemma 2.3 remains valid at the level β under the weaker assumption that (P.-S.)-sequences (u_m) for E such that $E(u_m) \to \beta$ are relatively compact. This observation will be useful when dealing with limiting cases for (P.-S.) in Chapter III.

2.5 Cerami's variant of (P.-S.). Cerami [1], [2; Teorema (*), p.166] has proposed the following variant of (P.-S.):

> (2.1)   Any sequence (u_m) such that $|E(u_m)| \leq c$ uniformly and $\|DE(u_m)\|(1 + \|u_m\|) \to 0$ $(m \to \infty)$ has a (strongly) convergent subsequence.

Condition (2.1) is slightly weaker than (P.-S.) while the most important implications of (P.-S.) are retained; see Cerami [1],[2] or Bartolo-Benci-Fortunato [1]. However, for most purposes it suffices to use the standard (P.-S.) condition. Therefore and in order to achieve a coherent and simple exposition consistent with the bulk of the literature in the field, in the following our presentation will be based on (P.-S.) rather than (2.1).

Variants of the Palais-Smale condition for non-differentiable functionals will be discussed in a later section. (See Section 10.)

3. A General Deformation Lemma

Besides the compactness condition, the second main ingredient in the proof of
Theorem 1.1 is the (local) gradient-line deformation Φ. Following Palais [4], we
will now construct a similar deformation for a general C^1-functional in a Banach
space. The construction may be carried out in the more general setting of C^1-
functionals on complete, regular $C^{1,1}$-Banach manifolds with Finsler structures;
see Palais [4]. However for most of our purposes it suffices to consider a Banach
space as the ambient space on which a functional is defined.

Pseudo-Gradient Flows on Banach Spaces

In the following we will at first assume that $E \in C^1(V)$ is a C^1-functional on
a Banach space V. Moreover, we denote

$$\tilde{V} = \{u \in V \; ; \; DE(u) \neq 0\}$$

the set of regular points of E. We define a notion to replace the gradient of a
functional on $\mathbb{R}^n$.

3.1 Definition. *A pseudo-gradient vector field for E is a locally Lipschitz con-*
tinuous vector field $v \colon \tilde{V} \to V$ such that the conditions
(1°) $\|v(u)\| < 2\min\{\|DE(u)\|, 1\}$,
(2°) $\langle v(u), DE(u) \rangle > \min\{\|DE(u)\|, 1\}\|DE(u)\|$
hold for all $u \in \tilde{V}$.

Note that we require v to be locally Lipschitz. Hence even in a Hilbert space
the following result is somewhat remarkable.

3.2 Lemma. *Any functional $E \in C^1(V)$ admits a pseudo-gradient vector field*
$v \colon \tilde{V} \to V$.

Proof. For $u \in \tilde{V}$ choose $w = w(u)$ such that

$$(3.1) \qquad \|w\| < 2\min\{\|DE(u)\|, 1\} ,$$

$$(3.2) \qquad \langle w, DE(u) \rangle > \min\{\|DE(u)\|, 1\}\|DE(u)\| .$$

By continuity, for any $u \in \tilde{V}$ there is a neighborhood $W(u)$ such that (3.1–
2) hold (with $w = w(u)$) for all $u' \in W(u)$. Since $\tilde{V} \subset V$ is metrizable and
hence paracompact, there exists a locally finite refinement $\{W_\iota\}_{\iota \in I}$ of the cover
$\{W(u)\}_{u \in \tilde{V}}$ of $\tilde{V}$, consisting of neighborhoods $W_\iota \subset W(u_\iota)$; see for instance
Kelley [1; Corollary 5.35, p.160].
    Choose a Lipschitz continuous partition of unity $\{\varphi_\iota\}_{\iota \in I}$ subordinate to
$\{W_\iota\}_{\iota \in I}$; that is, Lipschitz continuous functions $0 \leq \varphi_\iota \leq 1$ with support in
W_ι and such that $\sum_{\iota \in I} \varphi_\iota \equiv 1$ on $\tilde{V}$. For instance, following Palais [4; p. 205
f.], we may let

$$\rho_\iota(u) = dist(u, V \setminus W_\iota) = \inf\{\|u - v\| \; ; \; v \notin W_\iota\}$$

and define

$$\varphi_\iota(u) = \frac{\rho_\iota(u)}{\sum_{\iota' \in I} \rho_{\iota'}(u)} \; .$$

Clearly, $0 \leq \varphi_\iota \leq 1$, $\varphi_\iota = 0$ outside W_ι, and $\sum_{\iota \in I} \varphi_\iota \equiv 1$. Moreover, since $\{W_\iota\}_{\iota \in I}$ is locally finite, for any $u \in \tilde{V}$ there exists a neighborhood W of u such that $W \cap W_{\iota'} \neq \emptyset$ for at most finitely many indices $\iota' \in I$, and Lipschitz continuity of φ_ι on W is immediate from Lipschitz continuity of the family $\{\rho_\iota\}_{\iota \in I}$.

Finally, we may let

$$v(u) = \sum_{\iota \in I} \varphi_\iota(u) \; w(u_\iota) \; .$$

Relations (3.1–2) being linear in w, these relations remain true for any convex linear combination, and v is a pseudo-gradient vector field for E, as required.

□

3.3 Remark. If E admits some compact group action G as symmetries, v may be constructed to be G-equivariant. In particular, if E is even, that is, $E(u) = E(-u)$, with symmetry group $\{id, -id\} \cong \mathbb{Z}_2$, we may choose

$$\tilde{v}(u) = \frac{1}{2}\left(v(u) - v(-u)\right) \; ,$$

where v is any pseudo-gradient vector field for E, to obtain a $\mathbb{Z}_2$-equivariant pseudo-gradient vector field v for E, satisfying $v(-u) = -v(u)$.

More generally, suppose G is a compact group acting on V; that is, suppose there is a group homomorphism of G onto a subgroup – indiscriminately denoted by G – of the group of linear isomorphisms of V such that the evaluation map

$$G \times V \to V \; ; \; (g, u) \mapsto gu$$

is continuous. Also suppose that G leaves E invariant

$$E(gu) = E(u), \qquad \forall(g, u) \in G \times V \; .$$

Then it suffices to let

$$\tilde{v}(u) = \int_G g^{-1} v(gu) \, dg$$

be the average of any pseudo-gradient vector field v for E with respect to an invariant Haar's measure dg on G in order to obtain a G-equivariant pseudo-gradient vector field $\tilde{v}$, satisfying

$$\tilde{v}(gu) = g\tilde{v}(u)$$

for all g and u.

We are now ready to state the main theorem in this section.

3.4 Theorem (Deformation Lemma). *Suppose $E \in C^1(V)$ satisfies (P.-S.). Let $\beta \in \mathbb{R}$, $\bar{\varepsilon} > 0$ be given and let N be any neighborhood of K_β. Then there exist a number $\varepsilon \in]0, \bar{\varepsilon}[$ and a continuous 1-parameter family of homeomorphisms $\Phi(\cdot, t)$ of V, $0 \le t < \infty$, with the properties*
($1°$) $\Phi(u, t) = u$, if $t = 0$, or $DE(u) = 0$, or $|E(u) - \beta| \ge \bar{\varepsilon}$;
($2°$) $E(\Phi(u, t))$ is non-increasing in t for any $u \in V$;
($3°$) $\Phi(E_{\beta+\varepsilon} \setminus N, 1) \subset E_{\beta-\varepsilon}$, and $\Phi(E_{\beta+\varepsilon}, 1) \subset E_{\beta-\varepsilon} \cup N$.
Moreover, $\Phi: V \times [0, \infty[\to V$ has the semi-group property; that is, $\Phi(\cdot, t) \circ \Phi(\cdot, s) = \Phi(\cdot, s + t)$ for all $s, t \ge 0$.

Proof. Lemma 2.3 permits to choose numbers $\delta, \rho > 0$ such that

$$N \supset U_{\beta, 2\rho} \supset U_{\beta, \rho} \supset N_{\beta, \delta} .$$

Let η be a Lipschitz continuous function on V such that $0 \le \eta \le 1$, $\eta \equiv 1$ outside $N_{\beta, \delta}$, $\eta \equiv 0$ in $N_{\beta, \delta/2}$. Also let φ be a Lipschitz continuous function on $\mathbb{R}$ such that $0 \le \varphi \le 1$, $\varphi(s) \equiv 0$, if $|\beta - s| \ge \min\{\bar{\varepsilon}, \delta/4\}$, $\varphi(s) \equiv 1$, if $|\beta - s| \le \min\{\bar{\varepsilon}/2, \delta/8\}$. Finally, let $v: \tilde{V} \to V$ be a pseudo-gradient vector field for E. Define

$$e(u) = \begin{cases} -\eta(u)\, \varphi(E(u))\, v(u), & \text{if } u \in \tilde{V} \\ 0, & \text{else} \end{cases} .$$

By choice of φ and η, the vector field e vanishes identically (and therefore is Lipschitz continuous) near critical points u of E. Hence e is locally Lipschitz continuous throughout V. Moreover, since $\|v\| < 2$ uniformly, also $\|e\| \le 2$ is uniformly bounded. Hence there exists a global solution $\Phi: V \times \mathbb{R} \to V$ of the initial value problem

$$\frac{\partial}{\partial t} \Phi(u, t) = e(\Phi(u, t))$$

$$\Phi(u, 0) = u .$$

Φ is continuous in u, differentiable in t and has the semi-group property $\Phi(\cdot, s) \circ \Phi(\cdot, t) = \Phi(\cdot, s + t)$, for any $s, t \in \mathbb{R}$. In particular, for any $t \in \mathbb{R}$ the map $\Phi(\cdot, t)$ is a homeomorphism of V.

Properties ($1°$) and ($2°$) are trivially satisfied by construction and the properties of v. Moreover, for $\varepsilon \le \min\{\bar{\varepsilon}/2, \delta/8\}$ and $u \in E_{\beta+\varepsilon}$, if $E(\Phi(u, 1)) \ge \beta - \varepsilon$ it follows from ($2°$) that $|E(\Phi(u, t)) - \beta| \le \varepsilon$ and hence that $\varphi(E(\Phi(u, t))) = 1$ for all $t \in [0, 1]$.

Differentiating, by the chain rule we thus obtain

$$E(\Phi(u, 1)) = E(u) + \int_0^1 \frac{d}{dt} E(\Phi(u, t))\, dt$$

(3.3)
$$< \beta + \varepsilon - \int_0^1 \eta(\Phi(u, t)) \langle v(\Phi(u, t)), DE(\Phi(u, t)) \rangle\, dt$$

$$< \beta + \varepsilon - \int_{\{t; \Phi(u, t) \notin N_{\beta, \delta}\}} \langle v(\Phi(u, t)), DE(\Phi(u, t)) \rangle\, dt$$

$$\le \beta + \varepsilon - \mathcal{L}^1\left(\{t; \Phi(u, t) \notin N_{\beta, \delta}\}\right) \cdot \delta^2 .$$

But if either $u \notin N$ or $\Phi(u,1) \notin N$, by uniform boundedness $\|e\| \leq 2$ and since $V \setminus N$ and $N_{\beta,\delta}$ are separated by the "annulus" $U_{\beta,2\rho} \setminus U_{\beta,\rho}$ of width ρ, certainly

$$(3.4) \qquad \mathcal{L}^1\left(\{t \; ; \; \Phi(u,t) \notin N_{\beta,\delta}\}\right) \geq \frac{\rho}{2} \; .$$

Hence, if we choose $\varepsilon \leq \frac{\delta^2 \rho}{4}$, estimate (3.3) gives

$$E\left(\Phi(u,1)\right) \leq \beta + \varepsilon - \frac{\rho\delta^2}{2} \leq \beta - \varepsilon \; ,$$

and $(3°)$ follows. □

3.5 Remarks.
$(1°)$ Since the deformation $\Phi \colon V \times [0,\infty[\to B$ is obtained by integrating a suitably truncated pseudo-gradient vector field, Φ will be called a (local) pseudo-gradient flow.
$(2°)$ If $K_\beta = \emptyset$, we may choose $N = \emptyset$ and hence obtain a uniform reduction of energy near β in this case.
$(3°)$ The Palais-Smale condition was only used to obtain estimate (3.4). For this it is enough to assume that (P.-S.) holds at the level β, see Remark 2.4.$(2°)$. In particular, if $N = K_\beta = \emptyset$ the conclusion of Theorem 3.4 remains valid if condition (P.-S.)is replaced by the assumption that $N_{\beta,\delta} = \emptyset$ for some $\delta > 0$.
$(4°)$ If E is invariant under a compact group action G, as in Remark 3.3 we can achieve that Φ is G-equivariant in the sense that there holds

$$\Phi(gu,t) = g\,\Phi(u,t) \qquad \text{for all } u \in V, \; g \in G, \; t \geq 0 \; .$$

3.6 Comparison with gradient flows.
It may be of interest to consider the special case of a C^2-functional E on a real Hilbert space H with scalar product $(\cdot,\cdot)$ and induced norm $\|\cdot\|$. In this case, a gradient vector field $\nabla E \colon H \to H$ is defined as in the finite dimensional case by letting $\nabla E(u)$ at any $u \in H$ be the unique vector in H such that $\left(\nabla E(u), v\right) = DE(u)v$ for all $v \in H$, equivalently characterized by

$$(3.5) \qquad \|\nabla E(u)\| = \|DE(u)\|, \quad \langle \nabla E(u), DE(u) \rangle = \|DE(u)\|^2 \; .$$

Moreover, since

$$\|\nabla E(u) - \nabla E(v)\| = \|DE(u) - DE(v)\| \; ,$$

if $E \in C^2$, ∇E is of class C^1 and defines a local gradient flow Φ by letting

$$\frac{\partial}{\partial t}\Phi(u,t) = -\nabla E\left(\Phi(u,t)\right)$$

$$\Phi(u,0) = u \; .$$

To interpret Φ we identify E with its graph $\mathcal{G}(E) = \{(u, E(u)) \in H \times \mathbb{R}\}$. Then in the picture of Section 1.2 the flow-lines of Φ become paths of steepest descent, and the rest points of Φ are precisely the critical points of E where $\mathcal{G}(E)$ has a horizontal tangent plane.

In a Banach space V as ambient space, note that in general by (3.5) a gradient vector need not be uniquely determined unless V is uniformly locally convex. Moreover, also in this case, the duality map $j\colon V^* \to V$, which maps $v \in V^*$ to $w = j(v) \in V$ satisfying $\langle w, v \rangle = \|v\|^2 = \|w\|^2$, in general is only uniformly continuous on bounded sets but may fail to be Lipschitz.

Fortunately, it is not at all necessary to deform along lines of "steepest" descent to obtain existence results for saddle points, "steep enough" suffices. The notions of pseudo-gradient vector field and pseudo-gradient flow allow for the necessary flexibility.

Pseudo-Gradient Flows on Manifolds

The above constructions of pseudo-gradient vector fields and pseudo-gradient flows can easily be conveyed to C^1-functionals on complete $C^{1,1}$-Finsler manifolds. We basically follow Palais [4].

3.7 Finsler manifolds. Let F be a Banach space bundle over a space M and let $\| \cdot \|$ be a continuous real valued function on F such that the restriction $\| \cdot \|_u$ of $\| \cdot \|$ to each fiber F_u is an admissible norm for F_u. If we trivialize F in a neighborhood of a point $u_0 \in M$, using F_{u_0} as the standard fiber, then for each u near u_0 the norm $\| \cdot \|_u$ becomes a norm on F_{u_0}. We say $\| \cdot \|$ is a Finsler structure for the bundle F if for any $\varepsilon > 0$, each $u_0 \in M$, and each such trivialization in an atlas defining the bundle structure of E

$$\sup_{v \in F_{u_0} \setminus \{0\}} \frac{\|v\|_u}{\|v\|_{u_o}} \; , \quad \sup_{v \in F \setminus \{0\}} \frac{\|v\|_{u_0}}{\|v\|_u} < 1 + \varepsilon \; ,$$

if u is sufficiently near u_0.

Recall that a topological space M is regular if for each point $x \in M$ and any neighborhood U of u there is a closed neighborhood $\tilde{U}$ of u such that $\tilde{U} \subset U$. Now, a Finsler manifold of class C^r, $r \geq 1$, is a regular C^r-Banach manifold M, modelled on a Banach space V, together with a Finsler structure $\| \cdot \|$ on the tangent bundle TM. Then also the co-tangent space T^*M carries a natural Finsler structure, indiscriminately denoted by $\| \cdot \|$, characterized by letting

$$\|v^*\|_u = \sup\{|\langle v, v^* \rangle| \; ; \; v \in T_u M, \; \|v\|_u \leq 1\}$$

for any $v^* \in T_u^* M$. Finally, $\| \cdot \|$ induces a metric

$$(3.5) \qquad d(u, v) = \inf_p \int_0^1 \left\| \frac{d}{dt} p(t) \right\|_{p(t)} dt \; ,$$

where the infimum is taken over all C^1-paths $p\colon [0, 1] \to M$ joining $p(0) = u$, $p(1) = v$; see Palais [4; p. 208 ff.]. For reasons which will become clear in a moment, we will assume that M is complete with respect to the metric d.

As an example, we may consider any complete C^r-submanifold M of a Banach space V, with $T_u M$ carrying the norm induced by the inclusion $T_u M \subset T_u V \cong V$.

3.8 Definition. *Let M be a $C^{1,1}$ Finsler manifold modelled on a Banach space V, $E \in C^1(M)$, $\tilde{M} = \{u \in M \ ; \ DE(u) \neq 0\}$ the set of regular points of E. A pseudo-gradient vector field for E is a Lipschitz continuous vector field (section) $v \colon \tilde{M} \mapsto TM$ in the tangent bundle TM with the property that $v(u) \in T_u M$ and*
$(1°)$ $\|v(u)\|_u < 2 \min\{\|DE(u)\|_u, 1\}$,
$(2°)$ $\langle v(u), DE(u) \rangle > \min\{\|DE(u)\|_u, 1\} \|DE(u)\|_u$,
for all $u \in \tilde{M}$, where $\|\cdot\|_u$ denotes the norm in the tangent space $T_u M \cong V$ to M at the point u.

As in the proof of Lemma 3.2, for any $u \in \overset{\circ}{\tilde{M}}$ there exists a (constant) pseudo-gradient vector field in a suitable coordinate neighborhood (chart) $W \subset \tilde{M}$ around u. Since a $C^{1,1}$-Finsler manifold is paracompact, see Palais [4; p. 203], and a paracompact $C^{1,1}$-Banach manifold always admits locally Lipschitz partitions of unity, see Palais [4; p. 205 f.], a family of local pseudo-gradient vector fields as above may be patched together to yield a pseudo-gradient vector field $v \colon \tilde{M} \to TM$ just as in the "flat" case $M = V$. Thus we obtain (Palais [4; 3.3. p. 206]):

3.9 Lemma. *Any functional $E \in C^1(M)$ on a $C^{1,1}$-Finsler manifold M admits a pseudo-gradient vector field $v \colon \tilde{M} \to TM$.*

3.10 Remark. Again, if G is a compact group acting (smoothly) on M and if E is G-invariant, v may be constructed to be G-equivariant in the sense that

$$v\big(g(u)\big) = Dg(u)\, v(u) \ , \quad \forall u \in M, \ g \in G \ .$$

(Note that in the "flat" case $M = V$, with G acting through linear isomorphisms, the tangent map $Dg(u) \colon T_u M \cong V \to T_{g(u)} M \cong V$ may be identified with the map g itself.) Indeed, we may let

$$\tilde{v}(u) = \int_G \big(Dg(u)\big)^{-1} v\,(g(u))\ dg$$

with respect to an invariant measure dg on G.

For $\beta \in \mathbb{R}$, $\delta, \rho > 0$ define K_β, $N_{\beta,\delta}$, $U_{\beta,\delta}$ as in the flat case using the Finsler structure to define norm $\|DE(u)\|_u$ and distance (3.5).

The Palais-Smale condition can now be stated as in Section 2, and Lemma 2.3 remains true for $E \in C^1(M)$ on a $C^{1,1}$-Finsler manifold M. Then the construction in the proof of Theorem 3.4 can be conveyed to obtain a local pseudo-gradient flow $\Phi \colon D(\Phi) \subset M \times [0, \infty[\to M$ for E. Note that for Φ to be defined globally on $M \times [0, \infty[$ we also need to assume that M is complete with respect to the metric (3.5). This yields the following result.

3.11 Theorem. *Suppose M is a complete $C^{1,1}$-Finsler manifold and $E \in C^1(M)$ satisfies (P.-S.). Let $\beta \in \mathbb{R}$, $\bar{\varepsilon} > 0$ be given and let N be any neighborhood of K_β. Then there exist a number $\varepsilon \in \,]0, \bar{\varepsilon}[$ and a continuous 1-parameter family of homeomorphisms $\Phi(\cdot, t)$ of M, $0 \le t < \infty$, with the properties*
(1°) $\Phi(u, t) = u$, if $t = 0$, or $DE(u) = 0$, or $|E(u) - \beta| \ge \bar{\varepsilon}$;
(2°) $E(\Phi(u, t))$ is non-increasing in t for any $u \in M$;
(3°) $\Phi(E_{\beta+\varepsilon} \setminus N, 1) \subset E_{\beta-\varepsilon}$, and $\Phi(E_{\beta+\varepsilon}, 1) \subset E_{\beta-\varepsilon} \cup N$.
Moreover, Φ has the semi-group property $\Phi(\cdot, s) \circ \Phi(\cdot, t) = \Phi(\cdot, s + t)$, $\forall s, t \ge 0$. If M admits a compact group of symmetries G and if E is G-invariant, Φ can be constructed to be G-equivariant, that is, such that $\Phi(g(u), t) = g(\Phi(u, t))$ for all $g \in G$, $u \in M, t \ge 0$.

3.12 Remarks. (1°)It suffices to assume that (P.-S.) is satisfied at the level β, see Remark 2.4.(2°). In particular, if $N = K_\beta = \emptyset$, condition (P.-S.) may be replaced by the assumption that $N_{\beta, \delta} = \emptyset$ for some $\delta > 0$, see Remark 3.5.(3°). (2°)Completeness of M is only needed to ensure that the trajectories of the pseudo-gradient flow Φ are complete in forward direction.

4. The Minimax Principle

The "deformation lemma" Theorem 3.4 is a powerful tool for proving the existence of saddle points of functionals under suitable hypotheses on the topology of the manifold M or the sub-level sets E_α of E.

We proceed to state a very general result in this direction: the generalized minimax principle of Palais [4; p.210]. Later in this section we sketch some applications of this useful result.

4.1 Definition. *Let $\Phi: M \times [0, \infty[\to M$ be a semi-flow on a manifold M. A family $\mathcal{F}$ of subsets of M is called (positively) Φ-invariant if and only if $\Phi(F, t) \in \mathcal{F}$ for all $F \in \mathcal{F}$, $t \ge 0$.*

4.2 Theorem. *Suppose M is a complete Finsler manifold of class $C^{1,1}$ and $E \in C^1(M)$ satisfies (P.-S.). Also suppose $\mathcal{F} \subset \mathcal{P}(M)$ is a collection of sets which is invariant with respect to any continuous semi-flow $\Phi: M \times [0, \infty[\to M$ such that $\Phi(\cdot, 0) = id|_M$, $\Phi(\cdot, t)$ is a homeomorphism of M for any $t \ge 0$, and $E(\Phi(u, t))$ is non-increasing in t for any $u \in M$. Then, if*

$$\beta = \inf_{F \in \mathcal{F}} \sup_{u \in F} E(u)$$

is finite, β is a critical value of E.

Proof. Assume by contradiction that $\beta \in \mathbb{R}$ is a regular value of E. Choose $\bar{\varepsilon} = 1$, $N = \emptyset$ and let $\varepsilon > 0$, $\Phi: M \times [0, \infty[\to M$ be determined according to Theorem 3.11. By definition of β there exists $F \in \mathcal{F}$ such that

$$\sup_{u \in F} E(u) < \beta + \varepsilon \; ;$$

that is, $F \subset E_{\beta+\varepsilon}$. By property 3.11.($3°$) of Φ and invariance of $\mathcal{F}$, if we let $F_1 = \Phi(F, 1)$, we have $F_1 \in \mathcal{F}$ and $F_1 \subset E_{\beta-\varepsilon}$; that is,

$$\sup_{u \in F_1} E(u) \leq \beta - \varepsilon \;,$$

which contradicts the definition of β. □

Of course, it would be sufficient to assume condition (P.-S.) is satisfied at the level β and that $\mathcal{F}$ is forwardly invariant only with respect to the pseudo-gradient flow.

In the above form, the minimax principle can be most easily applied if E is a functional on a manifold M with a rich topology. But also in the "flat" case $E \in C^1(V)$, V a Banach space, such topological structure may be hidden in the sub-level sets E_γ of E.

4.3 Examples. (Palais, [3; p.190 f.]) Suppose M is a complete Finsler manifold of class $C^{1,1}$, and $E \in C^1(M)$ satisfies (P.-S.).
($1°$) Let $\mathcal{F} = \{M\}$. Then, if

$$\beta = \inf_{F \in \mathcal{F}} \sup_{u \in F} E(u) = \sup_{u \in M} E(u)$$

is finite, $\beta = \max_{u \in M} E(u)$ is attained at a critical point of E.
($2°$) Let $\mathcal{F} = \{\{u\}; u \in M\}$. Then if

$$\beta = \inf_{F \in \mathcal{F}} \sup_{u \in F} E(u) = \inf_{u \in M} E(u)$$

is finite, $\beta = \min_{u \in M} E(u)$ is attained at a critical point of E.
($3°$) Let X be any topological space, and let $[X, M]$ denote the set of homotopy classes $[f]$ of continuous maps $f \colon X \to M$. For given $[f] \in [X, M]$ let

$$\mathcal{F} = \{g(X) \; ; \; g \in [f]\} \;.$$

Since $[\Phi \circ f] = [f]$ for any homeomorphism Φ of M homotopic to the identity, the family $\mathcal{F}$ is invariant under such mappings Φ. Hence if

$$\beta = \inf_{F \in \mathcal{F}} \sup_{u \in F} E(u)$$

is finite, β is a critical value.
($4°$) Let $H_k(M)$ denote the k-dimensional homology of M (with arbitrary co-efficients). Given a non-trivial

$$f \in H_k(M), \quad f \neq 0 \;,$$

denote $\mathcal{F}$ the collection of all $F \subset M$ such that f is in the image of

$$H_k(i_F)\colon H_k(F) \to H_k(M) \, ,$$

where $H_k(i_F)$ is the homomorphism induced by the inclusion $i_F\colon F \hookrightarrow M$. Then $\mathcal{F}$ is invariant under any homeomorphism $\varPhi$ homotopic to the identity, and by Theorem 4.2, if

$$\beta = \inf_{F \in \mathcal{F}} \sup_{u \in F} E(u)$$

is finite, then β is a critical value.

There is a "dual version" of $(4°)$:

$(5°)$ If H^k is any k-dimensional cohomology functor, f a nontrivial element

$$f \in H^k(M), \qquad f \neq 0 \, ,$$

let $\mathcal{F}$ denote the family of subsets $F \subset M$ such that f is not annihilated by the restriction map

$$H^k(i_F)\colon H^k(M) \to H^k(F) \, .$$

Then, if

$$\beta = \inf_{F \in \mathcal{F}} \sup_{u \in F} E(u)$$

is finite, β is a critical value.

In the more restricted setting of a functional $E \in C^1(V)$, similar results are valid if M is replaced by any sub-level set E_γ, $\gamma \in \mathbb{R}$. We leave it to the reader to find the analogous variants of $(3°)$, $(4°)$, and $(5°)$. See also Ghoussoub [1].

Closed Geodesics on Spheres

Still a different construction of a flow-invariant family is at the basis of the next famous result: We assume the notion of geodesic to be familiar from differential geometry. Otherwise the reader may regard (4.2) below as a definition.

4.4 Theorem (Birkhoff [1]). On any compact surface S in $\mathbb{R}^3$ which is C^3-diffeomorphic to the standard sphere, there exists a non-constant closed geodesic.

Proof. Denote $\dot{u} = \frac{d}{dt}u$. Define

$$H^{1,2}\left(\mathbb{R}/2\pi; S\right) = \left\{u \in H^{1,2}_{loc}\left(\mathbb{R}; \mathbb{R}^3\right) \; ; \; u(t) = u(t + 2\pi), \right.$$
$$\left. u(t) \in S \text{ for almost every } t \in \mathbb{R}\right\}$$

the space of closed curves $u\colon \mathbb{R}/2\pi \to S$ with finite energy

$$E(u) = \frac{1}{2} \int_0^{2\pi} |\dot{u}|^2 \, dt \, .$$

By Hölder's inequality

$$(4.1) \quad |u(s) - u(t)| \leq \int_s^t |\dot{u}| \, d\tau \leq \left(|t - s| \int_s^t |\dot{u}|^2 \, d\tau\right)^{1/2} \leq (2|t - s|E(u))^{1/2} \, ,$$

functions $u \in H^{1,2}(\mathbb{R}/2\pi; S)$ with $E(u) \leq \gamma$ will be uniformly Hölder continuous with Hölder exponent $1/2$ and Hölder norm bounded by $\sqrt{2\pi\gamma}$. Hence, if $S \in C^3$, $H^{1,2}(\mathbb{R}/2\pi; S)$ becomes a complete C^2-submanifold of the Hilbert space $H^{1,2}(\mathbb{R}/2\pi; \mathbb{R}^3)$ with tangent space

$$T_u H^{1,2}(\mathbb{R}/2\pi; S) =$$
$$= \{\varphi \in H^{1,2}(\mathbb{R}/2\pi; \mathbb{R}^3) \ ; \ \varphi(t) \in T_{u(t)}S \cong \mathbb{R}^2\};$$

see for instance Klingenberg [1; Theorem 1.2.9].

By (4.1), if $E(u)$ is sufficiently small so that the image of u is covered by a single coordinate chart, of course

$$T_u H^{1,2}(\mathbb{R}/2\pi; S) \cong H^{1,2}(\mathbb{R}/2\pi; \mathbb{R}^2) \ .$$

Moreover, E is analytic on $H^{1,2}(\mathbb{R}/2\pi; \mathbb{R}^3)$, hence as smooth as $H^{1,2}(\mathbb{R}/2\pi; S)$ when restricted to that space. At a critical point $u \in C^2$, upon integrating by parts

$$(4.2) \qquad \int_0^{2\pi} \dot{u}\,\dot{\varphi}\,dt = \int_0^{2\pi} -\ddot{u}\,\varphi\,dt = 0 \ , \quad \forall \varphi \in T_u H^1(\mathbb{R}/2\pi; S) \ ;$$

that is, $\ddot{u}(t) \perp T_{u(t)}S$ for all t, which is equivalent to the assertion that u is a geodesic, parametrized by arc length.

More generally, at a critical point $u \in H^{1,2}(\mathbb{R}/2\pi; S)$, if $n: S \to \mathbb{R}^3$ denotes a (C^2-) unit normal vector field on S, for any $\varphi \in H^{1,2}(\mathbb{R}/2\pi; \mathbb{R}^3)$ we have

$$\varphi - n(u)\big(n(u) \cdot \varphi\big) \in T_u H^{1,2}(\mathbb{R}/2\pi; S) \ .$$

Inserting this into (4.2) and observing that

$$(4.3) \qquad \ddot{u} \cdot n(u) = \big(\dot{u} \cdot n(u)\big)^{\cdot} - \dot{u} \cdot Dn(u)\,\dot{u} = -\dot{u} \cdot Dn(u)\,\dot{u}$$

in the distribution sense, we obtain that

$$(4.4) \qquad \ddot{u} + \big(\dot{u} \cdot Dn(u)\,\dot{u}\big)n(u) = 0 \ .$$

From (4.4) we now obtain that $u \in H^{2,1}\big([0, 2\pi]\big) \hookrightarrow C^1\big([0, 2\pi]\big)$. Hence, by iteration, $u \in C^2\big([0, 2\pi]\big)$. Thus closed geodesics on S and critical points of E on $H^{1,2}(\mathbb{R}/2\pi; S)$ coincide.

Moreover, E satisfies the Palais-Smale condition on $H^{1,2}(\mathbb{R}/2\pi; S)$: If (u_m) is a sequence in $H^{1,2}(\mathbb{R}/2\pi; S)$ such that $E(u_m) \leq c < \infty$ and

$$\|DE(u_m)\| = \sup_{\substack{\varphi \in T_u H^{1,2}(\mathbb{R}/2\pi; S) \\ \|\varphi\|_{1,2} \leq 1}} \left| \int_0^{2\pi} \dot{u}_m\,\dot{\varphi}\,dt \right| \to 0 \ ,$$

then (u_m) contains a strongly convergent subsequence.

Proof of (P.-S.). Since $E(u_m) \leq c$ uniformly, by (4.1) the sequence (u_m) is equi-continuous. But S is compact, in particular bounded; hence (u_m) is equi-bounded. Thus, by Arzéla-Ascoli's theorem we may assume that $u_m \to u$ uniformly and weakly in $H^{1,2}(\mathbb{R}/2\pi; \mathbb{R}^3)$. It follows that $u \in H^{1,2}(\mathbb{R}/2\pi; S)$.

Via the unit normal vector field n we can define the projection $\pi_u \colon H^{1,2}(\mathbb{R}/2\pi; \mathbb{R}^3) \to T_u H^{1,2}(\mathbb{R}/2\pi; S)$ by letting

$$\varphi \mapsto \pi_u \varphi(t) = \varphi(t) - n\left(u(t)\right)\left(n\left(u(t)\right) \cdot \varphi(t)\right)$$

as above. Since $n \in C^2$, π_u is bounded on bounded sets and weakly continuous in $H^{1,2}$, for any (fixed) $u \in H^{1,2}(\mathbb{R}/2\pi; S)$. In particular, we have

$$\varphi_m := \pi_u(u_m - u) \in T_u H^{1,2}(\mathbb{R}/2\pi; S) \ ,$$

and (φ_m) is bounded in $H^{1,2}$. Thus, $\langle \varphi_m, DE(u_m) \rangle \to 0$. Moreover, since $u_m \to u$ weakly in $H^{1,2}$ and uniformly, it follows that also $\varphi_m \to 0$ weakly in $H^{1,2}$ and uniformly. Consequently, we have

$$\langle \varphi_m, DE(u_m) \rangle = \int_0^{2\pi} \dot{u}_m \dot{\varphi}_m \, dt$$

$$= \int_0^{2\pi} (\dot{u}_m - \dot{u}) \dot{\varphi}_m \, dt + o(1)$$

$$= \int_0^{2\pi} |\dot{u}_m - \dot{u}|^2 - (\dot{u}_m - \dot{u}) \cdot \frac{d}{dt}\left[n(u)\left(n(u) \cdot (u_m - u)\right)\right] dt + o(1)$$

$$= \int_0^{2\pi} |\dot{u}_m - \dot{u}|^2 - \left((\dot{u}_m - \dot{u}) \cdot n(u)\right)\left(n(u) \cdot (\dot{u}_m - \dot{u})\right) dt + o(1) \ ,$$

where $o(1) \to 0$ as $m \to \infty$. But

$$n(u) \cdot \dot{u} = 0 = n(u_m) \cdot \dot{u}_m$$

almost everywhere. Hence

$$\int_0^{2\pi} |n(u)(\dot{u}_m - \dot{u})|^2 \, dt = \int_0^{2\pi} |\left(n(u_m) - n(u)\right) \dot{u}_m|^2 \, dt$$

$$\leq 2\|n(u_m) - n(u)\|_\infty^2 E(u_m) \to 0 \quad (m \to \infty) \ ,$$

and $u_m \to u$ strongly. $\qquad\qquad\square$

In order to construct a flow-invariant family $\mathcal{F}$ we now proceed as follows:

By assumption, there exists a C^3-diffeomorphism $\Psi \colon S \to S^2$ from S onto the standard sphere $S^2 \subset \mathbb{R}^3$. Let $p \colon [-\frac{\pi}{2}, \frac{\pi}{2}] \to H^{1,2}(\mathbb{R}/2\pi; S)$ be a 1-parameter family of closed curves $u = p(\theta)$ on S, such that $p(\pm \frac{\pi}{2}) \equiv p_\pm$ are constant "curves" in S. Via Ψ we associate with p a map $\tilde{p} \in C^0(S^2; S^2)$ by representing $S^2 \cong [-\frac{\pi}{2}, \frac{\pi}{2}] \times \mathbb{R}/2\pi$ in polar coordinates with $\{-\frac{\pi}{2}\} \times \mathbb{R}/2\pi$, respectively $\{\frac{\pi}{2}\} \times \mathbb{R}/2\pi$ collapsed to points. Then let

$$\tilde{p}(\theta, \varphi) = \Psi\left(p(\theta)(\varphi)\right) .$$

Consider now the collection

$$P = \{p \in C^0\left([-\frac{\pi}{2}, \frac{\pi}{2}]; H^{1,2}(\mathbb{R}/2\pi; S)\right) ; \ p(\pm\frac{\pi}{2}) \equiv const. \in S\}$$

and let

$$\mathcal{F} = \{p \in P ; \ \tilde{p} \text{ is homotopic to } id|_{S^2}\} .$$

Choosing for $p(\theta)$ the pre-image under Ψ of a family of equilateral circles covering S^2, we find that $\mathcal{F} \neq \emptyset$. Also note that the map $P \ni p \mapsto \tilde{p} \in C^0(S^2; S^2)$ is continuous. Hence $\mathcal{F}$ is Φ-invariant under any homeomorphism Φ of $H^{1,2}(\mathbb{R}/2\pi; S)$ homotopic to the identity, and which maps constant maps to constant maps. Note that, in particular, any Φ which does not increase E will have this latter property.

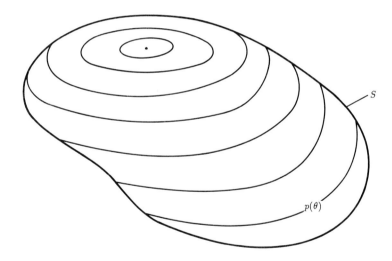

Fig. 4.1. An admissible comparison path $p \in P$

Finally, by Theorem 4.2

$$\beta = \inf_{p \in \mathcal{F}} \sup_{u \in p} E(u)$$

is critical.

This almost completes the proof of Theorem 4.4. However, it remains to rule out the possibility that $\beta = 0$: the energy of trivial (constant) "closed geodesics".

4.5 Lemma. $\beta > 0$.

Proof. There exists $\delta > 0$ such that for any x at distance $dist(x, S) \le \delta$ from S there is an unique nearest neighbor $\pi(x) \in S$, characterized by

$$|\pi(x) - x| = \inf_{y \in S} |x - y| \ ,$$

and $\pi(x)$ depends continuously on x. Moreover, π is C^2 if S is of class C^3. By (4.1) there exists $\gamma > 0$ such that for $u \in H^{1,2}(\mathbb{R}/2\pi; S)$ with $E(u) \le \gamma$ there holds

$$(4.5) \qquad \mathrm{diam}(u) = \sup_{0 \le \varphi, \psi \le 2\pi} |u(\varphi) - u(\psi)| < \delta \ .$$

Now suppose $\beta < \gamma$, and let $p \in \mathcal{F}$ be such that $E(u) \le \gamma$, for any $u = p(\theta) \in p$. By (4.5), if we fix $\varphi_0 \in [0, 2\pi]$, we can continuously contract any such curve u to $u(\varphi_0)$ in the δ-neighborhood of S by letting

$$u_s(\varphi) = (1 - s)u(\varphi) + s\, u(\varphi_0) \ .$$

Composing with π, we obtain a homotopy

$$p_s(\theta, \varphi) = \pi\big((1 - s)p(\theta)(\varphi) + sp(\theta)(\varphi_0)\big)$$

to a curve $p_1 \in P$ consisting entirely of constant loops $p_1(\theta) \equiv p(\theta)(\varphi_0)$ for all θ. Composing p_s with $\Psi \colon S \to S^2$, we also obtain a homotopy of $\tilde{p} \sim id|_{S^2}$ to $\tilde{p}_1(\theta, \varphi) = \Psi\left(p_1(\theta)(\varphi_0)\right)$. Finally, letting

$$\tilde{p}_{1,r}(\theta, \varphi) = \tilde{p}_1(r\theta, \varphi) = \Psi\big(p_1(r\theta)(\varphi_0)\big) \ , \ 0 \le r \le 1 \ ,$$

shows that $\tilde{p}_1$ is 0-homotopic, contradicting the definition of $\mathcal{F}$. □

4.6 Notes. Birkhoff's result of 1917 and a later extension to spheres of arbitrary dimension (Birkhoff [2]) mark the beginning of the calculus of variations in the large. A major advance then came with the celebrated work of Lusternik-Schnirelmann [1] of 1929 who – by variational techniques – established the existence of three geometrically distinct closed geodesics free of self-intersections on any compact surface of genus 0. (Full proofs were published by Lyusternik [1] in 1947.)

For recent developments in the theory of closed geodesics, see for instance Klingenberg [1] and Bangert [1].

5. Index Theory

In most cases the topology of the space M where a functional E is defined will be rather poor. However, if E is invariant under a compact group G acting on M, this may change drastically if we can pass to the quotient M/G with respect to the symmetry group. Often this space will have a richer topological structure which we may hope to exploit in order to obtain multiple critical points.

However, in general if the group G does not act freely on M, the quotient space will be singular, in particular no longer a manifold, and the above results no longer can be applied.

A nice way around this difficulty is to consider flow-invariant families $\mathcal{F}$ in Theorem 4.2 which are also invariant under the group action. Since by Remark 3.5.($4°$) we may choose our pseudo-gradient flows Φ to be equivariant if E is, this approach is promising. Moreover, at least for special kinds of group actions G, the topological complexity of the elements of such equivariant families can be easily measured or estimated in terms of an "index" which then may be used to distinguish different critical points.

Krasnoselskii Genus

The concept of an index theory is most easily explained for an even functional E on some Banach space V, with symmetry group $G = \mathbb{Z}_2 = \{id, -id\}$. Define

$$\mathcal{A} = \{A \subset V \mid A \text{ closed}, \ A = -A\}$$

to be the class of closed symmetric subsets of V.

5.1 Definition. For $A \in \mathcal{A}$, $A \neq \emptyset$, following Coffman [1] let

$$\gamma(A) = \begin{cases} \inf \left\{ m \ ; \ \exists h \in C^0(A; \mathbb{R}^m \setminus \{0\}), \ h(-u) = -h(u) \right\} \\ \infty, \quad \text{if } \{..\} = \emptyset, \text{ in particular, if } A \ni 0 \ , \end{cases}$$

and define $\gamma(\emptyset) = 0$. Remark that for any $A \in \mathcal{A}$ by the Tietze extension theorem any odd map $h \in C^0(A; \mathbb{R}^m)$ may be extended to a map $\tilde{h} \in C^0(V; \mathbb{R}^m)$. Letting $h(u) = \frac{1}{2} \left(\tilde{h}(u) - \tilde{h}(-u) \right)$ the extension can be chosen to be symmetric.

$\gamma(A)$ is called the Krasnoselskii genus of A. (The equivalence of Coffman's definition above with Krasnoselskii's [1] original definition – see also Krasnoselskii-Zabreiko [1; p.385 ff.] – was established by Rabinowitz [1; Lemma 3.6].) A notion of coindex with related properties was introduced by Connor-Floyd [1].

The notion of genus generalizes the notion of dimension of a linear space:

5.2 Proposition. *For any bounded symmetric neighborhood Ω of the origin in $\mathbb{R}^m$ there holds: $\gamma(\partial\Omega) = m$.*

Proof. Trivially, $\gamma(\partial\Omega) \leq m$. (Choose $h = id$.) Let $\gamma(\partial\Omega) = k$ and let $h \in C^0(\mathbb{R}^m; \mathbb{R}^k)$ be an odd map such that $h(\partial\Omega) \not\ni 0$. We may consider $\mathbb{R}^k \subset \mathbb{R}^m$. But then the topological degree of $h: \mathbb{R}^m \to \mathbb{R}^k \subset \mathbb{R}^m$ on Ω with respect to 0 is well-defined (see Deimling [1; Definition 1.2.3]). In fact, since h is odd, by the Borsuk-Ulam theorem (see Deimling [1; Theorem 1.4.1]) we have

$$\deg(h, \Omega, 0) = 1 .$$

Hence by continuity of the degree also

$$\deg(h, \Omega, y) = 1 \neq 0$$

for $y \in \mathbb{R}^m$ close to 0 and thus, by the solution property of the degree, h covers a neighborhood of the origin in $\mathbb{R}^m$; see Deimling [1; Theorem 1.3.1]. But then $k = m$ as claimed. □

Proposition 5.2 has a converse:

5.3 Proposition. *Suppose $A \subset V$ is a compact symmetric subset of a Hilbert space V with inner product $(\cdot, \cdot)_V$, and suppose $\gamma(A) = m < \infty$. Then A contains at least m mutually orthogonal vectors u_k, $1 \leq k \leq m$, $(u_k, u_l)_V = 0$ $(k \neq l)$.*

Proof. Let $u_1, \ldots, u_l$ be a maximal set of mutually orthogonal vectors in A, and denote $W = \text{span}\{u_1, \ldots, u_l\} \cong \mathbb{R}^l$, $\pi: V \to W$ orthogonal projection onto W. Then $\pi(A) \not\ni 0$, and π defines an odd continuous map $h = \pi|_A: A \to \mathbb{R}^l \setminus \{0\}$. By definition of $\gamma(A) = m$ we conclude that $l \geq m$, as claimed. □

Moreover, the genus has the following properties:

5.4 Proposition. *Let $A, A_1, A_2 \in \mathcal{A}$, $h \in C^0(V; V)$ an odd map. Then the following hold:*
($1°$) $\gamma(A) \geq 0$, $\gamma(A) = 0$ $\Leftrightarrow$ $A = \emptyset$.
($2°$) $A_1 \subset A_2 \Rightarrow \gamma(A_1) \leq \gamma(A_2)$.
($3°$) $\gamma(A_1 \cup A_2) \leq \gamma(A_1) + \gamma(A_2)$.
($4°$) $\gamma(A) \leq \gamma\left(\overline{h(A)}\right)$.
($5°$) If $A \in \mathcal{A}$ is compact and $0 \notin A$, then $\gamma(A) < \infty$ and there is a neighborhood N of A in V such that $\overline{N} \in \mathcal{A}$ and $\gamma(A) = \gamma(\overline{N})$.
That is, γ is a definite, monotone, sub-additive, supervariant and "continuous" map $\gamma: \mathcal{A} \to \mathbb{N}_0 \cup \{\infty\}$.

Proof. ($1°$) follows by definition.

(2°) If $\gamma(A_2) = \infty$ we are done. Otherwise, suppose $\gamma(A_2) = m$. By definition there exists $h \in C^0(A_2; \mathbb{R}^m \setminus \{0\})$, $h(-u) = -h(u)$. Restricting h to A_1 yields an odd map $h|_{A_1} \in C^0(A_1; \mathbb{R}^m \setminus \{0\})$, whence $\gamma(A_1) \leq \gamma(A_2)$.

(3°) Again we may suppose that both $\gamma(A_1) = m_1$, $\gamma(A_2) = m_2$ are finite, and we may let h_1, h_2 be odd maps $h_i \in C^0(A_i; \mathbb{R}^{m_i} \setminus \{0\})$, $i = 1, 2$, as in the definition of the genus. As remarked above, we may extend h_1, h_2 to odd maps $h_i \in C^0(V; \mathbb{R}^{m_i})$, $i = 1, 2$. But then letting $h(u) = (h_1(u), h_2(u))$ defines an odd map $h \in C^0(V; \mathbb{R}^{m_1 + m_2})$ which does not vanish for $u \in A_1 \cup A_2$, and claim (3°) follows.

(4°) Any odd map $\tilde{h} \in C^0\left(\overline{h(A)}; \mathbb{R}^m \setminus \{0\}\right)$ induces an odd map $\tilde{h} \circ h \in C^0(A; \mathbb{R}^m \setminus \{0\})$, and (4°) is immediate.

(5°) If A is compact and $0 \notin A$ there is $\rho > 0$ such that $A \cap B_\rho(0) = \emptyset$. The cover $\left\{ \tilde{B}_\rho(u) = B_\rho(u) \cup B_\rho(-u) \right\}_{u \in A}$ of A admits a finite sub-cover $\{\tilde{B}_\rho(u_1), \dots \tilde{B}_\rho(u_m)\}$. Let $\{\varphi_j\}_{1 \leq j \leq m}$ be a partition of unity on A subordinate to $\{\tilde{B}_\rho(u_j)\}_{1 \leq j \leq m}$; that is, let $\varphi_j \in C^0\left(\tilde{B}_\rho(u_j)\right)$ with support in $\tilde{B}_\rho(u_j)$ satisfy $0 \leq \varphi_j \leq 1$, $\sum_{j=1}^m \varphi_j(u) = 1$, for all $u \in A$. Replacing φ_j by $\overline{\varphi}_j(u) = \frac{1}{2}\left(\varphi_j(u) + \varphi_j(-u)\right)$ if necessary, we may assume that each φ_j is even, $1 \leq j \leq m$. By choice of ρ for any j the neighborhoods $B_\rho(u_j)$, $B_\rho(-u_j)$ are disjoint.

Hence the map $h: V \to \mathbb{R}^m$ with j-th component

$$h_j(u) = \begin{cases} \varphi_j(u), & \text{if } u \in B_\rho(u_j) \\ -\varphi_j(u), & \text{if } u \in B_\rho(-u_j) \end{cases}$$

is continuous, odd, and does not vanish on A.

Finally, assume that A is compact, $0 \notin A$, $\gamma(A) = m < \infty$ and let $h \in C^0(A; \mathbb{R}^m \setminus \{0\})$ be as in the definition of $\gamma(A)$. We may assume $h \in C^0(V; \mathbb{R}^m)$. Moreover, A being compact, also $h(A)$ is compact, and there exists a symmetric open neighborhood $\tilde{N}$ of $h(A)$ compactly contained in $\mathbb{R}^m \setminus \{0\}$. Choosing $N = h^{-1}(\tilde{N})$, by construction $h(\overline{N}) \not\ni 0$ and $\gamma(\overline{N}) \leq m$. On the other hand, $A \subset N$. Hence $\gamma(\overline{N}) = \gamma(A)$ by monotonicity of γ, property (2°). $\qquad\Box$

5.5 Observation. It is easy to see that if A is a finite collection of antipodal pairs $u_i, -u_i$ ($u_i \neq 0$), then $\gamma(A) = 1$.

Minimax Principles for Even Functionals

Suppose E is a functional of class C^1 on a closed symmetric $C^{1,1}$-submanifold M of a Banach space V and satisfies (P.-S.). Moreover, suppose that E is even, that is, $E(u) = E(-u)$ for all u. Also let $\mathcal{A}$ be as above. Then for any $k \leq \gamma(M) \leq \infty$ by Proposition 5.4.(4°) the family

$$\mathcal{F}_k = \{A \in \mathcal{A} ;\ A \subset M,\ \gamma(A) \geq k\}$$

is invariant under any odd and continuous map and non-empty. Hence, analogous to Theorem 4.2, for any $k \leq \gamma(M)$, if

$$\beta_k = \inf_{A \in \mathcal{F}_k} \sup_{u \in A} E(u)$$

is finite, then β_k is a critical value of E; see Theorem 5.7 below.

It is instructive to compare this result with the well-known Courant-Fischer minimax principle for linear eigenvalue problems. Recall that on $\mathbb{R}^n$ with scalar product $(\cdot, \cdot)$ the k-th eigenvalue of a symmetric linear map $K: \mathbb{R}^n \to \mathbb{R}^n$ is given by the formula

$$\lambda_k = \min_{\substack{V' \subset V, \\ \dim V' = k}} \max_{\substack{u \in V' \\ \|u\| = 1}} (Ku, u) .$$

Translated into the above setting we may likewise determine λ_k by considering the functional

$$E(u) = (Ku, u)$$

on the unit sphere $M = S^{n-1}$ and computing β_k as above. Trivially, E satisfies (P.-S.); moreover, it is easy to see that $\beta_k = \lambda_k$ for all k. (By Proposition 5.2 the inequality $\beta_k \leq \lambda_k$ is immediate. The reverse inequality follows by Proposition 5.3 and linearity of K.)

In the linear case now it is clear that if successive eigenvalues $\lambda_k = \lambda_{k+1} = \ldots = \lambda_{k+l-1} = \lambda$ coincide, then K has an l-dimensional eigenspace of eigenvectors $u \in V$ satisfying $Ku = \lambda u$. Is there a similar result in the non-linear setting? Actually, there is. For this we again assume that $E \in C^1(M)$ is an even functional on a closed, symmetric $C^{1,1}$-submanifold M in V, satisfying (P.-S.). Let β_k, $k \leq \gamma(M)$, be defined as above.

5.6 Lemma. *Suppose for some k, l there holds*

$$-\infty < \beta_k = \beta_{k+1} = \ldots = \beta_{k+l-1} = \beta < \infty .$$

Then $\gamma(K_\beta) \geq l$. By Observation 5.5, in particular, if $l > 1$, K_β is infinite.

Proof. By (P.-S.) the set K_β is compact and symmetric. Hence $\gamma(K_\beta)$ is well-defined and by Proposition 5.4.($5°$) there exists a symmetric neighborhood N of K_β in M such that $\gamma(\overline{N}) = \gamma(K_\beta)$. For $\bar{\varepsilon} = 1$, N, and β as above let $\varepsilon > 0$ and Φ be determined according to Theorem 3.11. We may assume Φ is odd. Choose $A \subset M$ such that $\gamma(A) \geq k + l - 1$ and $E(u) < \beta + \varepsilon$ for $u \in A$.

Let $\overline{\Phi(A, 1)} = \tilde{A} \in \mathcal{A}$. By property ($3°$) of Φ in Theorem 3.11

$$\tilde{A} \subset \overline{(E_{\beta - \varepsilon} \cup N)} .$$

Moreover, by definition of $\beta = \beta_k$ it follows that

$$\gamma\left(\overline{E_{\beta - \varepsilon}}\right) < k .$$

Thus by Proposition 5.4.($2°$)–($4°$)

$$\gamma(\overline{N}) \geq \gamma\left(\overline{E_{\beta-\varepsilon} \cup N}\right) - \gamma\left(\overline{E_{\beta-\varepsilon}}\right)$$
$$> \gamma(\tilde{A}) - k \;\geq\; \gamma(A) - k$$
$$\geq k + l - 1 - k = l - 1 \; ;$$

that is, $\gamma(\overline{N}) = \gamma(K_\beta) \geq l$, as claimed. $\qquad\qquad\qquad\qquad\qquad\square$

In consequence, we note

5.7 Theorem. *Suppose $E \in C^1(M)$ is an even functional on a complete symmetric $C^{1,1}$-manifold $M \subset V \setminus \{0\}$ in some Banach space V and suppose E satisfies (P.-S.) and is bounded from below on M. Let $\hat{\gamma}(M) = \sup\{\gamma(K) \; ; \; K \subset M$ compact and symmetric$\}$. Then the functional E possesses at least $\hat{\gamma}(M) \leq \infty$ pairs of critical points.*

Remarks. Note that the definition of $\hat{\gamma}(M)$ assures that for $k \leq \hat{\gamma}(M)$ the numbers β_k are finite.

Completeness of M can be replaced by the assumption that the flow defined by any pseudo-gradient vector field on M exists for all positive time.

Applications to Semilinear Elliptic Problems

As a particular case, Theorem 5.7 includes the following classical result of Ljusternik-Schnirelmann [2]: Any even function $E \in C^1(\mathbb{R}^n)$ admits at least n distinct pairs of critical points when restricted to S^{n-1}. In infinite dimensions, Theorem 5.7 and suitable variants of it have been applied to the solution of nonlinear partial differential equations and nonlinear eigenvalue problems for partial differential equations. See for instance Amann [1], Clark [1], Coffman [1], Hempel [1], Rabinowitz [7], Thews [1], and the surveys and lecture notes by Browder [2], Rabinowitz [7], [11].

Here we present only a simple example of this kind for which we return to the setting of problem

$(I.2.1)$ $\qquad\qquad\qquad -\Delta u + \lambda u = |u|^p u \qquad$ in Ω ,

$(I.2.3)$ $\qquad\qquad\qquad\qquad\quad u = 0 \qquad\qquad$ on $\partial\Omega$,

considered earlier. This time, however, we also admit solutions of varying sign.

5.8 Theorem. *Let Ω be a bounded domain in $\mathbb{R}^n$, $2 < p$; if $n \geq 3$ we assume in addition $p < \frac{2n}{n-2}$. Then for any $\lambda \geq 0$ problem (I.2.1), (I.2.3) admits infinitely many distinct pairs of solutions.*

Proof. By Theorem 5.7 the even functional

$$E(u) = \frac{1}{2} \int_\Omega \left(|\nabla u|^2 + \lambda|u|^2\right) dx$$

admits infinitely many distinct pairs of critical points on the sphere $S = \{u \in H_0^{1,2} \; ; \; \|u\|_{L^p} = 1\}$, for any $\lambda \geq 0$. Scaling suitably, we obtain infinitely many destinct pairs of solutions for (I.2.1), (I.2.3). □

General Index Theories

The concept of index can be generalized. Our presentation is based on Rabinowitz [11]. Suppose M is a complete $C^{1,1}$-Finsler manifold with a compact group action G. Let

$$\mathcal{A} = \{A \subset M \; ; \; A \text{ is closed}, \; g(A) = A \text{ for all } g \in G\}$$

be the set of G-invariant subsets of V, and let

$$\Gamma = \{h \in C^0(M; M) \; ; \; h \circ g = g \circ h \text{ for all } g \in G\}$$

be the class of G-equivariant mappings of M. (Since our main objective is that the flow $\Phi(\,\cdot\,, t)$ constructed in Theorem 3.11 be in Γ, we might also restrict Γ for instance to the class of G-equivariant homeomorphisms of M.) Finally, if $G \neq \{id\}$ denote

$$\text{Fix } G = \{u \in M \; ; \; gu = u \text{ for all } g \in G\}$$

the set of fixed points of G.

5.9 Definition. *An index for $(G, \mathcal{A}, \Gamma)$ is a mapping $i: \mathcal{A} \to \mathbb{N}_0 \cup \{\infty\}$ such that for all $A, B \in \mathcal{A}$, $h \in \Gamma$ there holds*
($1°$) (definiteness:) $i(A) \geq 0$, $i(A) = 0 \iff A = \emptyset$.
($2°$) (monotonicity:) $A \subset B \Rightarrow i(A) \leq i(B)$.
($3°$) (sub-additivity:) $i(A \cup B) \leq i(A) + i(B)$.
($4°$) (supervariance:) $i(A) \leq i\left(\overline{h(A)}\right)$.
($5°$) (continuity:) If A is compact and $A \cap Fix \, G = \emptyset$, then $i(A) < \infty$ and there is a G-invariant neighborhood N of A such that $i(\overline{N}) = i(A)$.
($6°$) (normalization): If $u \notin Fix \, G$, then $i\left(\bigcup_{g \in G} gu\right) = 1$.

5.10 Remarks and Examples. If $A \in \mathcal{A}$ and $A \cap Fix \, G \neq \emptyset$ then $i(A) = \sup_{B \in \mathcal{A}} i(B)$; indeed, by monotonicity, for $u_0 \in A \cap Fix \, G$ there holds $i(\{u_o\}) \leq i(A) \leq \sup_{B \in \mathcal{A}} i(B)$. On the other hand, for any $B \in \mathcal{A}$ the map $h: B \to \{u_0\}$, given by $h(u) = u_0$ for all $u \in B$, is continuous and equivariant, whence $i(B) \leq i(\{u_0\})$ by supervariance of the index. Hence, in general nothing will be lost if we define $i(A) = \infty$ for $A \in \mathcal{A}$ such that $A \cap Fix \, G \neq \emptyset$.

By Example 5.1 the Krasnoselskii genus γ is an index for $G = \{id, -id\}$, the class $\mathcal{A}$ of closed, symmetric subsets, and Γ the family of odd, continuous maps.

Analogous to Theorem 5.7 we have the following general existence result for variational problems that admit an index theory.

5.11 Theorem. *Let $E \in C^1(M)$ be a functional on a complete $C^{1,1}$-Finsler manifold M and suppose E is bounded from below and satisfies (P.-S.). Suppose G is a compact group acting on M without fixed points and let $\mathcal{A}$ be the set of closed G-invariant subsets of M, Γ be the group of G-equivariant homeomorphisms of M. Suppose i is an index for $(G, \mathcal{A}, \Gamma)$, and let $\hat{i}(M) = \sup\{i(K) ; K \subset M$ is compact and $G - invariant\} \leq \infty$.*
Then E admits at least $\hat{i}(M)$ critical points which are distinct modulo G.

The proof is the same as that of Theorem 5.7 and Lemma 5.6. Again note that completeness of M can be replaced by the assumption that any pseudo-gradient flow on M is complete in forward time.

Ljusternik-Schnirelman Category

The first example of an index was introduced in Ljusternik-Schnirelmann [2].

5.12 Definition. *Let M be a topological space and consider a closed subset $A \subset M$. We say that A has category k relative to M $(\mathrm{cat}_M(A) = k)$ if A is covered by k closed sets A_j, $1 \leq j \leq k$, which are contractible in M, and if k is minimal with this property. If no such finite covering exists we let $\mathrm{cat}_M(A) = \infty$.*

This notion fits in the frame of Definition 5.9 if we let

$$G = \{id\}$$
$$\mathcal{A} = \{A \subset M ; A \text{ closed }\},$$
$$\Gamma = \{h \in C^0(M; M) ; h \text{ is a homeomorphism }\}.$$

Then we have

5.13 Proposition. cat_M *is an index for* $(G, \mathcal{A}, \Gamma)$.

Proof. (1°)–(3°) of Definition 5.9 are immediate. (4°) is also clear, since a homeomorphism h preserves the topological properties of any sets A_j covering A. (5°) Any open cover of a compact set A by open sets O whose closure is contractible has a finite subcover $\{O_j , 1 \leq j \leq k\}$. Set $N = \bigcup O_j$. (6°) is obvious. ☐

5.14 Categories of some standard sets.
(1°) If $M = T^m = \mathbb{R}^m / \mathbb{Z}^m$ is the m-torus, then $\mathrm{cat}_{T^m}(T^m) = m + 1$, see Ljusternik-Schnirelman [2] or Schwartz [2; Lemma 5.15, p.161]. Thus, any functional $E \in C^1(T^m)$ possesses at least $m + 1$ distinct critical points. In particular, if $m = 2$, any C^1-functional on the standard torus, besides an absolute minimum and maximum must possess at least one additional critical point.

(2°) For the m-sphere $S^m \subset \mathbb{R}^{m+1}$ we have $\mathrm{cat}_{S^m}(S^m) = 2$. (Take A_1, A_2 slightly overlapping northern and southern hemispheres.)

(3°) For the unit sphere S in an infinite dimensional Banach space we have $\mathrm{cat}_S(S) = 1$. (S is contractible in itself.)

(4°) For real or complex m-dimensional projective space P^m we have $\mathrm{cat}_{P^m}(P^m) = m + 1 \ (m \leq \infty)$.

Since real projective $P^m = S^m/\mathbb{Z}_2$, we may ask whether, in the presence of a $\mathbb{Z}_2$-symmetry $u \to -u$, the category and Krasnoselskii genus of symmetric sets are always related as in the above example 5.14.(4°). This is indeed the case, see (Rabinowitz [1; Theorem 3.7]):

5.15 Proposition. *Suppose $A \subset \mathbb{R}^m \setminus \{0\}$ is compact and symmetric, and let $\tilde{A} = A/\mathbb{Z}_2$ with antipodal points collapsed. Then $\gamma(A) = \mathrm{cat}_{\mathbb{R}^m \setminus \{0\}/\mathbb{Z}_2}(\tilde{A})$.*

Using the notion of category, results in the spirit of Theorem 5.8 have been established by Browder [1], [3] and Schwartz [1], for example.

With index theories offering a very convenient means to characterize different critical points of functionals possessing certain symmetries it is not surprising that, besides the classical examples treated above, a variety of other index theories have been developed. Confer the papers by Fadell-Husseini [1], Fadell-Husseini-Rabinowitz [1], Fadell-Rabinowitz [1] on cohomological index theories – a very early paper in this regard is due to Yang [1]. Relative or pseudo-indices were introduced by Benci [3] and Bartolo-Benci-Fortunato [1].

For our final model problem in this section it will suffice to consider the S^1-index of Benci [2] as another particular case.

A Geometrical S^1-Index

If M is a complete $C^{1,1}$-Finsler manifold with an S^1-action (in particular, if M is a complex Hilbert space with $S^1 = \{e^{i\phi} \ ; \ 0 \leq \phi \leq 2\pi\}$ acting through scalar multiplication) we may define an index for this action as follows; see Benci [2].

5.16 Definition. Let $\mathcal{A}$ be the family of closed, S^1-invariant subsets of M, and Γ the family of S^1-equivariant maps (or homeomorphisms). For $A \neq \emptyset$, define

$$\tau(A) = \begin{cases} \inf\{m \ ; \ \exists h \in C^0(A; \mathbb{C}^m \setminus \{0\}), \ l \in \mathbb{N} : \\ \qquad h \circ g = g^l \circ h \text{ for all } g \in S^1\}, \\ \infty, \qquad\qquad\qquad\qquad\qquad\qquad \text{if } \{\ldots\} = \emptyset, \end{cases}$$

and let $\tau(\emptyset) = 0$. (Note the similarity with the Krasnoselskii index γ.) Here, S^1 acts on $\mathbb{C}^m$ by component-wise complex multiplication.

5.17 Proposition. τ *is an index for* $(S^1, \mathcal{A}, \Gamma)$.

Proof. It is easy to see that τ satisfies properties (1°) and (2°) of Definition 5.9. To see (3°) we may assume that $\tau(A_i) = m_i < \infty$, $i = 1, 2$, and we may choose h_i, l_i as in the definition of τ such that $h_i \in C^0(A_i; \mathbb{C}^{m_i} \setminus \{0\})$ satisfies $h_i \circ g = g^{l_i} \circ h_i$ for all $g \in S^1$, $i = 1, 2$.

Extending h_i to M and averaging

$$\tilde{h}_i(u) = \int_{S^1} g^{-l_i} h_i(gu) \, dg$$

with respect to an invariant measure (arc-length) on S^1, we may assume that $h_i \in C^0(M; \mathbb{C}^m)$, $i = 1, 2$. But then the map

$$h(u) = \left(\left(h_1(u) \right)^{l_2}, \left(h_2(u) \right)^{l_1} \right) ,$$

where for $(z_1, \ldots, z_m) \in \mathbb{C}^m$, $l \in \mathbb{N}$ we let $(z_1, \ldots, z_m)^l := (z_1^l, \ldots, z_m^l)$, defines a map

$$h \in C^0\left(A_1 \cup A_2 \; ; \; \mathbb{C}^{m_1 + m_2} \setminus \{0\} \right)$$

such that $h \circ g = \left(\left(g^{l_1} \circ h_1 \right)^{l_2}, \left(g^{l_2} \circ h_2 \right)^{l_1} \right) = g^l \circ h$ for all $g \in S^1$ with $l = l_1 l_2$.

To see (6°), for an element $u_0 \notin Fix(S^1)$ let

$$G_0 = \{g \in S^1 \; ; \; g u_0 = u_0\}$$

be the subgroup of S^1 fixing u_0. Since $u_0 \notin Fix(S^1)$, G_0 is discrete, hence represented by

$$G_0 = \left\{ e^{2\pi i k/l} \; ; \; 0 \leq k < l \right\}$$

for some $l \in \mathbb{N}$.

For $u = g u_0$ now let $h(u) = g^l \in S^1 \subset \mathbb{C} \setminus \{0\}$. Then h is well-defined and continuous along the S^1-orbit $S^1 u_0 = \{g u_0 \; ; \; g \in S^1\}$ of u_0. Extending h equivariantly, we see that $\tau(S^1 u_0) = 1$, which proves (6°).

Finally, to see (5°), suppose A is S^1-invariant, compact, and $A \cap \mathrm{Fix}(S^1) = \emptyset$. For any $u_0 \in A$ let h be constructed as in the proof of (6°) and let $O(u_0)$ be an S^1-invariant neighborhood of $S^1 u_0$ such that $h\left(O(u_0) \right) \not\ni 0$.

By compactness of A finitely many such neighborhoods $\{O(u_i)\}_{1 \leq i \leq m}$ cover A, whence $\tau(A) < \infty$ by sub-additivity. The remainder of the proof of (5°) is the same as in Proposition 5.4. $\qquad\square$

As in the case of the Krasnoselskii genus, Proposition 5.2, the S^1-index may be interpreted as a measure of the dimension of a closed S^1-invariant set; see Benci [2; Proposition 2.6]. However, we will not pursue this.

Instead, we observe that in case of a free S^1-action on a manifold M, a simpler variant of Benci's S^1-index can be defined as follows. (Recall that a group G acts freely on a manifold M if only the identity element in G fixes points in M.)

5.16' Definition. Suppose S^1 acts freely on a manifold M. Let $\mathcal{A}$ be the family of closed, S^1-invariant subsets of M, and Γ the family of S^1-equivariant maps (or homeomorphisms). For $A \neq \emptyset$, define

$$\tilde{\gamma}(A) = \begin{cases} \inf\{m \; ; \; \exists h \in C^0(A; \mathbb{C}^m \setminus \{0\}) : \\ \qquad h \circ g = g \circ h \text{ for all } g \in S^1\}, \\ \infty, \qquad\qquad\qquad\qquad\qquad \text{if } \{\ldots\} = \emptyset, \end{cases}$$

and let $\tilde{\gamma}(\emptyset) = 0$.

Proposition 5.17 and its proof may be carried over easily.

Multiple Periodic Orbits of Hamiltonian Systems

As an application, we present the following theorem on the existence of "many" periodic solutions of Hamiltonian systems, due to Ekeland and Lasry [1]:

5.18 Theorem. *Suppose $H \in C^1(\mathbb{R}^{2n}; \mathbb{R})$, and for some $\beta > 0$ assume that $C = \{x \in \mathbb{R}^{2n} \; ; \; H(x) \leq \beta\}$ is strictly convex, with boundary $S = \{x \in \mathbb{R}^{2n} \; ; \; H(x) = \beta\}$ satisfying $x \cdot \nabla H(x) > 0$ for $x \in S$. Suppose that for numbers $r, R > 0$ with*

$$r < R < \sqrt{2}\, r$$

we have

$$B_r(0) \subset C \subset B_R(0) \; .$$

Then there exist at least n distinct periodic solutions of the equation

(5.1) $$\dot{x} = \mathcal{J}\nabla H(x)$$

on S. ($\mathcal{J}$ is defined on p. 57.)

Theorem 5.18 provides a "global" analogue of a result by Weinstein [1] on the existence of periodic orbits of Hamiltonian systems near an equilibrium. Further extensions and generalizations of Theorem 5.18 were given by Ambrosetti-Mancini [2] and Berestycki-Lasry-Mancini-Ruf [1] who allow for energy surfaces "pinched" between ellipsoids rather than spheres. Moreover, Ekeland-Lassoued [1] have been able to show that any strictly convex energy surface carries at least two distinct periodic orbits of (5.1). It is conjectured that a result like Theorem 5.18 holds true in general on such surfaces; the proof of this conjecture, however, remains open.

Proof of Theorem 5.18. We follow Ambrosetti and Mancini [2]. As observed in Chapter I.6 we may assume that

(5.2) $$H(sx) = s^q H(x)$$

is homogeneous of degree q, $1 < q < 2$, and strictly convex. Moreover, dividing H by βq we may assume that $\beta = \frac{1}{q}$. By our assumption on S and (5.2), finally, we have

$$(5.3) \qquad \frac{1}{qR^q}|x|^q \le H(x) \le \frac{1}{qr^q}|x|^q .$$

Let H^* be the Legendre-Fenchel transform of H. Note that $H^* \in C^1$, $H^*(sy) = s^p H^*(y)$ with $p = \frac{q}{q-1} > 2$, and (5.3) translates into the condition

$$(5.4) \qquad \frac{1}{p}r^p|y|^p \le H^*(y) \le \frac{1}{p}R^p|y|^p .$$

Also let K be the integral operator

$$(Ky)(t) = \int_0^t \mathcal{J}y \, dt$$

on

$$L_0^p = \left\{ y \in L_{loc}^p(\mathbb{R}; \mathbb{R}^{2n}) \; ; \; y(t + 2\pi) = y(t), \int_0^{2\pi} y \, dt = 0 \right\} .$$

Then – as described in detail in I.6 – $y \in L_0^p \setminus \{0\}$ solves the equation

$$(5.5) \qquad Ky = \nabla H^*(y)$$

if and only if there is $x_0 \in \mathbb{R}^{2n}$ such that $x = Ky + x_0$ solves (5.1) with $H(x(t)) =: h/q > 0$, which in turn implies that $\tilde{x}(t) = h^{-1/q}x\left(h^{2/q-1}t\right)$ solves (5.1) with $H(\tilde{x}(t)) \equiv \beta = \frac{1}{q}$.

Suppose we can exhibit n distinct solutions y_k of (5.5) with minimal period 2π corresponding to distinct solutions x_k of (5.1) with energies $H(x_k(t)) = h_k$. Then either $h_j = h_k$ and the corresponding $\tilde{x}_j \ne \tilde{x}_k$ (since the solutions x_k are distinct). Or $h_j \ne h_k$ and $\tilde{x}_j$, $\tilde{x}_k$ will have different minimal period and hence be distinct. In any event we will have achieved the proof of the theorem.

Denote $E: L_0^p \to \mathbb{R}$ the dual variational integral corresponding to (5.5), given by

$$E(y) = \int_0^{2\pi} \left(H^*(y) - \frac{1}{2}\langle y, Ky \rangle \right) dt .$$

Note that we have an S^1-action on L_0^p, via

$$(\tau, y) \mapsto y_\tau(t) = y(t + \tau) , \qquad \text{for all } \gamma = e^{i\tau} \in S^1 .$$

This action leaves E invariant. Moreover, y has minimal period 2π if and only if $y_\tau = y$ precisely for $\tau \in 2\pi\mathbb{Z}$. Denote

$$m = \inf \left\{ E(y) \; ; \; y \in L_0^p \right\}$$

and let

$$m^* = \inf \left\{ E(y) \; ; \; y \in L_0^p, \exists \tau \notin 2\pi\mathbb{Z} : y_\tau = y \right\} .$$

Observe that, since $p > 2$ and since the spectrum of K contains positive eigenvalues, we have $m \leq m^* < 0$. Also note that the S^1-action will be free on the set

$$M = \{y \in L_0^p \; ; \; m \leq E(y) < m^*\} \; .$$

In particular, any $y \in M$ will have minimal period 2π.

Finally, E satisfies (P.-S.). Indeed, since E is coercive on L_0^p, any (P.-S.)-sequence (y_m) for E is bounded in L_0^p. Thus we may assume that $y_m \rightharpoonup y$ weakly in L_0^p and $Ky_m \to Ky$ strongly in $L^q([0, 2\pi]; \mathbb{R}^{2n})$. Recall that by strict convexity and homogeneity of H^* the differential ∇H^* is strongly monotone. (Confer Lemma I.6.3 and the discussion following it.) Hence we obtain

$$\langle y_m - y, DE(y_m) - DE(y) \rangle = \int_0^{2\pi} \langle y_m - y, \nabla H^*(y_m) - \nabla H^*(y) \rangle -$$
$$- \langle y_m - y, Ky_m - Ky \rangle \, dt$$
$$\geq \alpha \int_0^{2\pi} |y_m - y|^p \, dt - o(1) \; ,$$

where $o(1) \to 0$ as $m \to \infty$, and $y_m \to y$ strongly in L_0^p, as claimed. Thus, by Theorem 5.11 the proof will be complete if we can show that the (simplified) S^1-index of a compact S^1-invariant subset of M is $\geq n$. (From the definition of M it is clear that any pseudo-gradient flow on M will be complete in forward time.)

Note that since E is weakly lower semi-continuous and coercive on L_0^p and since in both cases the set of comparison functions is weakly closed, m and m^* will be attained in their corresponding classes. Let $y^* \in L_0^p$ satisfy

$$E(y^*) = m^*, \; y_\tau^* = y^* \text{ for some } \tau \notin 2\pi \mathbb{Z} \; .$$

By minimality of y^*

$$\langle y^*, DE(y^*) \rangle = p \int_0^{2\pi} H^*(y^*) \, dt - \int_0^{2\pi} \langle y^*, Ky^* \rangle \, dt = 0 \; ,$$

whence in particular

$$m^* = \left(\frac{1}{p} - \frac{1}{2} \right) \int_0^{2\pi} \langle y^*, Ky^* \rangle \, dt \; .$$

We may assume that $\tau = \frac{2\pi}{k}$ for some $k \in \mathbb{N}$, $k > 1$. Hence we obtain as comparison function

$$y(t) = y^* \left(\frac{t}{k} \right) \in L_0^p \; ,$$

and

$$m \leq \inf_{s>0} E(sy) = \inf_{s>0} \left(s^p \int_0^{2\pi} H^*(y) \, dt - \frac{s^2}{2} \int_0^{2\pi} \langle y, Ky \rangle \, dt \right) \; .$$

But

$$\int_0^{2\pi} H^*(y)\, dt = \int_0^{2\pi} H^*(y^*)\, dt = \frac{1}{p}\int_0^{2\pi} \langle y^*, Ky^*\rangle\, dt \ ,$$

while

(5.6)
$$\int_0^{2\pi} \langle y, Ky\rangle\, dt = k\int_0^{2\pi} \langle y^*, Ky^*\rangle\, dt \ .$$

Hence

(5.7)
$$\begin{aligned}
m &\le \inf_{s>0}\left(\frac{s^p}{p} - \frac{s^2}{2}k\right)\int_0^{2\pi}\langle y^*, Ky^*\rangle\, dt\\
&= k^{\frac{p}{p-2}}\left(\frac{1}{p} - \frac{1}{2}\right)\int_0^{2\pi}\langle y^*, Ky^*\rangle\, dt\\
&\le 2^{\frac{p}{p-2}} m^* < 0 \ .
\end{aligned}$$

To obtain a lower bound on m, let $y \in L_0^p$ satisfy

$$E(y) = m \ .$$

Then

(5.8)
$$\langle y, DE(y)\rangle = p\int_0^{2\pi} H^*(y)\, dt - \int_0^{2\pi}\langle y, Ky\rangle\, dt = 0 \ ,$$

and hence

(5.9)
$$m = \left(\frac{1}{p} - \frac{1}{2}\right)\int_0^{2\pi}\langle y, Ky\rangle\, dt = \left(1 - \frac{p}{2}\right)\int_0^{2\pi} H^*(y)\, dt \ .$$

Note that by (5.4) for $z \in L_0^p$ we have

(5.10)
$$\frac{1}{2\pi}\int_0^{2\pi} H^*(z)\, dt \ge \frac{r^p}{2\pi p}\int_0^{2\pi}|z|^p\, dx \ge \frac{r^p}{p}\left(\frac{1}{2\pi}\int_0^{2\pi}|z|^2\, dx\right)^{p/2} \ .$$

Let $\Sigma = \{z \in L_0^p \ ; \ \frac{1}{2\pi}\int_0^{2\pi}|z|^2\, dt = 1\}$ and denote

$$b = \sup\left\{\int_0^{2\pi}\langle z, Kz\rangle\, dt \ ; \ z \in \Sigma\right\} \ .$$

Then, if we let $y = \lambda z$, $z \in \Sigma$, by (5.8), (5.10) we have

(5.11)
$$\begin{aligned}
\lambda^{2-p}b &\ge \lambda^{2-p}\int_0^{2\pi}\langle z, Kz\rangle\, dt = p\int_0^{2\pi} H^*(z)\, dt\\
&\ge 2\pi r^p \ .
\end{aligned}$$

Since $p > 2$ this implies an estimate from above for λ. Moreover, by

$$m = \left(\frac{1}{p} - \frac{1}{2}\right)\lambda^2 \int_0^{2\pi} \langle z, Kz\rangle \, dt \geq \left(\frac{1}{p} - \frac{1}{2}\right)\lambda^2 b$$

(5.12)

$$\geq \left(\frac{1}{p} - \frac{1}{2}\right)(2\pi)^{-2/(p-2)}b^{p/(p-2)}r^{-2p/(p-2)} =: c_0 r^{-2p/(p-2)} \; ,$$

the latter translates into a lower bound for m, and hence for m^*, by (5.7). Now let

$$\Sigma_n = \left\{ z(t) = e^{Jt}(\xi, \eta) = (\xi\cos t + \eta\sin t, -\xi\sin t + \eta\cos t) \; ; \right.$$

$$\left. (\xi, \eta) \in \mathbb{R}^{2n} \text{ with } |\xi|^2 + |\eta|^2 = 1 \right\} \subset \Sigma \; .$$

Clearly Σ_n is S^1-invariant. Moreover, note that for $z \in \Sigma_n$ we have $H^*(z(t)) \leq R^p/p$ for all t, whence

(5.13)
$$\frac{1}{2\pi}\int_0^{2\pi} H^*(z) \, dt \leq R^p/p \; .$$

Now, we have

5.19 Lemma. $\int_0^{2\pi}\langle z, Kz\rangle \, dt = b$ for any $z \in \Sigma_n$.

Postponing the proof of Lemma 5.19 we conclude the proof of Theorem 5.18 as follows. For $z \in \Sigma_n$ let $\lambda = \lambda(z) > 0$ satisfy

(5.14)
$$p\lambda^p \int_0^{2\pi} H^*(z) \, dt - \lambda^2 \int_0^{2\pi} \langle z, Kz\rangle \, dt = 0 \; .$$

By (5.10) the map $z \mapsto \lambda(z)z$ is an S^1-equivariant C^1-embedding of Σ_n into L_0^p, mapping Σ_n onto an S^1-invariant set $\tilde{\Sigma}_n$ diffeomorphic to Σ_n by radial projection; in particular, the (simplified) S^1-index $\tilde{\tau}(\tilde{\Sigma}_n) \geq \tilde{\tau}(\Sigma_n)$.

From (5.13), (5.14) and Lemma 5.19, as in (5.11) we now obtain that for $z \in \Sigma_n$, $\lambda = \lambda(z)$ there holds

$$\lambda^{2-p}b = \lambda^{2-p}\int_0^{2\pi}\langle z, Kz\rangle \, dt = p\int_0^{2\pi} H^*(z) \, dt \leq 2\pi R^p \; .$$

Hence analogous to (5.12) we obtain

$$\sup_{y \in \tilde{\Sigma}_n} E(y) \leq c_0 R^{-2p/p-2} \; .$$

Since by assumption $R < \sqrt{2}\, r$, and in view of (5.7), this implies that

$$\sup_{y \in \tilde{\Sigma}_n} E(y) < \frac{1}{2^{p/p-2}}c_0 r^{-2p/p-2} \leq \frac{1}{2^{p/p-2}}m \leq m^* \; ;$$

that is, $\tilde{\Sigma}_n \subset M$. Hence the proof of Theorem 5.18 is complete if we show that $\tilde{\tau}(\Sigma_n) \geq n$.

But any S^1-equivariant map $h: \Sigma_n \to \mathbb{C}^m \setminus \{0\}$ with $h \circ g = g \circ h$ induces an odd map of $S^{2n-1} = \{(\xi, \eta) \in \mathbb{R}^{2n} \; ; \; |\xi|^2 + |\eta|^2 = 1\}$ into $\mathbb{C}^m \setminus \{0\} \cong \mathbb{R}^{2m} \setminus \{0\}$, given by

$$(\xi, \eta) \mapsto h\big(e^{\mathcal{J}t}(\xi, \eta)\big) \ .$$

By Proposition 5.2 we conclude that $2m \geq 2n$, whence $\tilde{\tau}(\Sigma_n) \geq n$. (Since the map

$$z = z(t) = e^{\mathcal{J}t}(\xi, \eta) \mapsto z(0) = (\xi, \eta) \in \mathbb{R}^{2n} \setminus \{0\} \cong \mathbb{C}^n \setminus \{0\}$$

is S^1-equivariant, we actually have equality $\tilde{\tau}(\Sigma_n) = n$.) $\square$

Proof of Lemma 5.19. Since K is compact there exists $z \in \Sigma$ satisfying

$$\int_0^{2\pi} \langle z, Kz \rangle \, dt = b \ .$$

By (5.6) z must have minimal period 2π.

Moreover, z satisfies

$$Kz + x_0 = \lambda z$$

for some $\lambda \in \mathbb{R}$, where $x_0 = \frac{1}{2\pi} \int_0^{2\pi} Kz \, dt$. Setting

$$x = Kz + x_0,$$

equivalently, x solves

$$\lambda \dot{x} = \mathcal{J}x \ .$$

Hence, $|x| \equiv const$. Moreover, the properties of z imply that $x \neq 0$ and has minimal period 2π. A scalar multiple of x (and hence also z) thus belongs to Σ_n.

By S^1-invariance of $I(z) = \int_0^{2\pi} \langle z, Kz \rangle \, dt$ and invariance of I under rotations in $\mathbb{R}^{2n}$, if $I(z) = b$ for *some* $z \in \Sigma_n$, it follows that $I(z) = b$ for *all* $z \in \Sigma_n$. The proof is complete. $\square$

6. The Mountain Pass Lemma and its Variants

The minimax principle and its variants essentially cover all possibilities how existence results for saddle points can be drawn from information about the topology of the sub-level sets of a functional E.

However, unless the domain of E itself has a rich topology, finding the right notion of flow-invariant family may be quite tiresome. Fortunately, there are existence results for saddle points taylor-made for applications. These are the famous (infinite-dimensional) mountain pass lemma and its variants, due to Ambrosetti and Rabinowitz [1]. In its simplest form this result reads as follows.

6.1 Theorem. *Suppose $E \in C^1(V)$ satisfies (P.-S.). Suppose*
(1°) $E(0) = 0$;
(2°) $\exists \rho > 0$, $\alpha > 0$: $\|u\| = \rho \Rightarrow E(u) \geq \alpha$;
(3°) $\exists u_1 \in V$: $\|u_1\| \geq \rho$ and $E(u_1) < \alpha$.
Define

$$P = \left\{ p \in C^0\left([0,1]; V\right) \ ; \ p(0) = 0, \ p(1) = u_1 \right\} .$$

Then

$$\beta = \inf_{p \in P} \ \sup_{u \in p} E(u)$$

is a critical value.

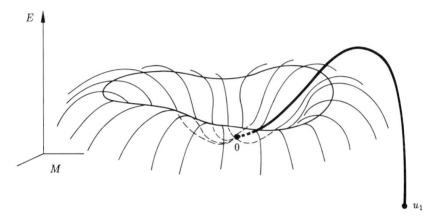

Fig. 6.1. On the mountain pass lemma of Ambrosetti and Rabinowitz

Proof. Suppose by contradiction that $K_\beta = \emptyset$, and let $\bar{\varepsilon} = \min\{\alpha, \alpha - E(u_1)\}$, $N = \emptyset$. Determine $\varepsilon > 0$ and a deformation Φ as in Theorem 3.4. By definition of β, there exists $p \in P$ such that

$$\sup_{u \in p} E(u) < \beta + \varepsilon .$$

Consider $p_1 = \Phi(p, 1)$. Note that by choice of $\bar{\varepsilon}$ the deformation $\Phi(\cdot, 1)$ leaves u_0 and u_1 fixed. Hence $p_1 \in P$. Moreover, $\Phi(E_{\beta+\varepsilon}, 1) \subset E_{\beta-\varepsilon}$; that is,

$$\sup_{u \in p_1} E(u) = \sup_{u \in p} E\left(\Phi(u, 1)\right) < \beta - \varepsilon ,$$

by choice of p. But this contradicts the definition of β, and the proof is complete.
□

Remark. Observe that by Remark 3.5.(3°) the proof, and hence the assertion of Theorem 6.1 remain true if we only require (P.-S.) at the level β.

Applications to Semilinear Elliptic Boundary Value Problems

Theorem 6.1 would permit to give an alternative proof of Theorem I.2.1. However, Theorem 6.1 can also be applied to more general problems of the type

$$(6.1) \qquad\qquad -\Delta u = g(\cdot, u) \qquad \text{in } \Omega ,$$

$$(6.2) \qquad\qquad u = 0 \qquad \text{on } \partial\Omega ,$$

which cannot be solved by a constrained minimization method.

6.2 Theorem. *Let Ω be a smooth, bounded domain in $\mathbb{R}^n$, $n \geq 3$, and let $g\colon \Omega \times \mathbb{R} \to \mathbb{R}$ be a Carathéodory function with primitive $G(x, u) = \int_0^u g(x, v) \, dv$. Suppose the following conditions hold:*

(1°) $\limsup_{u \to 0} \frac{g(x,u)}{u} \leq 0$, uniformly in $x \in \Omega$;

(2°) $\exists p < 2^* = \frac{2n}{n-2}, \ C\colon |g(x, u)| \leq C(1 + |u|^{p-1})$, for almost every $x \in \Omega$, $u \in \mathbb{R}$;

(3°) $\exists q > 2, \ R_0 : 0 < q\, G(x, u) \leq g(x, u)\, u$, for almost every $x \in \Omega$, if $|u| \geq R_0$. Then problem (6.1), (6.2) admits non-trivial solutions $u^+ \geq 0 \geq u_-$. If, in addition, g is Hölder continuous in both variables, then $u^+ > 0 > u_-$ in Ω.

Remark. An analogous result is valid for $n = 2$. However, notation is simpler if we consider only $n \geq 3$. Wang [1], under (essentially) the assumptions of Theorem 6.2, recently has established the existence of even a third non-trivial solution.

Proof. For simplicity, at first we establish only the existence of a single non-trivial solution of (6.1), (6.2).

Problem (6.1), (6.2) corresponds to the Euler-Lagrange equation of the functional

$$E(u) = \frac{1}{2} \int_\Omega |\nabla u|^2 \, dx - \int_\Omega G(x, u) \, dx$$

on the space $H_0^{1,2}(\Omega)$. Assumption (2°) implies that E is of class C^1.

To see that E verifies (P.-S.) first note that by (2°) the map $u \mapsto g(\cdot, u)$ takes bounded sets in $L^p(\Omega)$ into bounded sets in $L^{p/(p-1)}(\Omega) \subset H^{-1}(\Omega)$. There-
fore, and since for $p < \frac{2n}{n-2}$ by Rellich's theorem the space $H_0^{1,2}(\Omega)$ embeds into $L^p(\Omega)$ compactly, the map $K\colon H_0^{1,2}(\Omega) \to H^{-1}(\Omega)$, given by $K(u) = g(\cdot, u)$, is compact. Now $DE(u) = -\Delta u - g(\cdot, u)$, and hence by Proposition 2.2 it suffices to show that any (P.-S.)-sequence (u_m) for E is bounded in $H_0^{1,2}(\Omega)$.

Let (u_m) be a (P.-S.)-sequence. Then we obtain

$$C + o(1)\|u_m\|_{H_0^{1,2}} \geq q\, E(u_m) - \langle u_m, DE(u_m) \rangle =$$

$$(6.3) \qquad = \frac{q-2}{2} \int_\Omega |\nabla u_m|^2 \, dx + \int_\Omega \big(g(x, u_m)u_m - qG(x, u_m)\big) \, dx$$

$$\geq \frac{q-2}{2}\|u_m\|_{H_0^{1,2}}^2 + \mathcal{L}^n(\Omega) \cdot \operatorname*{ess\,inf}_{x \in \Omega, v \in \mathbb{R}} \big(g(x, v)v - qG(x, v)\big) ,$$

where $o(1) \to 0 \ (m \to \infty)$.

But by $(2°)$ and $(3°)$ the last term is finite and the desired conclusion follows.

Next observe that $E(0) = 0$; moreover, by $(1°)$, $(2°)$ for any $\varepsilon > 0$ there exists $C(\varepsilon)$ such that

$$G(x, u) \leq \varepsilon |u|^2 + C(\varepsilon)|u|^p$$

for all $u \in \mathbb{R}$ and almost every $x \in \Omega$. It follows that

$$E(u) \geq \frac{1}{2} \int_\Omega |\nabla u|^2 \, dx - \varepsilon \int_\Omega |u|^2 \, dx - C(\varepsilon) \int_\Omega |u|^p \, dx$$

$$\geq \left(\frac{1}{2} - \frac{\varepsilon}{\lambda_1} - C(\varepsilon)\|u\|_{H_0^{1,2}}^{p-2} \right) \|u\|_{H_0^{1,2}}^2 \geq \alpha > 0 \ ,$$

provided $\|u\|_{H_0^{1,2}} = \rho$ is sufficiently small. Here λ_1 denotes the first eigenvalue of $-\Delta$ in Ω with homogeneous Dirichlet boundary conditions, given by the Rayleigh-Ritz quotient $(I.2.4)$. Moreover, we have used the Sobolev embedding $H_0^{1,2}(\Omega) \hookrightarrow L^p(\Omega)$; see Theorem A.5 of the appendix.

Finally, condition $(3°)$ can be restated as a differential inequality for the function G, of the form

$$u|u|^q \frac{d}{du}\left(|u|^{-q}G(x, u)\right) \geq 0 \ , \quad \text{for } |u| \geq R_0 \ .$$

Upon integration we infer that for $|u| \geq R_0$ we have

(6.4)
$$G(x, u) \geq \gamma_0(x)|u|^q$$

with $\gamma_0(x) = R_0^{-q} \min \left\{ G(x, R_0), \, G(x, -R_0) \right\} > 0$. Hence, if $u \in H_0^{1,2}(\Omega)$ does not vanish identically, we obtain that for some constant $c(u) > 0$ there holds

(6.5)
$$E(\lambda u) = \frac{\lambda^2}{2} \int_\Omega |\nabla u|^2 \, dx - \int_\Omega G(x, \lambda u) \, dx$$

$$\leq C(u)\lambda^2 - c(u)\lambda^q + \mathcal{L}^n(\Omega) \operatorname*{ess\,inf}_{x \in \Omega, |v| \leq R_0} |G(x, v)| \ ,$$

$$\to -\infty \qquad \text{as } \lambda \to \infty \ .$$

But then, for fixed $u \neq 0$ and sufficiently large $\lambda > 0$ we may let $u_1 = \lambda u$, and from Theorem 6.1 we obtain the existence of a non-trivial solution u to (6.1), (6.2).

In order to obtain solutions $u^+ \geq 0 \geq u_-$, we may truncate g above or below $u = 0$, replacing g by 0, if $\pm u \leq 0$. Denote the truncated functions by $g^\pm(x, u)$, with primitive $G^\pm(x, u) = \int_0^u g^\pm(x, v) \, dv$. Note that $(1°)$, $(2°)$ remain valid for $g^\pm$ while $(3°)$ will hold for $\pm u \geq R_0$, almost everywhere in Ω. Moreover, for $\pm u \leq 0$ all terms in $(3°)$ vanish.

Denote

$$E^\pm(u) = \frac{1}{2} \int_\Omega |\nabla u|^2 \, dx - \int_\Omega G^\pm(x, u) \, dx$$

the functional related with $g^\pm$. Then as above $E^\pm \in C^1(H_0^{1,2}(\Omega))$ satisfies (P.-S.), $E^\pm(0) = 0$, and condition (2°) of Theorem 6.1 holds. Moreover, choosing a comparison function $u > 0$, and letting $u_1 = \lambda u$ for large positive, respectively negative λ, by (6.4), (6.5) also condition (3°) of Theorem 6.1 is satisfied. Our former reasoning then yields a non-trivial critical point $u^\pm$ of $E^\pm$, weakly solving the equation

$$-\Delta u^\pm = g^\pm(\cdot, u^\pm) \quad \text{in } \Omega .$$

Since $g^\pm(u) = 0$ for $\pm u \leq 0$, by the weak maximum principle $\pm u^\pm \geq 0$; see Theorem B.5 of the appendix. Hence the functions $u^\pm$ in fact are weak solutions of the original equation (6.1).

Finally, if g is Hölder continuous, by (2°) and Lemma B.3 of the appendix and the remarks following it, $\pm u \in C^{2,\alpha}(\overline{\Omega})$ for some $\alpha > 0$. Hence the strong maximum principle gives $u^+ > 0 > u_-$, as desired. (For a related problem a similar reasoning was used by Ambrosetti-Lupo [1].) □

The Symmetric Mountain Pass Lemma

More generally, for problems which are invariant under the involution $u \mapsto -u$, we expect the existence of infinitely many solutions, as in the case of problem (I.2.1), (I.2.3); see Theorem 5.8. However, if a problem does not exhibit the particular homogeneity of problem (I.2.1), (I.2.3), in general it cannot be reduced to a variational problem on the unit sphere in L^p, and a global method is needed. Fortunately, there is a "higher-dimensional" version of Theorem 6.1, especially adapted to functionals with a $\mathbb{Z}_2$-symmetry, the symmetric mountain pass lemma – again due to Ambrosetti-Rabinowitz [1]:

6.3 Theorem. *Suppose $E \in C^1(V)$ is even, that is $E(u) = E(-u)$, and satisfies (P.-S.). Let $V^+, V^- \subset V$ be closed subspaces of V with* codim $V^+ \leq$ dim $V^- < \infty$ *and suppose there holds*
(1°) $E(0) = 0$,
(2°) $\exists \alpha > 0$, $\rho > 0$ $\forall u \in V^+ : \|u\| = \rho \Rightarrow E(u) \geq \alpha$.
(3°) $\exists R > 0$ $\forall u \in V^- : \|u\| \geq R \Rightarrow E(u) \leq 0$.
Then for each j, $1 \leq j \leq k = $ dim V^- $-$ codim V^+ the numbers

$$\beta_j = \inf_{h \in \Gamma} \sup_{u \in V_j} E(h(u))$$

are critical, where

$$\Gamma = \{h \in C^0(V; V); h \text{ is odd}, \ h(u) = u \text{ if } u \in V^- \text{ and } \|u\| \geq R\} ,$$

and where $V_j \subset V^-$ is a fixed subspace of dimension

$$\dim \ V_j = \text{codim } V^+ + j .$$

Moreover, $\beta_k \geq \beta_{k-1} \geq \ldots \geq \beta_1 \geq \alpha$.

For the proof of Theorem 6.3 we need a topological lemma.

6.4 Intersection Lemma. *Let V, V^+, V^-, Γ, V_j, R be as in Theorem 6.3. Then for any $\rho > 0$, any $h \in \Gamma$ there holds*

$$\gamma\big(h(V_j) \cap S_\rho \cap V^+\big) = j \ ,$$

where S_ρ denotes the ρ-sphere $S_\rho = \{u \in V; \|u\| = \rho\}$, and γ denotes the Krasnoselskii genus introduced in Section 5.1; in particular $h(V_j) \cap S_\rho \cap V^+ \neq \emptyset$.

Proof. Denote $S_\rho^+ = S_\rho \cap V^+$. For any $h \in \Gamma$ the set $A = h(V_j) \cap S_\rho^+$ is symmetric, compact, and $0 \notin A$. Thus by Proposition 5.4.($5°$) there exists a neighborhood U of A such that $\gamma(\overline{U}) = \gamma(A)$. Then

$$\gamma\big(h(V_j) \cap S_\rho^+\big) \geq \gamma\big(h(V_j) \cap S_\rho \cap \overline{U}\big)$$
$$\geq \gamma\big(h(V_j) \cap S_\rho\big) - \gamma\big(h(V_j) \cap S_\rho \setminus U\big) \ .$$

But, if π denotes projection onto V^-, since U is a neighborhood of $h(V_j) \cap S_\rho^+$ we have $\pi\big(h(V_j) \cap S_\rho \setminus U\big) \not\ni 0$, and hence

$$\gamma\big(h(V_j) \cap S_\rho \setminus U\big) \leq \dim V^- < \infty \ .$$

On the other hand, by Proposition 5.4.($2°$), ($4°$)

$$\gamma\big(h(V_j) \cap S_\rho\big) \geq \gamma\big(h^{-1}(S_\rho) \cap V_j\big) \ .$$

But $h(0) = 0$, $h = id$ on $V_j \setminus B_R(0)$; hence $h^{-1}(S_\rho) \cap V_j$ bounds a symmetric neighborhood of the origin in V_j. Thus, by Proposition 5.2 we conclude that

$$\gamma\big(h(V_j) \cap S_\rho\big) \geq \dim V_j = \dim V^- + j \ ,$$

which implies the lemma. □

Proof of Theorem 6.3. By Lemma 6.4 we have $\beta_j \geq \alpha$ for all $j \in \{1, \ldots, k\}$. Moreover, clearly $\beta_{j+1} \geq \beta_j$ for all j. Suppose by contradiction that β_j is regular for some $j \in \{1, \ldots, k\}$. For $\beta = \beta_j$, $N = \emptyset$, and $\overline{\varepsilon} = \alpha$ let $\varepsilon > 0$ and Φ be as constructed in Theorem 3.4. By Remark 3.5.($4°$) we may assume that $\Phi(\cdot, t)$ is odd for any $t \geq 0$.

Note that by choice of $\overline{\varepsilon}$ we have $\Phi(\cdot, t) \circ h \in \Gamma$ for any $h \in \Gamma$. Choose $h \in \Gamma$ such that there holds $E(h(u)) \leq \beta + \varepsilon$ for all $u \in V_j$. Then $h_1 = \Phi(\cdot, 1) \circ h \in \Gamma$, and for all $u \in V_j$ there holds

$$E(h_1(u)) = E(\Phi(h(u), 1)) < \beta - \varepsilon \ ,$$

contradicting the definition of β. □

Note that in contrast to Theorem 5.7 from Theorem 6.3 we do not obtain optimal multiplicity results in the case of degenerate critical values $\beta_j = \beta_{j+1} = \ldots = \beta_k$. However, for most applications this defect does not matter. (Actually, by using a notion of pseudo-index, Bartolo-Benci-Fortunato [1; Theorem 2.4] have been able to prove optimal multiplicity results also in the case of degenerate critical values.) The following result is typical.

6.5 Theorem. *Suppose V is an infinite dimensional Banach space and suppose $E \in C^1(V)$ satisfies (P.-S.), $E(u) = E(-u)$ for all u, and $E(0) = 0$. Suppose $V = V^- \oplus V^+$, where V^- is finite dimensional, and assume the following conditions:*

($1°$) $\exists \alpha > 0$, $\rho > 0 : \|u\| = \rho$, $u \in V^+ \Rightarrow E(u) \geq \alpha$.

($2°$) For any finite dimensisonal subspace $W \subset V$ there is $R = R(W)$ such that $E(u) \leq 0$ for $u \in W$, $\|u\| \geq R$.

Then E possesses an unbounded sequence of critical values.

Proof. Choose a basis $\{\varphi_1, \ldots\}$ for V^+ and for $k \in \mathbb{N}$ let $W_k = V^- \oplus$ span $\{\varphi_1, \ldots, \varphi_k\}$, with $R_k = R(W_k)$. Since $W_k \subset W_{k+1}$, we may assume that $R_k \leq R_{k+1}$ for all k.

Define classes

$$\Gamma_k = \big\{ h \in C^0(V; V) \; ; \; h \text{ is odd,}$$

$$\forall j \leq k, \ u \in W_j \; : \; \|u\| \geq R_j \Rightarrow h(u) = u \big\}$$

and let

$$\beta_k = \inf_{h \in \Gamma_k} \sup_{u \in W_k} E\big(h(u)\big) \ .$$

By Theorem 6.3, choosing $V_k = W_k$, each of these numbers defines a critical value $\beta_k \geq \alpha$ of E.

The proof therefore will be complete when we show the following:

Assertion. *The sequence (β_k) is unbounded.*

First observe that, since $\Gamma_k \supset \Gamma_{k+1}$ while $W_k \subset W_{k+1}$ for all k, the sequence (β_k) is non-decreasing. Suppose by contradiction that $\sup_k \beta_k = \beta < \infty$. Note that $K = K_\beta$ is compact and symmetric. Moreover, since $\beta \geq \alpha > 0$ we also have $0 \notin K$. By Proposition 5.4.($5°$) then $\gamma(K) < \infty$, and there exists a symmetric neighborhood N of K such that $\gamma(\overline{N}) = \gamma(K_\beta)$.

Choose $\bar{\varepsilon} = \alpha > 0$, and let $\varepsilon \in]0, \bar{\varepsilon}[$ and Φ be determined according to Theorem 3.4 corresponding to $\bar{\varepsilon}, \beta$, and N. Observe that Φ may be chosen to be odd, see Remark 3.5.($4°$) , and that by choice of $\bar{\varepsilon}$ for any $j \in \mathbb{N}$ we have $\Phi(u, t) = u$ for all $t \geq 0$ and any $u \in W_j$ such that $\|u\| \geq R_j$.

Let $\beta = \beta^{(0)}$, $\varepsilon = \varepsilon^{(0)}$, $\Phi = \Phi_{\beta^{(0)}}$. We iterate the above procedure. For each number $\beta' \in [\alpha, \beta - \varepsilon]$ also the set $K' = K_{\beta'}$ is compact, has finite genus $\gamma(K_{\beta'})$, and possesses a neighborhood $N' = N_{\beta'}$ with $\gamma(\overline{N}') = \gamma(K')$. Moreover, for each such β', with $\bar{\varepsilon} = \alpha$ and $N = N_{\beta'}$ we may let $\varepsilon_{\beta'} \in]0, \bar{\varepsilon}[$, $\Phi_{\beta'}$ be determined according to Theorem 3.4. Finitely many neighborhoods $]\beta^{(l)} - \varepsilon^{(l)}, \beta^{(l)} + \varepsilon^{(l)}[$, $l = 1, \ldots, L$, where $\beta^{(l)} \in [\alpha, \beta - \varepsilon]$, $\varepsilon^{(l)} = \varepsilon_{\beta^{(l)}}$, cover $[\alpha, \beta - \varepsilon]$. Clearly, we may assume

$$\beta = \beta^{(0)} \geq \beta^{(1)} \geq \ldots \geq \beta^{(l)} \geq \beta^{(l+1)} \geq \ldots \geq \beta^{(L)} \geq \alpha$$

and also that

$$\beta^{l-1} > \beta^l + \varepsilon^l > \beta^{l-1} - \varepsilon^{l-1} > \beta^l$$

for all l, $1 \le l \le L$.

For any $k \in \mathbb{N}$, with $\varepsilon = \varepsilon^{(0)} > 0$ as above, now choose $h \in \Gamma_k$ satisfying

$$\sup_{u \in W_k} E\left(h(u)\right) \le \beta_k + \varepsilon \le \beta + \varepsilon .$$

Composing h with $\Phi = \Phi_{\beta^{(0)}}$ yields a map $\Phi(\cdot, 1) \circ h =: h' \in \Gamma_k$ with

$$h'(W_k) \subset E_{\beta - \varepsilon} \cup N .$$

Consider the compositions

$$H^{(m,m)} = id, \ H^{m,m-1} = \Phi_{\beta^{(m-1)}}(\cdot, 1) ,$$

$$H^{(m,l)} = \Phi_{\beta^{(m-1)}}(\cdot, 1) \circ \ldots \circ \Phi_{\beta^{(l)}}(\cdot, 1), \qquad 0 \le l < m \le L+1$$

and let $h^{(0)} = h$, $h^{(l)} = H^{(l,0)} \circ h$, for $l = 1, \ldots, L+1$. Note that $h' = h^{(1)}$ in this notation. Moreover, by Theorem 3.4 each $H^{(m,l)}$ is the composition of homeomorphisms, hence a homeomorphism itself.

By induction, letting $N^{(0)} = N$, $N^{(l)} = N_{\beta^{(l)}}$, $l = 1, \ldots, L$, we have

$$h^{(m+1)}(W_k) \subset E_{\beta^{(m)} - \varepsilon^{(m)}} \bigcup_{0 \le l \le m} H^{(m,l)}\left(N^{(l)}\right)$$

$$\subset E_{\beta^{(m+1)} + \varepsilon^{(m+1)}} \bigcup_{0 \le l \le m} H^{(m,l)}\left(N^{(l)}\right) ,$$

for all m, $0 \le m \le L$. Moreover, $h^{(m)} \subset \Gamma_k$ for all m, $0 \le m \le L+1$. In particular, $h^{(L+1)} \in \Gamma_k$ and

$$h^{(L+1)}(W_k) \subset E_\alpha \bigcup_{0 \le l \le L} H^{(L,l)}\left(N^{(l)}\right) .$$

Let $S_\rho^+ = \partial B_\rho(0; V^+)$. Then by assumption (1°) we have

$$h^{(L+1)}(W_k) \cap S_\rho^+ \subset \bigcup_{0 \le l \le L} \left(H^{(L,l)}\left(N^{(l)}\right) \cap S_\rho^+\right) \subset \bigcup_{0 \le l \le L} H^{(L,l)}\left(N^{(l)}\right) .$$

By monotonicity and sub-additivity of the genus γ, see Proposition 5.4.(2°), (3°), and since each map $H^{(L,l)}$ is a homeomorphism, this implies that

$$\gamma\left(h^{(L+1)}(W_k) \cap S_\rho^+\right) \le \sum_{0 \le l \le L} \gamma\left(H^{(L,l)}\left(\overline{N^{(l)}}\right)\right) =$$

(6.6)

$$= \sum_{0 \le l \le L} \gamma\left(\overline{(N^{(l)})}\right) = \sum_{0 \le l \le L} \gamma\left(K_{\beta^{(l)}}\right) =: k_0 < \infty ,$$

with k_0 independent of k.

On the other hand, by the Intersection Lemma 6.4, for any $k \in \mathbb{N}$ we have

$$(6.7) \qquad\qquad \gamma\left(h(W_k) \cap S_\rho^+\right) \geq k \ .$$

This holds for any $h \in \Gamma_k$, in particular, for $h = h^{(L+1)}$.

Since k is arbitrary, this contradicts (6.6); hence the proof of Theorem 6.5 is complete. $\qquad\square$

Applications to Semilinear Equations with Symmetry

As an application we state the following existence result for problems of the type (6.1), (6.2) involving odd nonlinearities g. Results of this kind are well-known; see Section 5. Note that the symmetry assumption allows to remove any conditions on g near $u = 0$.

6.6 Theorem. *Let Ω be a smoothly bounded domain in $\mathbb{R}^n$, $n \geq 3$, and let $g \colon \Omega \times \mathbb{R} \to \mathbb{R}$ be a Carathéodory function with primitive $G(\cdot, u) = \int_0^u g(\cdot, v) \, dv$. Suppose:*
(1°) g is odd: $g(x, -u) = -g(x, u)$,
and conditions $(2^\circ),(3^\circ)$ of Theorem 6.2 are satisfied, that is
(2°) $\exists p < 2^* = \frac{2n}{n-2}$, $C \colon |g(x, u)| \leq C\left(1 + |u|^{p-1}\right)$ almost everywhere,
(3°) $\exists q > 2$, $R_0 \colon 0 < q\, G(x, u) \leq g(x, u)u$ for almost every x, $|u| \geq R_0$.
Then problem (6.1), (6.2) admits an unbounded sequence (u_k) of solutions $u_k \in H_0^{1,2}(\Omega)$.

Proof. As in the proof of Theorem 6.2, define

$$E(u) = \frac{1}{2}\int_\Omega |\nabla u|^2 \, dx - \int_\Omega G(u) \, dx \ .$$

Hypothesis (2°) implies that E is Fréchet differentiable on $H_0^{1,2}(\Omega)$. The assertion of the theorem is equivalent to the assertion that E admits an unbounded sequence of critical points.

To prove the latter we invoke Theorem 6.5. Note that by the proof of Theorem 6.2 the functional E satisfies (P.-S.); see (6.3). Moreover, since g is odd, E is even: $E(u) = E(-u)$. Finally, $E(0) = 0$.

Denote $0 < \lambda_1 < \lambda_2 \leq \lambda_3 \leq \dots$ the eigenvalues of $-\Delta$ on Ω with homogeneous Dirichlet data, as usual, and let φ_j be the corresponding eigenfunctions.

We claim that for k_0 sufficiently large there exist $\rho > 0$, $\alpha > 0$ such that for all $u \in V^+ := \text{span}\{\varphi_k \ ; \ k \geq k_0\}$ with $\|u\|_{H_0^{1,2}} = \rho$ there holds $E(u) \geq \alpha$. Indeed, by (2°), Sobolev's embedding $H_0^{1,2}(\Omega) \hookrightarrow L^{2^*}(\Omega)$, and Hölder's inequality, for $u \in V^+$ we have (with constants C_1, C_2 independent of u)

$$
\begin{aligned}
E(u) &\geq \frac{1}{2}\int_\Omega |\nabla u|^2 \, dx - C\int_\Omega |u|^p \, dx - C \\
(6.8) \qquad &\geq \frac{1}{2}\|u\|_{H_0^{1,2}}^2 - C\, \|u\|_{L^2}^r \|u\|_{L^{2^*}}^{p-r} - C \\
&\geq \left(\frac{1}{2} - C_1 \lambda_{k_0}^{-r/2} \|u\|_{H_0^{1,2}}^{p-2}\right)\|u\|_{H_0^{1,2}}^2 - C_2
\end{aligned}
$$

where $\frac{r}{2} + \frac{p-r}{2^*} = 1$. In particular, $r = n(1 - \frac{p}{2^*}) > 0$, and we may let $\rho = 2\sqrt{(C_2 + 1)}$ and choose $k_0 \in \mathbb{N}$ such that

$$C_1 \lambda_{k_0}^{-r/2} \rho^{p-2} \le \frac{1}{4}$$

to achieve that

$$E(u) \ge 1 =: \alpha , \text{ for all } u \in V^+ \text{ with } \|u\|_{H_0^{1,2}} = \rho .$$

Now fix V^+ as above and denote $V^- = \operatorname{span}\{\varphi_j ; j < k_0\}$ its orthogonal complement.

Finally, on any finite dimensional subspace $W \subset H_0^{1,2}$, by $(3°)$ and (6.4–5) there exist constants $C_i = C_i(W) > 0$ such that

(6.9) $$\sup_{u \in \partial B_R(0;W)} E(u) \le C_1 R^2 - C_2 R^q + C_3 \to -\infty$$

as $R \to \infty$.

Hence, Theorem 6.5 guarantees the existence of an unbounded sequence of critical values

$$\beta_k = \inf_{h \in \Gamma_k} \sup_{u \in W_k} E(h(u)), \quad k \ge k_0 ,$$

where $W_k = \operatorname{span}\{\varphi_j ; j \le k\}$ and Γ_k is defined as in the proof of Theorem 6.5. The proof is complete.   □

6.7 Remark. We can estimate the rate at which the sequence (β_k) diverges. Indeed, by the Intersection Lemma 6.4 , letting $V_k = \operatorname{span}\{\varphi_j ; j \ge k\}$, $W_k = \operatorname{span}\{\varphi_1 \ldots, \varphi_k\}$ as in the proof above, we have

$$h(W_k) \cap S_\rho \cap V_k \neq \emptyset$$

for any $\rho > 0$, any $k \in \mathbb{N}$. Hence, by (6.8), with $r = n\left(1 - \frac{p}{2^*}\right) = p - \frac{n(p-2)}{2}$ we obtain

$$\beta_k \ge \sup_{\rho > 0} \inf_{u \in S_\rho \cap V_k} E(u)$$

$$\ge \sup_{\rho > 0} \left(\frac{1}{2} - C_1 \lambda_k^{-r/2} \rho^{p-2}\right) \rho^2 - C_2$$

$$\ge c\lambda_k^{\frac{r}{p-2}} = c\lambda_k^{\left(\frac{p}{p-2} - \frac{n}{2}\right)} ,$$

for k large. Finally, by the asymptotic formula (see Weyl [1], or Edmunds-Moscatelli [1])

$$\lambda_k k^{-2/n} \to c > 0 \qquad (k \to \infty) ,$$

and it follows that with a constant $c > 0$ we have

(6.10) $$\beta_k \ge c k^{\frac{2p}{n(p-2)} - 1}$$

for large k. Estimate (6.10) will prove useful in the next section.

7. Perturbation Theory

A natural question which even today is not adequately settled is whether the symmetry of the functional is important for results like Theorem 6.5 to hold.

A partial answer for problems of the kind studied in Theorem 6.6 was independently obtained by Bahri-Berestycki [1] and Struwe [1] in 1980.

In abstract form, the variational principle underlying these results has been phrased by Rabinowitz [10]. His result provides an analogue for Theorem 6.5 in a non-symmetric setting:

7.1 Theorem. *Suppose $E \in C^1(V)$ satisfies (P.-S.). Let $W \subset V$ be a finite-dimensional subspace of V, $w^* \in V \setminus W$, and let $W^* = W \oplus \mathrm{span}\{w^*\}$; also let*

$$W_+^* = \{w + tw^* \; ; \; w \in W, \; t \geq 0\}$$

denote the "upper half-space" in W^*. Suppose
(1°) $E(0) = 0$,
(2°) $\exists R > 0 \; \forall u \in W : \; \|u\| \geq R \Rightarrow E(u) \leq 0$,
(3°) $\exists R^* \geq R \; \forall u \in W^* : \; \|u\| \geq R^* \Rightarrow E(u) \leq 0$,
and let

$$\Gamma = \{h \in C^0(V, V) \; ; \; h \text{ is odd}, \; h(u) = u \; \text{ if } \max\{E(u), E(-u)\} \leq 0 ,$$
$$\text{in particular, if } u \in W, \|u\| \geq R, \quad \text{or if } u \in W^*, \|u\| \geq R^*\} .$$

Then, if

$$\beta^* = \inf_{h \in \Gamma} \; \sup_{u \in W_+^*} \; E(h^*(u)) > \beta = \inf_{h \in \Gamma} \; \sup_{u \in W} \; E(h(u)) \geq 0 ,$$

the functional E possesses a critical value $\geq \beta^*$.

Proof. For $\gamma \in]\beta, \beta^*[$ let

$$\Lambda = \{h \in \Gamma \; ; \; E(h(u)) \leq \gamma \text{ for } u \in W\} .$$

By definition of β, clearly $\Lambda \neq \emptyset$. Hence

$$\gamma^* = \inf_{h \in \Lambda} \; \sup_{u \in W_+^*} \; E(h(u)) \geq \beta^*$$

is well-defined. We contend that γ^* is critical.

Assume by contradiction that γ^* is regular and choose $\varepsilon > 0$, Φ according to Theorem 3.4, corresponding to γ^*, $\bar{\varepsilon} = \gamma^* - \gamma > 0$, $N = \emptyset$. Also select $h \in \Lambda$ with

$$\sup_{u \in W_+^*} \; E(h(u)) < \gamma^* + \varepsilon .$$

Define an odd map $h' \colon W^* \to V$

$$h'(u) = \begin{cases} \Phi\left(h(u), 1\right), & \text{if } u \in W_+^* \\ -\Phi\left(h(-u), 1\right), & \text{if } -u \in W_+^* . \end{cases}$$

Note that by choice of $\bar{\varepsilon}$ and since $h \in \Lambda$ we have

$$\Phi\left(h(-u), 1\right) = h(-u) = -h(u) = -\Phi\left(h(u), 1\right)$$

for $u \in W$. Hence h' is well-defined, odd, and continuous. By the Tietze extension theorem h' may be extended to an odd map $h' \in C^0(V; V)$. Moreover, since $0 \le \beta < \gamma < \gamma^* - \bar{\varepsilon}$, the map $\Phi(\cdot, 1)$ fixes any point u that satisfies $E(u) \le 0$ and $E(-u) \le 0$. By definition of Λ so does h, and hence the composition $\Phi(\cdot, 1) \circ h$. It follows that $h' \in \Lambda$. But now the estimate

$$\sup_{u \in W_+^*} E\left(h'(u)\right) = \sup_{u \in W_+^*} E\left(\Phi\left(h(u), 1\right)\right) < \gamma^* - \varepsilon$$

yields the desired contradiction. □

Theorem 7.1 suggests to compare a non-symmetric functional $\tilde{E}$ satisfying the remaining hypotheses (1°), (2°) of Theorem 6.5 with its symmetrization $E(u) = \frac{1}{2}\left(\tilde{E}(u) + \tilde{E}(-u)\right)$. Then Theorem 6.5 applies to E and we obtain an unbounded sequence (β_k) of critical values of E. Playing the speed of divergence $\beta_k \to \infty$ off against the perturbation from symmetry $E - \tilde{E}$, from Theorem 7.1 we can glean the existence of infinitely many critical points for functionals perturbed from symmetry.

Applications to Semilinear Elliptic Equations

Stating a general theorem of this kind, however, seems to involve so many technical conditions that any reader would doubtfully ask himself if these conditions can ever be met in practice. Therefore we prefer to present such an application immediately. Moreover, we return to the setting of problem (6.1),(6.2) under (essentially) the assumptions (1°)–(3°) of Theorem 6.6 studied previously. However, in order to make estimate (6.4) uniform in $x \in \Omega$, we assume that g is continuous in all its variables. For convenience we restate also the remaining conditions.

7.2 Theorem. *Let Ω be a smoothly bounded domain in $\mathbb{R}^n, n \ge 3$. Suppose*
(1°) $g: \Omega \times \mathbb{R} \to \mathbb{R}$ is continuous and odd with primitive $G(x, u) = \int_0^u g(x, v)\, dv$;
(2°) $\exists p < 2^* = \frac{2n}{n-2}$, $C: |g(x, u)| \le C\left(1 + |u|^{p-1}\right)$ almost everywhere;
(3°) $\exists q > 2$, $R_0 : 0 < q\, G(x, u) \le g(x, u)u$ for almost every x, $|u| \ge R_0$.
Moreover, suppose that

$$\frac{2p}{n(p-2)} - 1 > \frac{q}{q-1} .$$

Then for any $f \in L^2(\Omega)$ the problem

$$-\Delta u = g(\cdot, u) + f \qquad in \ \Omega \ ,$$
$$u = 0 \qquad on \ \partial\Omega \ ,$$

has an unbounded sequence of solutions $u_k \in H_0^{1,2}(\Omega)$, $k \in \mathbb{N}$.

Proof. The assertion of the theorem is equivalent to the claim that the functional

$$\tilde{E}(u) = \frac{1}{2} \int_\Omega |\nabla u|^2 \ dx - \int_\Omega G(\cdot, u) \ dx - \int_\Omega fu \ dx$$

possesses an unbounded sequence of critical points in $H_0^{1,2}(\Omega)$.

Clearly, $\tilde{E}(0) = 0$, and on any finite-dimensional subspace $W \subset H_0^{1,2}(\Omega)$ by (3°) and (6.4–5) the estimate

$$(7.1) \qquad \sup_{u \in \partial B_\rho(0;W)} \tilde{E}(u) \leq C_1 \rho^2 - C_2 \rho^q + C_3 \rho + C_4 \to -\infty \qquad (\rho \to \infty)$$

holds with constants $C_i = C_i(W) > 0$. (The term $\int_\Omega fu \ dx$ is simply bounded by using Hölder's inequality.) To see (P.-S.), as in (6.3) in the proof of Theorem 6.2, for a (P.-S.) sequence (u_m) consider the estimate

$$C + o(1) \|u_m\|_{H_0^{1,2}} \geq q\tilde{E}(u_m) - \langle u_m, D\tilde{E}(u_m)\rangle$$
$$= q \, E(u_m) - \langle u_m, DE(u_m)\rangle - (q-1) \int_\Omega fu \ dx$$
$$\geq \frac{q-2}{2} \|u_m\|_{H_0^{1,2}}^2 - C\|u_m\|_{H_0^{1,2}} - C \ ,$$

from which boundedness of (u_m) in $H_0^{1,2}(\Omega)$ follows at once. (Note that the even symmetrization $E(u) = \frac{1}{2}\big(\tilde{E}(u) + \tilde{E}(-u)\big)$ of $\tilde{E}$ equals the functional E considered in Theorem 6.6.) Using Proposition 2.2, the existence of a strongly convergent subsequence of (u_m) now is established exactly as in the proof of Theorem 6.2.

Let λ_j, φ_j, $V^+ = \text{span}\{\varphi_k \; ; \; k \geq k_0\}$, $W_k = \text{span}\{\varphi_j \; ; \; j \leq k\}$ as in the proof of Theorem 6.6. By (7.1) we may choose a non-decreasing sequence (R_k) such that $\tilde{E}(u) \leq 0$ for all $u \in W_k$ with $\|u\|_{H_0^{1,2}} \geq R_k$; hence also $E(u) \leq 0$ for all such u. With this sequence (R_k) let (Γ_k) be defined as in the proof of Theorem 6.5 and for $k \geq k_0$ let

$$(7.2) \qquad \beta_k = \inf_{h \in \Gamma_k} \sup_{u \in W_k} E\left(h(u)\right) \geq c \, k^{\frac{2p}{n(p-2)} - 1} - \beta_0$$

be the sequence of critical values of the symmetrized functional E constructed in Theorem 6.6, the lower estimate following from (6.10). Suppose that $\tilde{E}$ does not have any critical values larger than some number $\overline{\beta} - 1$. Fix $\overline{k} \geq k_0$ such that $\beta_k \geq \overline{\beta}$ for $k \geq \overline{k}$. We will use Theorem 7.1 to derive a uniform bound

$$\beta_k \leq C, \qquad \text{for all } k \geq \overline{k} \ ,$$

which will yield the desired contradiction to (7.2).

For $k \geq \bar{k}$ let $W = W_k$, $w^* = \varphi_{k+1}$, $W^* = W_{k+1}$, and let Γ be defined as in Theorem 7.1, with $R = R_k$, $R^* = R_{k+1}$. Note that $\Gamma \subset \Gamma_{k+1} \subset \Gamma_k$. Since by assumption there are no critical values $\beta > \bar{\beta} - 1$ for $\tilde{E}$, from Theorem 7.1 we conclude that

$$\tilde{\beta}^* := \tilde{\beta}_k^* = \inf_{h^* \in \Gamma} \sup_{u \in W_+^*} \tilde{E}\left(h^*(u)\right) =$$

$$= \tilde{\beta} := \tilde{\beta}_k = \inf_{h \in \Gamma} \sup_{u \in W} \tilde{E}\left(h(u)\right) \ .$$

From oddness of any map $h \in \Gamma$ and the estimate

$$E(u) = \frac{1}{2}\left(\tilde{E}(u) + \tilde{E}(-u)\right) \leq \max\{\tilde{E}(u), \tilde{E}(-u)\} \ ,$$

valid for all u, we also deduce that

$$(7.3) \qquad \tilde{\beta}_k \geq \inf_{h \in \Gamma} \sup_{u \in W} E\left(h(u)\right) \geq \beta_k \geq \bar{\beta} \ , \text{ for all } k \geq \bar{k} \ .$$

Given $\varepsilon > 0$ (say $\varepsilon = 1$) choose a map $h^* \subset \Gamma$ such that

$$\sup_{u \in W} \tilde{E}\left(h^*(u)\right) \leq \sup_{u \in W_+^*} \tilde{E}\left(h^*(u)\right) < \tilde{\beta}_k^* + \varepsilon = \tilde{\beta}_k + \varepsilon \ .$$

Let $\|\cdot\|$ denote the norm in $H_0^{1,2}(\Omega)$, respectively $H^{-1}(\Omega)$. Consider a pseudo-gradient vector field v for E such that

$$\|v\| \leq 2, \ \langle v(u), DE(u)\rangle \geq \min\{1, \|DE(u)\|\}\|DE(u)\| \ .$$

Since E is even, we may assume v to be odd. Note that

$$\langle v(u), D\tilde{E}(u)\rangle = \langle v(u), DE(u)\rangle + \langle v(u), D\tilde{E}(u) - DE(u)\rangle$$

$$\geq \min\{1, \|DE(u)\|\} \|DE(u)\| - 2\|D\tilde{E}(u) - DE(u)\|$$

$$\geq \min\{1, \|D\tilde{E}(u)\|\} \|D\tilde{E}(u)\| - 4 \|D\tilde{E}(u) - DE(u)\| \ .$$

That is, v also is a pseudo-gradient vector field for $\tilde{E}$ off $\tilde{N}_\delta$ for any $\delta \geq 8$, where for $\delta > 0$ the set $\tilde{N}_\delta$ is given by

$$\tilde{N}_\delta = \left\{u \in H_0^{1,2}(\Omega) \ ; \ \|D\tilde{E}(u) - DE(u)\| > \delta^{-1} \min\{1, \|D\tilde{E}(u)\|\}\|D\tilde{E}(u)\|\right\} \ .$$

Note that – unless $f \equiv 0$, which is trivial – $\tilde{N}_\delta$ is a neighborhood of the set of critical points of E and $\tilde{E}$, for any $\delta > 1$. Denote $N_\delta = \tilde{N}_\delta \cup \left(-\tilde{N}_\delta\right)$ the symmetrized sets $\tilde{N}_\delta$. Let η be Lipschitz, even, $0 \leq \eta \leq 1$ and satisfy $\eta(u) = 0$ in N_{10} and $\eta(u) = 1$ off $N = N_{20}$. Also let φ be Lipschitz, $0 \leq \varphi \leq 1$, $\varphi(s) = 0$ for $s \leq 0$, $\varphi(s) = 1$ for $s \geq 1$, and let

$$e(u) = -\varphi\left(\max\{\tilde{E}(u), \tilde{E}(-u)\}\right)\eta(u)v(u)$$

denote the truncated pseudo-gradient vector field on $H_0^{1,2}(\Omega)$.

Then, if Φ denotes the (odd) pseudo-gradient flow for E induced by e, Φ will also be a pseudo-gradient flow for $\tilde{E}$ and will strictly decrease $\tilde{E}$ off N with "speed"

$$\frac{d}{dt}\tilde{E}\big(\Phi(u,t)\big)|_{t=0} \leq \frac{1}{2}\min\big\{1,\|D\tilde{E}(u)\|^2\big\} , \qquad \text{if } \tilde{E}(u) \geq 1 .$$

Moreover, $\Phi(\cdot,t) \in \Gamma$ for all $t \geq 0$.

Composing h^* with $\Phi(\cdot,t)$, unless $\tilde{E}$ achieves its supremum on $\Phi\big(h^*(W^*),t\big)$ at a point $u \in N$, thus we obtain that

$$\frac{d}{dt}\sup_{u\in W^*}\tilde{E}\big(\Phi\big(h^*(W^*),t\big)\big) \leq -c < 0 ,$$

uniformly in $t \geq 0$. (Here, condition (P.-S.) was used; the constant c may depend on our initial choice for the map h.) Replacing h^* by $\Phi(\cdot,t) \circ h^*$ and choosing t large, if necessary, we hence may assume that $\tilde{E}$ achieves its supremum on $h^*(W^*)$ in N. That is, we may estimate

(7.4)
$$\begin{aligned}
\tilde{\beta}_k + \varepsilon = \tilde{\beta}_k^* + \varepsilon &\geq \sup_{u\in W^*}\tilde{E}\big(h^*(u)\big) - \sup_{\substack{u\in N,\\ \tilde{E}(u)\leq\tilde{\beta}_k+\varepsilon}}|\tilde{E}(u) - \tilde{E}(-u)| \\
&= \sup_{u\in W^*}\tilde{E}\big(h^*(u)\big) - \sup_{\substack{u\in N,\\ \tilde{E}(u)\leq\tilde{\beta}_k+\varepsilon}}\big|2\int_\Omega fu\,dx\big| \\
&\geq \tilde{\beta}_{k+1} - c\sup_{\substack{u\in N\\ \tilde{E}(u)\leq\tilde{\beta}_k+\varepsilon}}\|u\|_{L^2} .
\end{aligned}$$

But for $u \in N$ with $\tilde{E}(u) \leq \tilde{\beta}_k + \varepsilon$, with constants $c > 0$ from (3°) and $(6.4\text{--}5)$ we obtain

$$\begin{aligned}
\|u\|_{L^2}^q \leq c\,\|u\|_{L^q}^q &\leq c\left(\frac{q-2}{q}\right)\int_\Omega G(u)\,dx + C \\
&\leq c\int_\Omega\left(\frac{1}{2}g(u)u - G(u)\right)dx + C \\
&= c\left(\tilde{E}(u) - \frac{1}{2}\langle u, D\tilde{E}(u)\rangle + \frac{1}{2}\int_\Omega fu\,dx\right) + C \\
&\leq c\,\tilde{E}(u) + c\left(1 + \|D\tilde{E}(u) - D\tilde{E}(-u)\|\right)\|u\|_{H_0^{1,2}} + c\,\|u\|_{L^2} + C \\
&\leq c\,\tilde{\beta}_k + c\,\|u\|_{H_0^{1,2}} + C .
\end{aligned}$$

From this there follows

$$\begin{aligned}
\|u\|_{H_0^{1,2}}^2 &\leq 2\,\tilde{E}(u) + 2\int_\Omega G(u)\,dx + C\|u\|_{L^2} \\
&\leq c\,\tilde{E}(u) + c\,\|u\|_{H_0^{1,2}} + C .
\end{aligned}$$

Together with the former estimate this shows that for such u there holds

$$\|u\|_{L^2} \leq c\, \tilde{\beta}_k^{1/q} \ .$$

Inserting this estimate in (7.4) above yields the inequality

$$\tilde{\beta}_{k+1} \leq \tilde{\beta}_k + c\, \tilde{\beta}_k^{1/q} = \tilde{\beta}_k \left(1 + c\, \tilde{\beta}_k^{\frac{1-q}{q}} \right) ,$$

for all $k \geq k_0$, with an uniform constant c. By iteration therefore

$$\tilde{\beta}_{k_0+l} \leq \tilde{\beta}_{k_0} \prod_{k=k_0}^{k_0+l-1} \left(1 + c\, \tilde{\beta}_k^{\frac{1-q}{q}} \right)$$

$$\leq \tilde{\beta}_{k_0}\, exp \left(\sum_{k=k_0}^{k_0+l-1} ln \left(1 + c\, \tilde{\beta}_k^{\frac{1-q}{q}} \right) \right)$$

$$\leq \tilde{\beta}_{k_0}\, exp \left(c \sum_{k=k_0}^{k_0+l-1} \tilde{\beta}_k^{\frac{1-q}{q}} \right) .$$

But by (7.2), (7.3) we have

$$\tilde{\beta}_k \geq \beta_k \geq c\, k^{\frac{2p}{n(p-2)}-1} \ .$$

Hence, if

$$\mu = \frac{1-q}{q} \cdot \left(\frac{2p}{n(p-2)} - 1 \right) < -1 \ ,$$

we can uniformly estimate

$$\tilde{\beta}_{k_0+l} \leq \tilde{\beta}_{k_0}\, exp \left(c \sum_{k=k_0}^{k_0+l-1} \tilde{\beta}_k^{\frac{1-q}{q}} \right)$$

$$\leq \tilde{\beta}_{k_0}\, exp \left(c \sum_{k=k_0}^{\infty} k^{\mu} \right) < \infty \ ,$$

for all $l \in \mathbb{N}$, which yields the desired contradiction. Thus, $\tilde{E}$ possesses an unbounded sequence of critical values and the proof is complete. □

Remark 7.3. (1°) Recently, Bahri-Lions [1] have been able to improve the estimate for β_k as follows

$$\beta_k \geq c\, k^{\frac{2}{n} \cdot \frac{p}{p-2}}$$

which yields the improved bounds

$$\frac{q}{q-1} < \frac{2}{n} \cdot \frac{p}{p-2}$$

for p and q for which Theorem 7.2 is valid. Bahri [1] moreover has shown that for any $p \in]2, 2^*[$ there is a dense open set of $f \in H^{-1}(\Omega)$ for which the problem

$$-\Delta u = u \, |u|^{p-2} + f \qquad \text{in } \Omega$$
$$u = 0 \qquad \text{on } \partial\Omega$$

possesses infinitely many solutions.

Generally, it is expected that this should be true for *all* f (for instance in $L^2(\Omega)$), also for more general problems of type (6.1), (6.2) considered earlier. Indeed, this is true in the case of ordinary differential equtions – see for example Nehari [1], Struwe [2], or Turner [1] – and in the radially symmetric case, where different solutions of (6.1),(6.2) can be characterized by the nodal properties they possess; see Struwe [3], [4]. However, the proof of this conjecture in general remains open.

(2°) Applications of perturbation theory to Hamiltonian systems are given by Bahri-Berestycki [2], Rabinowitz [10], and – more recently – by Long [1], [2].

(3°) Finally, Tanaka [1] has obtained similar perturbation results for nonlinear wave equations, as in Section I.6.

8. Linking

In the preceding chapter we have seen that symmetry is not essential for a variational problem to have "many" solutions. In this chapter we will describe another method for dealing with non-symmetric functionals. This method was introduced by Benci [1], Ni [1], and Rabinowitz [8]. Subsequently, it was generalized to a setting possibly involving also "indefinite" functionals by Benci-Rabinowitz [1]. It is based on the topological notion of "linking".

8.1 Definition. *Let S be a closed subset of a Banach space V, Q a submanifold of V with relative boundary ∂Q. We say S and ∂Q link if*

(1°) $S \cap \partial Q = \emptyset$, and

(2°) for any map $h \in C^0(V; V)$ such that $h|_{\partial Q} = id$ there holds $h(Q) \cap S \neq \emptyset$. More generally, if S and Q are as above and if Γ is a subset of $C^0(V; V)$, then S and ∂Q will be said to link with respect to Γ if (1°) holds and (2°) is satisfied for any $h \in \Gamma$.

8.2 Example. Let $V = V_1 \oplus V_2$ be decomposed into closed subspaces V_1, V_2, where $\dim V_2 < \infty$. Let $S = V_1$, $Q = B_R(0; V_2)$ with relative boundary $\partial Q = \{u \in V_2 \; ; \; \|u\| = R\}$. Then S and ∂Q link.

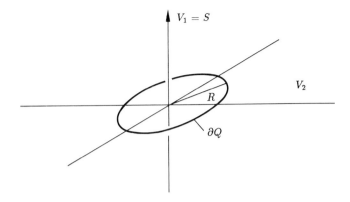

Fig. 8.1.

Proof. Let $\pi: V \to V_2$ be the (continuous) projection of V onto V_2, and let h be any continuous map such that $h|_{\partial Q} = id$. We have to show that $\pi\big(h(Q)\big) \ni 0$.

For $t \in [0,1]$, $u \in V_2$ define

$$h_t(u) = t\,\pi\big(h(u)\big) + (1-t)u \;.$$

Note that h_t defines a homotopy of $h_0 = id$ with $h_1 = \pi \circ h$. Moreover, $h_t|_{\partial Q} = id$ for all t. Hence the topological degree $\deg(h_t, Q, 0)$ is well-defined for all t. By homotopy invariance and normalization of the degree (see for instance Deimling [1; Theorem 1.3.1]), we have

$$\deg(\pi \circ h, Q, 0) = \deg(id, Q, 0) = 1 \;.$$

Hence $0 \in \pi \circ h(Q)$, as was to be shown. $\qquad\qquad\qquad\qquad\square$

8.3 Example. Let $V = V_1 \oplus V_2$ be decomposed into closed subspaces with $\dim V_2 < \infty$, and let $\underline{u} \in V_1$ with $\|\underline{u}\| = 1$ be given. Suppose $0 < \rho < R_1$, $0 < R_2$ and let

$$S = \{u \in V_1 \;;\; \|u\| = \rho\}\;,$$
$$Q = \{s\underline{u} + u_2 \;;\; 0 \le s \le R_1,\; u \in V_2, \|u_2\| \le R_2\}\;,$$

with relative boundary $\partial Q = \{s\underline{u} + u_2 \in Q \;;\; s \in \{0, R_1\} \text{ or } \|u_2\| = R_2\}$. Then S and ∂Q link.

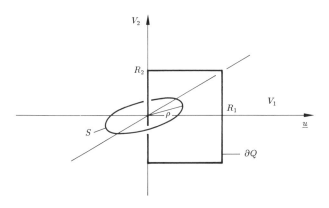

Fig. 8.2.

Proof. Let $\pi: V \to V_2$ denote the projection onto V_2, and let $h \in C^0(V; V)$ satisfy $h|_{\partial Q} = id$. We must show that there exists $u \in Q$ such that the relations $\|h(u)\| = \rho$ and $\pi\big(h(u)\big) = 0$ simultaneously hold. For $t \in [0,1]$, $s \in \mathbb{R}$, $u_2 \in V_2$ let

$$\overline{h}_t(s, u_2) = \big(t\|h(u) - \pi\big(h(u)\big)\| + (1 - t)s - \rho, \ t\pi\big(h(u)\big) + (1 - t)u_2\big) \ ,$$

where $u = s\underline{u} + u_2$. This defines a family of maps $\overline{h}_t: \mathbb{R} \times V_2 \to \mathbb{R} \times V_2$ depending continuously on $t \in [0,1]$. Moreover, if $u = s\underline{u} + u_2 \in \partial Q$, we have

$$\overline{h}_t(s, u_2) = \big(t\|u - u_2\| + (1 - t)s - \rho, u_2\big) = (s - \rho, u_2) \neq 0$$

for all $t \in [0,1]$. Hence, if we identify Q with a subset of $\mathbb{R} \times V_2$ via the decomposition $u = s\underline{u} + u_2$, the topological degree $\deg(\overline{h}_t, Q, 0)$ is well-defined and by homotopy invariance

$$\deg(\overline{h}_1, Q, 0) = \deg(\overline{h}_0, Q, 0) = 1 \ ,$$

where $\overline{h}_0(s, u_2) = (s - \rho, u_2)$. Thus, there exists $u = s\underline{u} + u_2 \in Q$ such that $\overline{h}_1(u) = 0$, which is equivalent to

$$\pi\big(h(u)\big) = 0 \qquad \text{and} \qquad \|h(u)\| = \rho \ ,$$

as desired. $\qquad\qquad\qquad\qquad\qquad\qquad\qquad\qquad\qquad\qquad\qquad\qquad\qquad\square$

8.4 Theorem. *Suppose $E \in C^1(V)$ satisfies (P.-S.). Consider a closed subset $S \subset V$ and a submanifold $Q \subset V$ with relative boundary ∂Q. Suppose*
(1°) S and ∂Q link,
(2°) $\alpha = \inf_{u \in S} E(u) > \sup_{u \in \partial Q} E(u) = \alpha_0$.
Let

$$\Gamma = \{h \in C^0(V; V) \ ; \ h|_{\partial Q} = id\} \ ,$$

Then the number

$$\beta = \inf_{h \in \Gamma} \sup_{u \in Q} E\big(h(u)\big)$$

defines a critical value $\beta \geq \alpha$ of E.

Proof. Suppose by contradiction that $K_\beta = \emptyset$. For $\bar\varepsilon = \alpha - \alpha_0 > 0$, $N = \emptyset$ let $\varepsilon > 0$ and $\Phi \colon V \times [0,1] \to V$ be the pseudo-gradient flow constructed in Theorem 3.4. Note that by choice of $\bar\varepsilon$ there holds $\Phi(\cdot,t)|_{\partial Q} = id$ for all t. Let $h \in \Gamma$ such that $E\big(h(u)\big) < \beta + \varepsilon$ for all $u \in Q$. Define $h' = \Phi(\cdot,t) \circ h$. Then $h' \in \Gamma$ and

$$\sup_{u \in Q} E\big(h(u)\big) < \beta - \varepsilon$$

by Theorem 3.4.($3°$), contradicting the definition of β.   □

Remark. By Example 8.3, letting $S = \partial B_\rho(0;V)$, $Q = \{t\underline{u} \ ; \ 0 \leq t \leq 1\}$ with relative boundary $\partial Q = \{0, \underline{u}\}$, $0 < \rho < \|\underline{u}\|$, Theorem 8.4 contains the mountain pass lemma Theorem 6.1 as a special case. In contrast to Theorem 6.1, Theorem 8.4 allows variations within a higher dimensional set Q, comparable to Theorem 6.3 in the symmetric case.

Applications to Semilinear Elliptic Equations

We demonstrate this with yet another variant of Theorem 6.2, allowing for linear growth of g near $u = 0$. Conceivably, condition ($1°$) below may be further weakened. However, we will not pursue this.

8.5 Theorem. *Let Ω be a smooth, bounded domain in $\mathbb{R}^n$, $n \geq 3$, and let $g \colon \Omega \times \mathbb{R} \to \mathbb{R}$ be measurable in $x \in \Omega$, differentiable in $u \in \mathbb{R}$ with derivative g_u, and with primitive G. Moreover, suppose that*
($1°$) $g(x,0) = 0$, and

$$\frac{g(x,u)}{u} \geq g_u(x,0), \quad \textit{for almost every } x \in \Omega, u \in \mathbb{R} \ ;$$

($2°$) $\exists p < \frac{2n}{n-2}$, $C \colon \big|g_u(x,u)\big| \leq C\big(1 + |u|^{p-2}\big)$, for almost every $x \in \Omega, u \in \mathbb{R}$;
($3°$) $\exists q > 2$, $R_0 \colon 0 < q\, G(x,u) \leq g(x,u)u$, for almost every $x \in \Omega$, if $|u| \geq R_0$.
Then the problem

(8.1)	$-\Delta u = g(\cdot, u)$	*in Ω*
(8.2)	$u = 0$	*on $\partial\Omega$*

admits a solution $u \not\equiv 0$.

Proof. For $u \in H_0^{1,2}(\Omega)$ let

$$E(u) = \frac{1}{2} \int_\Omega |\nabla u|^2 \, dx - \int_\Omega G(x,u) \, dx \ .$$

By assumptions $(2°),(3°)$ the functional E is of class C^2 and as in the proof of Theorem 6.2 satisfies (P.-S.) on $H_0^{1,2}(\Omega)$. It suffices to show that E admits a critical point $u \not\equiv 0$. Note that by assumption $(1°)$ problem (8.1),(8.2) admits $u \equiv 0$ as a trivial solution.

Let φ_k denote the eigenfunctions of the linearized equation

$$-\Delta \varphi_k = g_u(x,0)\varphi_k + \lambda_k \varphi_k \qquad \text{in } \Omega$$
$$\varphi_k = 0 \qquad \text{on } \partial\Omega$$

with eigenvalues $\lambda_1 < \lambda_2 \leq \lambda_3 \leq \dots$. Denote $k_0 = \min\{k \; ; \; \lambda_k > 0\}$ and let $V^+ = \text{span}\{\varphi_k \; ; \; k \geq k_0\}$, $V^- = \text{span}\{\varphi_1,\dots,\varphi_{k_0-1}\}$. Note that by $(1°)$ we have

$$G(x,u) = \int_0^u g(x,v)\,dv \geq \frac{1}{2}g_u(x,0)u^2 .$$

Hence there holds

(8.3) $$E(u) \leq \frac{1}{2}D^2 E(0)(u,u) \leq 0, \text{ for } u \in V^- ,$$

while by definition of V^+ clearly we have

$$D^2 E(0)(u,u) \geq \lambda_{k_o}\|u\|_{L^2}^2, \text{ for } u \in V^+ .$$

To strengthen the latter inequality note that by $(2°)$ the function $g_u(x,0)$ is essentially bounded. Thus for sufficiently large k_1 we have

$$D^2 E(0)(u,u) = \int_0 |\nabla u|^2 \, dx - \int_\Omega g_u(x,0)u^2 \, dx \geq \frac{1}{2}\|u\|_{H_0^{1,2}}^2$$

uniformly for $u \in \text{span}\{\varphi_k \; ; \; k \geq k_1\}$. Since the complement of this space in V^+ has finite dimension, we conclude that there exists $\lambda > 0$ such that

$$D^2 E(0)(u,u) \geq \lambda\|u\|_{H_0^{1,2}}^2 ,$$

uniformly for $u \in V^+$. But $E \in C^2\big(H_0^{1,2}(\Omega)\big)$; it follows that for sufficiently small $\rho > 0$ we have

(8.4) $$\inf_{u \in S_\rho^+} E(u) \geq \frac{1}{2}D^2 E(0)(u,u) - o\big(\|u\|_{H_0^{1,2}}^2\big) \geq \frac{\lambda\rho^2}{4} > 0 ,$$

where $S_\rho^+ = \{u \in V^+ \; ; \; \|u\|_{H_0^{1,2}} = \rho\}$ and where $o(s)/s \to 0 \ (s \to 0)$.

By $(3°)$ finally, as in the proof of Theorem 6.2 (see (6.4–5)), for any finite-dimensional subspace W we have

$$E(u) \to -\infty, \qquad \text{as } \|u\| \to \infty, \ u \in W .$$

Recalling (8.3), (8.4), we see that the assumptions of Theorem 8.4 are satisfied with $S = S_\rho^+$ and

$$Q = \{u^- + s\varphi_{k_0} \; ; \; u^- \in V^-, \; \|u^-\|_{H_0^{1,2}} \le R, \; 0 \le s \le R\} \, ,$$

for sufficiently large $R > 0$. Thus E admits a critical point u with $E(u) \ge \frac{\lambda\rho^2}{4}$. The proof is complete. $\qquad\qquad\qquad\qquad\qquad\qquad\qquad\qquad\qquad\qquad\qquad\qquad\square$

Further applications of Theorem 8.4 to semilinear elliptic boundary value problems are given in the survey notes by De Figueireido [1] or Rabinowitz [11; p. 25 ff.]. In particular, Theorem 8.4 offers a simple and unified approach to asymptotically linear equations, possibly "resonant" at $u = 0$ or at infinity, as in Ahmad-Lazer-Paul [1], Amann [4], Amann-Zehnder [1]. (See for instance Rabinowitz [11; Theorem 4.12], or Bartolo-Benci-Fortunato [1] and the references cited therein. See also Chang [1; p. 708].)

Applications to Hamiltonian Systems

A more refined application of linking is given in the next theorem due to Hofer and Zehnder [1], prompted by work of Viterbo [1]. Once again we deal with Hamiltonian systems

$$(8.5) \qquad\qquad\qquad\qquad \dot{x} = \mathcal{J}\nabla H(x) \, ,$$

where H is a given smooth Hamiltonian and $\mathcal{J}$ is the skew-symmetric matrix

$$\mathcal{J} = \begin{pmatrix} 0 & -id \\ id & 0 \end{pmatrix}$$

on $\mathbb{R}^{2n} = \mathbb{R}^n \times \mathbb{R}^n$.

8.7 Theorem. *Let $H \in C^2(\mathbb{R}^{2n}; \mathbb{R})$; suppose that 1 is a regular value of H and $S = S_1 = H^{-1}(1)$ is compact and connected. Then for any $\delta > 0$ there is a number $\beta \in]1 - \delta, 1 + \delta[$ such that $S_\beta = H^{-1}(\beta)$ carries a periodic solution of the Hamiltonian system (8.5).*

(Note that by the implicit function theorem and compactness of S there exists a number $\delta_0 > 0$ such that for any $\beta \in]1 - \delta_0, 1 + \delta_0[$ the hyper-surface S_β is of class C^2, compact, and diffeomorphic to S.)

Remarks. By Theorem 8.7, in order to obtain a periodic solution of (8.5) on the *fixed* surface $S = S_1$, it would suffice to obtain a priori bounds (in terms of the action integral) for periodic solutions to (8.5) on surfaces S_β near S. Benci-Hofer-Rabinowitz [1] have shown that this is indeed possible, provided certain geometric conditions are satisfied, including for instance the condition that S be strictly star-shaped with respect to the origin. However, also energy surfaces of Hamiltonians that are only convex either in the position or in the momentum variables are allowed.

The existence of periodic solutions to (8.5) on S likewise can be derived if S is of "contact type". This notion, introduced by Weinstein [3], allows to give an intrinsic interpretation of the existence results by Rabinowitz [5] and Weinstein [2] for closed trajectories of Hamiltonian systems on convex or strictly star-shaped energy hyper-surfaces. See Hofer-Zehnder [1], or Zehnder [2] for details.

Proof of Theorem 8.7. Observe that, as remarked earlier in the proof of Theorem I.6.5, the property that a level hypersurface $S_\beta = H^{-1}(\beta)$ carries a periodic solution of (8.5) is independent of the particular Hamiltonian H having S_β as a level surface. We now use this freedom to redefine H suitably. For $0 < \delta < \delta_0$ let $U = H^{-1}([1 - \delta, 1 + \delta]) \simeq S \times [-1, 1]$. Then $\mathbb{R}^{2n} \setminus U$ has two components. Indeed, by Alexander duality

$$\tilde{H}_0(\mathbb{R}^{2n} \setminus U \ ; \ \mathbb{Z}) \simeq \overline{H}^{2n-1}(U \ ; \ \mathbb{Z}) \simeq \overline{H}^{2n-1}(S \ ; \ \mathbb{Z}) \simeq \mathbb{Z} \ ,$$

in Spanier's [1] notation. Denote by A the unbounded component of $\mathbb{R}^{2n} \setminus U$ and by B the bounded component. Also let $\gamma = \text{diam } U > 0$. Fix numbers $r, b > 0$ such that

$$\gamma < r < 2\gamma$$

$$\frac{3}{2}\pi r^2 < b < 2\pi r^2$$

and choose a smooth function $f :] - \delta_0, \delta_0[\to \mathbb{R}$ such that $f|_{]-\delta_0,-\delta]} = 0$, $f|_{[\delta,\delta_0[} = b$, and such that $f'(s) > 0$ for $-\delta < s < \delta$. Also let $g : \mathbb{R} \to \mathbb{R}$ be a smooth function satisfying

(8.6) $\quad g(s) = b$ for $s \le r$, $g(s) \ge \frac{3}{2}\pi s^2$ for $s > r$, $g(s) = \frac{3}{2}\pi s^2$ for large s ,

and $0 < g'(s) \le 3\pi s$ for $s > r$.

Then define

$$\tilde{H}(x) = \begin{cases} 0, & \text{if } x \in B, \\ f(s), & \text{if } H(x) = 1 - s, \ -\delta \le s \le \delta, \\ b, & \text{if } x \in A, \ |x| \le r, \\ g(|x|), & \text{if } |x| > r \ . \end{cases}$$

Note that for $x \in \mathbb{R}^{2n}$ we can estimate

(8.7) $$-b + \frac{3}{2}\pi|x|^2 \le \tilde{H}(x) \le \frac{3}{2}\pi|x|^2 + b \ .$$

Now we look for solutions of (8.5) of period 1. As we have seen in Section I.6, such solutions uniquely correspond to the critical points of the functional

$$E(x) = \frac{1}{2}\int_0^1 \langle \dot{x}, \mathcal{J}x \rangle \, dt - \int_0^1 \tilde{H}(x) \, dt$$

on the space of 1-periodic (C^1-)maps $x : \mathbb{R} \to \mathbb{R}^{2n}$.

By the following lemma we are able to distinguish a periodic solution on a level-hypersurface S_β.

8.9 Lemma. *Suppose x is a 1-periodic C^1-solution of (8.5) which satisfies $E(x) > 0$. Then $x(t) \in S_\beta$ for some $\beta \in [1 - \delta, 1 + \delta]$, for all t.*

Proof. Since $\tilde{H} \geq 0$, any constant solution x of (8.5) satisfies $E(x) \leq 0$. Assume x is non-constant with $|x(t_0)| > r$. Then $g(|x(t_0)|) = \tilde{H}(x(t_0)) = \tilde{H}(x(t)) = g(|x(t)|)$ also for t close to t_0 and by (8.6) it follows that $|x(t)| \equiv |x(t_0)| = s_0$. In particular, x satisfies

$$\dot{x} = \mathcal{J} \frac{g'(s_0)\, x}{s_0}$$

and we compute

$$E(x) = \frac{1}{2} g'(s_0)\, s_0 - \tilde{H}(x(t_0))$$

$$\leq \frac{3}{2} \pi s_0^2 - \frac{3}{2} \pi s_0^2 = 0 \,,$$

by definition of $\tilde{H}$. The claim now follows. $\qquad\qquad\qquad\qquad\square$

In order to exhibit a 1-periodic solution x of (8.5) with $E(x) > 0$, a variational argument involving linking will be applied.

Denote by $V = H^{1/2,2}(S^1 \; ; \; \mathbb{R}^{2n})$ the Hilbert space of all 1-periodic functions

$$(8.8) \qquad x(t) = \sum_{k \in \mathbb{Z}} \exp\left(2\pi k t \mathcal{J}\right) x_k \in L^2\left([0,1] \; ; \; \mathbb{R}^{2n}\right) \,,$$

with Fourier coefficients $x_k \in \mathbb{R}^{2n}$ satisfying

$$\|x\|^2 = 2\pi \left(\sum_{k \in \mathbb{Z}} |k|\, |x_k|^2 \right) + |x_0|^2 < \infty \,.$$

The inner product on V is given by

$$(x, y)_V := 2\pi \left(\sum_{k \in \mathbb{Z}} |k| \langle x_k, y_k \rangle \right) + \langle x_0, y_0 \rangle$$

inducing the norm $\| \cdot \|$ above.

Split $V = V^- \oplus V^0 \oplus V^+$ orthogonally, where

$$V^- = \{x \in V \; ; \; x_k = 0 \text{ for } k \geq 0\} \,,$$
$$V^0 = \{x \in V \; ; \; x_k = 0 \text{ for } k \neq 0\} \,,$$
$$V^+ = \{x \in V \; ; \; x_k = 0 \text{ for } k \leq 0\}$$

with reference to the Fourier decomposition (8.8) of an element $x \in V$. Also denote P^-, P^0 and P^+ the corresponding projections. Thus, any $x \in V$ may be uniquely expressed $x = x^- + x^0 + x^+$, where $x^- = P^- x$, etc.

Note that the self-adjoint operator

$$Lx = -\mathcal{J}\dot{x}$$

has eigenspaces

$$V_k = \{\exp(2\pi kt\mathcal{J})x_k \ ; \ x_k \in \mathbb{R}^{2n}\}$$

with eigenvalues $2\pi k$, $k \in \mathbb{Z}$. Moreover, $\dim V_k = 2n$, for all k. Hence the related quadratic form

$$A(x) = \frac{1}{2}\int_0^1 \langle \dot{x}, \mathcal{J}x \rangle \, dt \quad \begin{cases} < 0, & \text{for } x \in V^-, \\ = 0, & \text{for } x \in V^0, \\ > 0, & \text{for } x \in V^+. \end{cases}$$

In fact, using the projections $P^\pm$, A can be written

$$A(x) = \frac{1}{2}\big((-P^- + P^+)x, x\big)_V \ .$$

Note that since $\tilde{H}$ is of class C^2 and behaves like $|x|^2$ for large $|x|$, its mean

$$G(x) = \int_0^1 \tilde{H}(x) \, dt$$

is of class C^2 on $L^2(S^1; \mathbb{R}^{2n})$. Hence also $E \in C^2(H^{1/2,2}(S^1))$. Restricting G to V, from compactness of the embedding $H^{1/2,2}(S^1) \hookrightarrow L^2(S^1)$ we also deduce that DG is a completely continuous map from V into its dual.

8.10 Lemma.   *There exist numbers $\alpha > 0$, $\rho \in]0,1[$ such that $E(x) \geq \alpha$ for $x \in V^+$, $\|x\| = \rho$.*

Proof. $\nabla\tilde{H}(0) = 0$, $\nabla^2\tilde{H}(0) = 0$ implies that $DG(0) = 0$, $D^2G(0) = 0$. Hence $D^2E(0) = -P^- + P^+$. Since $E \in C^2$, $E(0) = 0$, $DE(0) = 0$, the lemma follows. $\qquad\square$

Now define

$$Q = \{x = x^- + x^0 + s\,e \ \in V \ ; \ \|x^- + x^0\| \leq R, \text{ and } 0 \leq s \leq R\} \ ,$$

with $R > 1$ to be determined, and with a unit vector

$$e = \frac{1}{\sqrt{2\pi}}\exp\left(2\pi t\mathcal{J}\right)a \in V^+, \qquad |a| = 1 \ .$$

Denote ∂Q the relative boundary of Q in $V^- \oplus V^0 \oplus \mathbb{R} \cdot e$; that is,

$$\partial Q = \{x^- + x^0 + s\,e \in V \ ; \ \|x^- + x^0\| = R \text{ or } s \in \{0, R\}\} \ .$$

8.11 Lemma. *If R is sufficiently large, then $E|_{\partial Q} \leq 0$.*

Proof. Note that (8.7) implies that

$$G(x) = \int_0^1 \tilde{H}(x)\, dt \geq -b + \frac{3}{2}\pi\|x\|_{L^2}^2 \ .$$

Therefore, for $x = x^- + x^0 + se$ with $s = R$ or $\|x^- + x^0\|^2 = \|x^-\|^2 + \|x^0\|_{L^2}^2 = R$, we obtain

$$E(x) = \frac{1}{2}\big((-P^- + P^+)x, x\big)_V - G(x)$$

$$\leq -\frac{1}{2}\|x^-\|^2 + \frac{1}{2}s^2 + b - \frac{3}{2}\pi\big(\|x^-\|_{L^2}^2 + \|x^0\|_{L^2}^2 + \frac{s^2}{2\pi}\big)$$

$$\leq -\frac{1}{2}\|x^-\|^2 - \frac{1}{4}s^2 - \frac{3}{2}\pi\|x^0\|_{L^2}^2 + b$$

$$\leq 0 \ ,$$

if $R > 0$ is sufficiently large. Moreover, $\tilde{H} \geq 0$ implies $G \geq 0$ and therefore $E \leq 0$ on $V^- \oplus V^0$, that is at $s = 0$, which concludes the proof. □

Fix $0 < \rho < R$ as in Lemmas 8.10, 8.11. Denote $S_\rho^+ = \{x \in V^+ \ ; \ \|x\| = \rho\}$. Define a class Γ of maps $V \to V$ as follows:

Γ is the class of maps $h \in C^0(V; V)$ such that h is homotopic to the identity through a family of maps $h_t = L_t + K_t$, $0 \leq t \leq T$, where $L_0 = id$, $K_0 = 0$, and where for each t there holds:

$$L_t = (L_t^-, L_t^0, L_t^+) : V^- \oplus V^0 \oplus V^+ \to V^- \oplus V^0 \oplus V^+$$

is a linear isomorphism preserving sub-spaces, K_t is compact and $h_t(\partial Q) \cap S_\rho^+ = \emptyset$, for each t.

Following Benci-Rabinowitz [1], we now establish

8.12 Lemma. *∂Q and S_ρ^+ link with respect to Γ.*

Proof. Choose $h \in \Gamma$. We must show that $h(Q) \cap S_\rho^+ \neq \emptyset$, or equivalently that the equations

$$(8.9) \qquad\qquad \big((P^- + P^0) \circ h\big)(x) = 0, \qquad \|h(x)\| = \rho$$

are satisfied for some $x \in Q$.

Using a degree argument as in Example 8.3, we establish (8.9) for every h_t in the family defining $h \in \Gamma$. Consider $Q \subset V^- \oplus V^0 \oplus \mathbb{R} \cdot e =: W$ and represent $x = x^- + x^0 + se$. Applying L_t^{-1} to (8.9) these equations become

$$x^- + x^0 + (P^- + P^0)L_t^{-1}K_t(x) = 0 \ ,$$

$$\|h_t(x)\| = \rho \ .$$

Note that $k_t := (P^- + P^0)L_t^{-1}K_t: W \to V^- \oplus V^0$ is compact. Now define a map $T_t: W \to W$ by letting

$$T_t(x) = T_t(x^- + x^0 + se) = x^- + x^0 + k_t(x) + \|h_t(x)\|e$$
$$= x + k_t(x) + (\|h_t(x)\| - s)e \ .$$

Observe that T_t is of the form $id + compact$. Moreover, the condition $h_t(\partial Q) \cap S_\rho^+ = \emptyset$ translates into the condition $T_t(x) \neq \rho e$ for $x \in \partial Q$. Hence the Leray-Schauder degree (see for instance Deimling [1; 2.8.3, 2.8.4]) of T_t on Q with respect to ρe is well-defined and, in fact,

$$\deg(T_t, Q, \rho e) = \deg(T_0, Q, \rho e) = 1 \ ,$$

as desired. $\square$

Let ∇E denote the gradient of E with respect to the scalar product in V. Since ∇E has linear growth, the gradient flow $\Phi: V \times [0, \infty[\to V$ given by

$$\frac{\partial}{\partial t}\Phi(x, t) = -\nabla E(\Phi(x, t))$$
$$\Phi(x, 0) = x$$

exists globally. Note that E is non-increasing along flow-lines with

$$(8.10) \qquad \frac{d}{dt}E(\Phi(x, t)) = -\left\|\nabla E(\Phi(x, t))\right\|^2 \ .$$

8.13 Lemma. $\Phi(\cdot, T) \in \Gamma$ for any $T \geq 0$.

Proof. Clearly $h_t = \Phi(\cdot, t)$ for $0 \leq t \leq T$ is a homotopy of $\Phi(\cdot, T)$ to the identity; moreover by (8.10)

$$E(\Phi(x, t)) \leq E(x) \leq 0 \qquad \text{for } x \in \partial Q \ .$$

By Lemmata 8.10 and 8.11 therefore $h_t(\partial Q) \cap S_\rho^+ = \emptyset$ for all t. Finally, observe that

$$(8.11) \qquad \nabla E(x) = -x^- + x^+ + \nabla G(x) \ ,$$

where ∇G is compact. Thus the desired form of $\Phi(x, t)$ may be read off the variation of constant formula

$$\Phi(x, t) = e^t x^- + x^0 + e^{-t}x^+ + \int_0^1 (e^{t-s}P^- + P^0 + e^{-(t-s)}P^+)\nabla G(\Phi(x, s)) \ ds$$
$$=: L_t^- x^- + L_t^0 x^0 + L_t^+ x^+ + K_t(x) \ . \qquad \square$$

It remains to verify the Palais-Smale condition for E.

8.14 Lemma. *E satisfies (P.-S.) on V.*

Proof. Let (x_m) be a (P.-S.)sequence for E. By (8.11) and Proposition 2.2 it suffices to show that (x_m) is bounded in V. Suppose by contradiction that $\|x_m\| \to \infty$. Normalize $z_m = \frac{x_m}{\|x_m\|}$ and note that

$$(8.12) \qquad (-P^- + P^+)z_m - \frac{\nabla G(x_m)}{\|x_m\|} \to 0 \ .$$

By (8.6) we may estimate

$$\left(\frac{\nabla G(x_m)}{\|x_m\|}, \phi \right)_V = \left| \int_0^1 \frac{\langle \nabla \tilde{H}(x_m), \phi \rangle}{\|x_m\|} \, dt \right| \le c \left(1 + \left\| \frac{x_m}{\|x_m\|} \right\|_{L^2} \right) \|\phi\|_{L^2} \ ,$$

uniformly in $\phi \in H^{1/2,2}(S^1)$. Choosing $\phi = \phi_m = \left(\frac{\nabla G(x_m)}{\|x_m\|} \right)$ we infer that the sequence $y_m := \left(\frac{\nabla G(x_m)}{\|x_m\|} \right)$ is bounded, hence relatively weakly compact in $H^{1/2,2}$. Suppose that $y_m \to y$ weakly in $H^{1/2,2}$ and hence strongly in L^2. Choosing $\phi = \phi_m = y_m - y$ above, it then follows that $y_m \to y$ strongly in $H^{1/2,2}$ as $m \to \infty$. Hence, from (8.12) it follows that $z_m \to z$ in $H^{1/2,2}$ and almost everywhere in $[0,1]$. Since $\nabla \tilde{H}$ grows linearly, passing to the limit in the expression

$$\left\langle \frac{\phi}{\|x_m\|}, DE(x_m) \right\rangle = \int_0^1 \langle -\mathcal{J} \dot{z}_m, \phi \rangle - \frac{\langle \nabla \tilde{H}(\|x_m\|z_m), \phi \rangle}{\|x_m\|} \, dt$$

is allowed by Vitali's convergence theorem. From (8.6) we thus infer that z satisfies the equation

$$\dot{z} = 3\pi \mathcal{J} \, z \ .$$

However, 3π does not belong to the spectrum of $L = -\mathcal{J} \frac{d}{dt}$. Hence $z \equiv 0$. But since $z_m \to z$, and $\|z_m\| = 1$, this is impossible. Therefore the original sequence (x_m) must be bounded. □

In order to conclude we can now repeat the argument of Theorem 8.4.

Proof of Theorem 8.8. Define

$$\beta = \inf_{t \ge 0} \sup_{x \in Q} E\big(\Phi(x,t) \big) \ .$$

From Lemmata 8.12, 8.13 it follows that $\Phi(Q,t) \cap S_\rho^+ \ne \emptyset$ for all $t \ge 0$. Hence $\beta \ge \alpha > 0$. Suppose by contradiction that β is a regular value for E. By (P.-S.) there exists $\delta > 0$ such that

$$\|\nabla E(x)\| \ge \delta$$

for all x such that $|E(x) - \beta| < \delta$. Choose $t_0 > 0$ such that

$$\sup_{x \in Q} E\big(\Phi(x, t_0)\big) < \beta + \delta \ .$$

Then by definition of β and choice of δ, for $t \geq t_0$ the number

$$\sup_{x \in Q} E\big(\Phi(x, t)\big)$$

is achieved only at points where $\left\| \nabla E\big(\Phi(x, t)\big) \right\| \geq \delta.$

Hence by (8.10)

$$\frac{d}{dt} \left(\sup_{x \in Q} E\big(\Phi(x, t)\big) \right) \leq -\delta^2 \ ,$$

which gives a contradiction to the definition of β after time $T \geq t_0 + \frac{2}{\delta}$. The proof is complete.   □

9. Critical Points of Mountain Pass Type

Different critical points of functionals E sometimes can be distinguished by the topological type of their neighborhoods in the sub-level sets of E. This is the original idea of M. Morse which led to the development of what is now called Morse theory; see the Introduction. Hofer [3] has observed that such information about the topological type in some cases is available already from the minimax characterization of the corresponding critical value.

9.1 Definition. *Let $E \in C^1(V)$, and suppose u is a critical point of E with $E(u) = \beta$. We call u of mountain pass type if for any neighborhood N of u the set $N \cap E_\beta$ is disconnected.*

With this notion available we can strengthen the assertion of Theorem 6.1 as follows. For convenience we recall

6.1 Theorem. *Suppose $E \in C^1(V)$ satisfies (P.-S.). Suppose*
(1°) $E(0) = 0$.
(2°) $\exists \rho > 0, \ \alpha > 0: \ \|u\| = \rho \Rightarrow E(u) \geq \alpha.$
(3°) $\exists u_1 \in V: \ \|u_1\| \geq \rho \ and \ E(u_1) < \alpha.$
Define
$$\Gamma = \big\{ p \in C^0\left([0,1]; V\right) \ ; \ p(0) = 0, \ p(1) = u_1 \big\} \ .$$

Then
$$\beta = \inf_{p \in \Gamma} \ \sup_{u \in p} E(u)$$

is a critical value.

Now we assert:

9.2 Theorem. *Under the hypotheses of Theorem 6.1 the following holds:*
(1°) either E admits a relative minimizer $u \neq 0$ with $E(u) = \beta$, or
(2°) E admits a critical point u of mountain pass type with $E(u) = \beta$.

Remark. Simple examples on $\mathbb{R}$ show that in general also case (1°) occurs; see Figure 9.1.

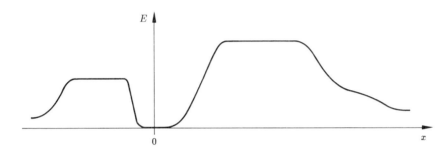

Fig. 9.1 . A function possessing no critical point of mountain pass type

Another variant of Theorem 6.1 is related to results of Chang [4] and Pucci-Serrin [1], [2].

9.3 Theorem. *Suppose $E \in C^1(V)$ satisfies (P.-S.). Suppose 0 is a relative minimizer of E with $E(0) = 0$, and suppose that E admits a second relative minimizer $u_1 \neq 0$. Then,*
(1°) Either there exists a critical point u of E which is not of minimum type, or
(2°) the origin and u_1 can be connected in any neighborhood of the set of relative minimizers u of E with $E(u) = 0$. Necessarily then, also $E(u_1) = 0$.

The proofs of these results are quite similar in spirit.

Proof of Theorem 9.3. Let Γ and β be defined as in Theorem 6.1. Suppose that K_β consists entirely of relative minimizers of E. Then for each $u \in K_\beta$ there exists a neighborhood $N(u)$ such that

$$E(u) = \beta \leq E(v) \qquad \text{for all } v \in N(u) .$$

Let $N_0 = \bigcup_{u \in K_\beta} N(u)$, and for any neighborhood $\tilde{N}$ of K_β let $\varepsilon > 0$ and Φ be determined according to Theorem 3.4 for $\bar{\varepsilon} = 1$ and $N = N_0 \cap \tilde{N}$. Choosing $p \in \Gamma$ such that $p([0,1]) \subset E_{\beta+\varepsilon}$, by (1°),(3°) of Theorem 3.4 the path $p' = \Phi(\cdot, 1) \circ p \in \Gamma$ satisfies

$$p'([0,1]) \subset \Phi(E_{\beta+\varepsilon}, 1) \subset N \cup E_{\beta-\varepsilon} \subset N_0 \cup E_{\beta-\varepsilon} \ .$$

But N_0 and $E_{\beta-\varepsilon}$ by construction are disjoint, hence disconnected. Thus either $p'([0,1]) \subset N$ or $p'([0,1]) \subset E_{\beta-\varepsilon}$. Since the latter contradicts the definition of β we conclude that $p'([0,1]) \subset N \subset \tilde{N}$, whence $p(0) = 0$ and $p(1) = u_1$ can be connected in any neighborhood $\tilde{N}$ of K_β, as claimed.   □

Proof of Theorem 9.2. Suppose, by contradiction, that K_β contains no relative minimizers nor critical points of mountain pass type. Then any $u \in K_\beta$ possesses a neighborhood $N(u)$ such that $N(u) \cap E_\beta$ is (path-) connected. Moreover, $K_\beta \subset \overline{E_\beta}$. Now we have the following topological lemma; see Hofer [3; Lemma 1]:

9.4 Lemma. *Let (M, d) be a metric space and let K and Λ be nonempty subsets of M such that K is compact, Λ is open, and $K \subset \overline{\Lambda}$, the closure of Λ. Suppose $\{N(u) \ ; \ u \in K\}$ is an open cover of K such that $u \in N(u)$ and $N(u) \cap \Lambda$ is connected for each $u \in K$.*
Then there exists a finite, disjoint open cover $\{U_1, \ldots, U_L\}$ of K such that $U_l \cap \Lambda$ for each l is contained in a connected component of Λ.

Proof. Choose $\delta > 0$ such that for any $u \in K$ we have

$$(9.1) \qquad\qquad B_\delta(u) \subset N(\overline{u}) \qquad \text{for some } \overline{u} \in K \ .$$

For instance, we may choose a finite subcover $\{N(\overline{u}_i) \ ; \ 1 \leq i \leq I\}$ of $\{N(u) \ ; \ u \in K\}$ and let

$$\delta = \min_{u \in K} \max_{1 \leq i \leq I} \big\{ \text{dist}(u, M \setminus N(\overline{u}_i)) \big\} > 0 \ .$$

Define an equivalence relation $\overset{*}{\sim}$ on K by letting

$$u \overset{*}{\sim} \overline{u} \ \Leftrightarrow \ \begin{array}{l} \text{There exist a number } m \in \mathbb{N} \text{ and points } u_i \in K, \ 0 \leq i \leq \\ m+1, \text{ such that } u_0 = u, \ u_{m+1} = \overline{u} \text{ and } d(u_i, u_{i+1}) < \delta \text{ for} \\ 0 \leq i \leq m. \end{array}$$

Since K is compact there are only finitely many equivalence classes, say $K_1, \ldots, K_L$. Let

$$U_l = \Big\{ x \in M \ ; \ \text{dist}(x, K_l) < \frac{\delta}{4} \Big\}, \qquad 1 \leq l \leq L \ .$$

Then it is immediate that $U_k \cap U_l = \emptyset$ if $k \neq l$ and $\bigcup_{l=1}^{L} U_l \supset K$. It remains to show that each $U_l^* = U_l \cap \Lambda$ is contained in a connected component of Λ.

Define another equivalence relation $\sim$ on Λ by letting

$$v \sim w \quad \Leftrightarrow \quad \begin{array}{l} v \text{ and } w \text{ belong to the same connected} \\ \text{component of } \Lambda. \end{array}$$

Fix l and let $v, w \in U_l \cap \Lambda$. We wish to show that $v \sim w$. By definition of U_l and K_l there exists a finite chain $u_i \in K_l$, $0 \leq i \leq m+1$, such that

$$d(v, u_0) < \frac{\delta}{4}, \quad d(w, u_{m+1}) < \frac{\delta}{4},$$

and

$$d(u_i, u_{i+1}) < \delta, \text{ for } 0 \leq i \leq m.$$

Set $\varepsilon = \delta - \max_{1 \leq i \leq m} d(u_i, u_{i+1}) > 0$. Since $K \subset \overline{\Lambda}$, for each $0 \leq i \leq m+1$ there exists $v_i \in \Lambda$ such that $d(u_i, v_i) < \frac{\varepsilon}{2} < \delta$, whence also

$$d(v_i, v_{i+1}) < \delta \qquad \text{for } 0 \leq i \leq m.$$

But now by (9.1) we have

$$v, v_0 \in \left(B_\delta(u_0) \cap \Lambda\right) \subset \left(N(\overline{u}_0) \cap \Lambda\right)$$

for some $\overline{u}_0 \in K$, and our assumption about the cover $\{N(u) \; ; \; u \in K\}$ implies that $v \sim v_0$. Similarly, $w \sim v_{m+1}$. Finally, for $0 \leq i \leq m$ we have

$$d(u_i, v_{i+1}) \leq d(u_i, u_{i+1}) + d(u_{i+1}, v_{i+1}) < \delta,$$

whence

$$v_i, v_{i+1} \in \left(B_\delta(u_i) \cap \Lambda\right) \subset \left(N(\overline{u}_i) \cap \Lambda\right),$$

and $v_i \sim v_{i+1}$ for all $i = 0, \ldots, m$. In conclusion

$$v \sim v_0 \sim v_1 \sim \ldots \sim v_{m+1} \sim w,$$

and the proof is complete $\square$

Proof of Theorem 9.2 (completed). Let $\{U_1, \ldots, U_L\}$ be a disjoint open cover of K_β as in Lemma 9.4 and set $N = \bigcup_{l=1}^{L} U_l$. Choose $\overline{\varepsilon} = \alpha > 0$ and let $\varepsilon > 0$, Φ be determined according to Theorem 3.4 corresponding to β, $\overline{\varepsilon}$, and N. Let $p \in \Gamma$ satisfy $p([0,1]) \subset E_{\beta+\varepsilon}$. Then $p' = \Phi(\cdot, 1) \circ p \in \Gamma$ and $p'([0,1]) \subset \Phi(E_{\beta+\varepsilon}, 1) \subset E_{\beta-\varepsilon} \cup N = E_{\beta-\varepsilon} \cup U_1 \cup \ldots \cup U_L$. Smoothing p' if necessary we may assume $p' \in C^1$. Choose $\gamma \in]\beta - \varepsilon, \beta[$ such that γ is a regular value of $E \circ p'$.

Let $0 < t_1 < t_2 < \ldots < t_{2k-1} < t_{2k} < 1$ denote the successive pre-images of γ under $E \circ p'$, and let $I_j = [t_{2j-1}, t_{2j}]$, $1 \leq j \leq k$. Note that $p'|_{I_j} \subset \bigcup_{l=1}^{L} U_l$. Since the latter is a union of disjoint sets, for any j there is $l \in \{1, \ldots, L\}$ such that $p'(I_j) \subset U_l$. But $E \circ p'(\partial I_j) = \gamma < \beta$ and $U_l \cap E_\beta$ is connected. Hence we may replace $p'|_{I_j}$ by a path $\tilde{p}: I_j \to U_l \cap E_\beta$ with endpoints $\tilde{p}|_{\partial I_j} = p|_{\partial I_j}$, for any $j = 1, \ldots k$. In this way we obtain a path $\tilde{p} \in \Gamma$ such that $\sup_{u \in \tilde{p}} E(u) < \beta$, which yields the desired contradiction. $\square$

Multiple Solutions of Coercive Elliptic Problems

We apply these results to a semi-linear elliptic problem. Let Ω be a bounded domain in $\mathbb{R}^n$, $g: \mathbb{R} \to \mathbb{R}$ a continuous function. For $\lambda \in \mathbb{R}$ consider the problem

$$(9.2) \qquad\qquad -\Delta u = \lambda u - g(u) \qquad \text{in } \Omega \text{ ,}$$
$$(9.3) \qquad\qquad u = 0 \qquad\qquad\qquad \text{on } \partial\Omega \text{ ,}$$

Let $0 < \lambda_1 < \lambda_2 \leq \lambda_3 \ldots$ denote the eigenvalues of $-\Delta: H_0^{1,2}(\Omega) \mapsto H^{-1}(\Omega)$. Then we may assert

9.5 Theorem. *Suppose g is Lipschitz and satisfies the condition*
$(1^\circ) \lim_{u \to o} \frac{g(u)}{u} = 0$, $\lim_{|u| \to \infty} \frac{g(u)}{u} = \infty$.
Then for any $\lambda > \lambda_2$ problem (9.2),(9.3) admits at least 3 distinct non-trivial solutions.

Remark that by (1°) problem (9.2),(9.3) for any $\lambda \in \mathbb{R}$ admits $u \equiv 0$ as (trivial) solution. Moreover, by (1°) the functional E related to (9.2), (9.3) is coercive. The latter stands in contrast with the examples studied earlier in this chapter. As a consequence, the existence of multiple solutions for problem (9.2), (9.3) heavily depends on the behavior of the functional E near $u = 0$, governed by the parameter λ, whereas in previous examples the behavior near ∞ had been responsible for the nice multiplicity results obtained.

Theorem 9.5 is essentially due to Struwe [5], improving an earlier result of Ambrosetti and Mancini [1]. For differentiable g a simple proof later was given by Ambrosetti-Lupo [1] and we shall basically follow their approach in the proof below. Here, by a further simplification of their argument, we also succeed in eliminating their differentiability assumption. See also Rabinowitz [11; Theorem 2.42], Chang [1; Theorem 3], and Hofer [1] for related results.

If g is odd, then for $\lambda_k < \lambda \leq \lambda_{k+1}$ problem (9.2), (9.3) possesses at least k pairs of distinct nontrivial solutions, see for example Ambrosetti [1]. (This can also be deduced from a variant of Theorem 5.7 above, applied to the sub-level set $M = E_0$ of the functional E related to (9.2), (9.3). Note that M is forwardly invariant under the pseudo-gradient flow for E and hence the trajectories of this flow are complete in forward time. Moreover, for $\lambda_k < \lambda \leq \lambda_{k+1}$ by (1°) it follows that the genus $\hat{\gamma}(M) \geq k$.)

Without any symmetry assumption optimal multiplicity results for (9.2), (9.3) are not known. However, results of Dancer [2] suggest that in general even for large λ one can expect no more than 4 non-trivial solutions.

Proof of Theorem 9.5. Let $u_\pm = \min\{u > 0 \;;\; \frac{g(\pm u)}{\pm u} \geq \lambda\}$. Then we may replace g by the truncated function

$$\hat{g}(u) = \begin{cases} g(u), & -u_- \leq u \leq u_+ \\ \lambda u, & u < -u_- \text{ or } u_+ < u. \end{cases}$$

Observe that if u satisfies

$$-\Delta u = \lambda u - \hat{g}(u) = \begin{cases} \lambda u - g(u), & \text{if } u \in [-u_-, u_+] \\ 0, & \text{else} \end{cases}$$

then by the weak maximum principle in fact u satisfies (9.2),(9.3).

Thus we may assume that $|g(x,u) - \lambda u| \leq c$ and $(\lambda u - g(u))u \geq 0$ for all $u \in \mathbb{R}$. Hence the functional

$$E(u) = \frac{1}{2} \int_\Omega \left(|\nabla u|^2 - \lambda |u|^2 \right) dx + \int_\Omega G(x,u)\, dx$$

is well-defined, and $E \in C^1\left(H_0^{1,2}(\Omega)\right)$. Moreover, (P.-S.) is satisfied. In fact, E is coercive, $DE(u) = id + compact$, and (P.-S.) follows from Proposition 2.2. In order to obtain a positive (negative) solution $\overline{u}$ (respectively $\underline{u}$) of (9.2),(9.3), we might further truncate the nonlinearity $\lambda u - g(u)$ below or above 0 and proceed as in the proof of Theorem 6.2. However, we can also use the trivial solution $u = 0$ as a sub-(super-) solution to problem (9.2),(9.3) and minimize in the cone of non-negative (non-positive) functions, as we did earlier in Sections I.2.3, I.2.4.

Let $\overline{u}, \underline{u}$ minimize E in $\overline{M} = \{u \in H_0^{1,2}(\Omega) \; ; \; u \geq 0\}$, respectively $\underline{M} = \{u \in H_0^{1,2}(\Omega) \; ; \; u \leq 0\}$. Then, since 0 is a trivial solution to (9.2)–(9.3), by Theorem I.2.4 the functions $\underline{u}, \overline{u}$ also solve (9.2)–(9.3). In particular, $\underline{u}, \overline{u} \in C^{2,\alpha}(\Omega)$ for some $\alpha > 0$; see Appendix B. Moreover, if $\lambda > \lambda_1$ and if $\varphi_1 > 0$, $\|\varphi_1\|_{L^2} = 1$, denotes a normalized eigenfunction corresponding ot λ_1, we have

$$E(\varepsilon \varphi_1) = \frac{1}{2}(\lambda_1 - \lambda)\varepsilon^2 - o(\varepsilon^2) \;,$$

where $o(s)/s \to 0$ as $s \to 0$, and this is < 0 for sufficiently small $|\varepsilon| \neq 0$; hence $\underline{u}, \overline{u} \not\equiv 0$, and in fact, by the strong maximum principle (Theorem B.4 of the appendix), the functions $\underline{u}, \overline{u}$ cannot have an interior zero.

We claim: $\underline{u}$ and $\overline{u}$ are relative minimizers of E in $H_0^{1,2}(\Omega)$. It suffices to show that for sufficiently small $\rho > 0$ we have

$$E(\overline{u}) \leq E(u) \text{ for all } u \in \overline{M}_\rho := \{u \in H_0^{1,2}(\Omega) \; ; \; \|(\overline{u} - u)_-\|_{H_0^{1,2}} < \rho\} \;,$$

where $(s)_\pm = \pm\max\{\pm s, 0\}$. Fix $\rho > 0$ and let $u_\rho \in \overline{M}_\rho$ be a minimizer of E in $\overline{M}_\rho$. Then

$$\langle v, DE(u_\rho) \rangle \geq 0$$

for all $v \geq 0$; that is, u_ρ is a (weak) supersolution to (9.2)–(9.3), satisfying

$$-\Delta u_\rho \geq \lambda u_\rho - g(u_\rho) \;.$$

Choosing $v = -(u_\rho)_- \geq 0$ as testing function, we obtain

$$\int_\Omega |\nabla (u_{\rho_-})|^2 - \lambda |u_{\rho_-}|^2 \, dx + \int_\Omega g(u_{\rho_-})u_{\rho_-} \, dx \leq 0 \;.$$

But by $(1°)$ there is a constant $c \geq 0$ such that $g(u)\,u \geq -c|u|^2$ for all u; thus

$$\int_\Omega |\nabla(u_{\rho_-})|^2 - (\lambda + c)|u_{\rho_-}|^2 \, dx \leq 0 \ .$$

Let $\Omega_{\rho_-} = \{x \in \Omega \ ; \ u_\rho < 0\}$. Then, since $\overline{u} > 0$ we have $\mathcal{L}^n(\Omega_{\rho_-}) \to 0$ as $\rho \to 0$. Now by Hölder's and Sobolev's inequalities we can estimate

$$\int_{\tilde{\Omega}} |u|^2 \, dx \leq \left(\mathcal{L}^n(\tilde{\Omega})\right)^{\frac{2}{n}} \left(\int_{\tilde{\Omega}} |u|^{\frac{2n}{n-2}} \, dx\right)^{\frac{n-2}{n}}$$

$$\leq C(n)\left(\mathcal{L}^n(\tilde{\Omega})\right)^{\frac{2}{n}} \int_{\tilde{\Omega}} |\nabla u|^2 \, dx$$

for all $u \in H_0^{1,2}(\tilde{\Omega})$, and for any $\tilde{\Omega} \subset \Omega$. Thus,

$$\lambda_{\rho_-} := \inf\left\{ \frac{\int_\Omega |\nabla u|^2 \, dx}{\int_\Omega |u|^2 \, dx} \ ; \ u \in H_0^{1,2}(\Omega_{\rho_-}) \setminus \{0\}\right\} \to \infty \text{ as } \rho \to 0 \ ,$$

and for $\rho > 0$ sufficiently small such that $\lambda_{\rho_-} > (\lambda + c)$ we obtain $u_{\rho_-} \equiv 0$; that is, $u_\rho \geq 0$. Hence $E(u_\rho) \geq E(\overline{u})$, and $\overline{u}$ is a relative minimizer of E in $\overline{M}_\rho$, whence also in $H_0^{1,2}(\Omega)$. The same conclusion is valid for $\underline{u}$.

Now let

$$\Gamma = \left\{ p \in C^0\left([0,1]; H_0^{1,2}(\Omega)\right) \ ; \ p(0) = \underline{u}, p(1) = \overline{u}\right\}$$

and denote

$$\beta = \inf_{p \in \Gamma} \sup_{u \in p} E(u) \ .$$

If $\beta \neq 0$ we are done, because Theorem 9.3 either guarantees the existence of infinitely many relative minimizers or the existence of a critical point u not of minimum type and thus distinct from $\underline{u}$, $\overline{u}$. Since $\beta \neq 0$, u must also be distinct from the trivial solution $u = 0$.

If $\beta = 0$, Theorem 9.3 may yield the third critical point $u = 0$. However; by Theorem 9.2 there exists a critical point u with $E(u) = \beta$ of minimum or mountain pass type. But for $\lambda > \lambda_2$ the set where the functional E is negative on a suitable deleted neighborhood of the origin is connected; thus $u = 0$ is not of one of these types, and the proof is complete. $\qquad\square$

Remark. Since in case $\beta = 0$ we only have to show that $K_\beta \neq \{0\}$, instead of appealing to Theorem 9.2 it would suffice to show that, if $u = 0$ were the only critical point besides $\underline{u}$ and $\overline{u}$, then 0 can be avoided by a path joining $\underline{u}$ with $\overline{u}$ without increasing energy. For $\lambda > \lambda_2$ this is easily shown by a direct construction in the spirit of Theorem 9.2.

Notes 9.6. (1°) In the context of Hamiltonian systems (8.5), Ekeland-Hofer [1] have applied Theorem 9.2 to obtain the existence of periodic solutions with prescribed minimal period for certain convex Hamiltonian functions H; see also Girardi-Matzeu [1], [2]. Similar applications to semilinear wave equations as considered in Section I.6.6 have recently been given by Salvatore [1].

(2°) Generalizations of Theorem 9.2 to higher-dimensional minimax methods like Theorem 6.3 have been obtained by Bahri-Lions [1], Lazer-Solimini [1], and Viterbo [2]. See Remark 7.3 for an application of these results. Recently, Ghoussoub [1] has presented a unified approach to results in the spirit of Theorems 9.2 and 9.3.

10. Non-Differentiable Functionals

Sometimes a functional $E\colon V \to \mathbb{R}\cup\{\pm\infty\}$ may fail to be Fréchet differentiable on V but may only be Gâteaux differentiable on its domain

$$\mathrm{Dom}(E) = \{u \in V \; ; \; E(u) < \infty\}$$

in direction of a dense space of "testing functions" $T \subset V$.

10.1 Nonlinear scalar field equations: The zero mass case. As an example we consider the problem

(10.1) $$-\Delta u = g(u) \quad \text{in } \mathbb{R}^n, \; n \geq 3 \;,$$

(10.2) $$u(x) \to 0 \quad (|x| \to \infty) \;,$$

with associated action integral

(10.3) $$E(u) = \frac{1}{2} \int_{\mathbb{R}^n} |\nabla u|^n \, dx - \int_{\mathbb{R}^n} G(u) \, dx \;,$$

studied earlier in Section I.4.1. Again, we denote

$$G(u) = \int_0^u g(v) \, dv$$

a primitive of g. Recall that in case of positive mass, that is

$$\limsup_{u \to 0} g(u)/u \leq -m < 0 \;,$$

problem (10.1–2) can be dealt with as a variational problem in $H^{1,2}(\mathbb{R}^n)$, where E is differentiable; see Berestycki-Lions [1]. In contrast to Section I.4.1, however, now we do not exclude the "0-mass case"

$$\frac{g(u)}{u} \to 0 \quad (u \to 0) \;.$$

Then a natural space on which to study E is the space $D^{1,2}(\mathbb{R}^n)$; that is, the closure of $C_0^\infty(\mathbb{R}^n)$ in the norm

$$\|u\|_{D^{1,2}}^2 = \int_{\mathbb{R}^n} |\nabla u|^2 \, dx \ .$$

By Sobolev's inequality, an equivalent characterization is

$$D^{1,2}(\mathbb{R}^n) = \left\{ u \in L^{\frac{2n}{n-2}}(\mathbb{R}^n) \ ; \ \nabla u \in L^2(\mathbb{R}^n) \right\} \ .$$

Note that, unless G satisfies the condition

$$|G(u)| \le c|u|^{\frac{2n}{n-2}} \ ,$$

the functional E may be infinite on a dense set of points in this space and hence cannot be Fréchet differentiable on $D^{1,2}(\mathbb{R}^n)$. In Struwe [6], [7] a variant of the Palais-Smale condition was introduced to handle precisely this difficulty. From this method the following result due to Berestycki and Lions [1], [2] can be derived easily.

10.2 Theorem. *Suppose g is continuous with primitive $G(u) = \int_0^u g(v) \, dv$ and satisfies the conditions*
(1°) $-\infty \le \limsup_{u \to 0} g(u)u/|u|^{\frac{2n}{n-2}} \le 0$,
(2°) $-\infty \le \limsup_{|u| \to \infty} g(u)u/|u|^{\frac{2n}{n-2}} \le 0$,
and suppose there exists a constant ξ_1 such that
(3°) $G(\xi_1) > 0$.
Moreover, assume that g is odd, that is,
(4°) $g(-u) = -g(u)$.
Then there exist infinitely many radially symmetric solutions $u_l \in D^{1,2}(\mathbb{R}^n)$ of $(10.1),(10.2)$ and $E(u_l) \to \infty$ as $l \to \infty$.

Remark. Observe that by the maximum principle we may replace g by the function $\tilde{g}$ given by

$$\tilde{g}(u) = \begin{cases} g(\zeta^*), & \text{if } u > \zeta^* \\ g(u), & \text{if } |u| \le \zeta^* \\ g(-\zeta^*), & \text{if } u < -\zeta^* \end{cases}$$

where $\zeta^* = \inf\{\zeta \ge \zeta_1 \ ; \ g(\zeta) \le 0\} \le +\infty$. Indeed, if u solves $(10.1),(10.2)$ for $\tilde{g}$, by the maximum principle u solves $(10.1),(10.2)$ for g. Hence we may assume that instead of (2°) g satisfies the stronger hypothesis
(5°) $\lim_{|u| \to \infty} g(u)u/|u|^{\frac{2n}{n-2}} = 0$
together with the assumption
(6°) $\exists \zeta_2 > 0 : G(\zeta) > 0$ for all $\zeta, |\zeta| > \zeta_2$.

For the proof of Theorem 10.2 we follow Struwe [6].

10.3 The abstract scheme. Suppose that

$$E: \mathrm{Dom}(E) \subset V \to \mathbb{R}$$

is a densely defined functional on a Banach space V with norm $\|\cdot\|$. Moreover, assume there is a family $(T_L)_{L\in\mathbb{N}}$ of Banach spaces

$$T_1 \subset \ldots \subset T_L \subset T_{L+1} \subset \ldots \subset V$$

with norms $\|\cdot\|_L$ such that

$$\|u\| \leq \ldots \leq \|u\|_{L+1} \leq \|u\|_L \qquad \text{for } u \in T_L \ ,$$

and whose union is dense in V:

(10.4) $$T := \bigcup_{L\in\mathbb{N}} T_L \overset{\text{dense}}{\subset} V \ .$$

(By default, all topological statements refer to the norm-topology of V.) Also suppose that for any $u \in \mathrm{Dom}(E)$ the restricted functional $E|_{\{u\}+T_L} \in C^1(T_L)$, for any $L \in \mathbb{N}$, and the partial derivative

$$D_L E: \mathrm{Dom}(E) \ni u \mapsto D_L E(u) \in T_L^*$$

is continuous in the topology of V for any $L \in \mathbb{N}$.

We define $u \in \mathrm{Dom}(E) \subset V$ to be critical if $D_L E(u) = 0$ for all $L \in \mathbb{N}$, and we denote

$$K_\beta = \{u \in \mathrm{Dom}(E) \ ; \ E(u) = \beta, \ D_L E(u) = 0, \ \forall L \in \mathbb{N}\}$$

the set of critical points with energy β.

Suppose that E satisfies the following variant of the Palais-Smale condition:

(P.-S.) Any sequence (u_m) in V such that $E(u_m) \to \beta$, while $D_L E(u_m) \to 0$ in T_L^* $(m \to \infty)$, for any $L \in \mathbb{N}$, possesses an accumulation point in K_β.

Note the following:

10.4 Lemma. *Suppose V satisfies (10.4) and $E: \mathrm{Dom}(E) \subset V \to \mathbb{R}$ satisfies $\widetilde{(P.\text{-}S.)}$. Then for any $\beta \in \mathbb{R}$ the set K_β is compact and any neighborhood N of K_β contains a member of the family*

$$N_{\beta,L} = \{u \in V \ ; \ |E(u) - \beta| < 1/L, \ \|D_L E(u)\|_L^* < 1/L\}, \ L > 0 \ .$$

Moreover, the system

$$U_{\beta,\rho} = \{u \in V \ ; \ \exists v \in K_\beta : \|u - v\| < \rho\}$$

is a fundamental system of neighborhoods of K_β.

Proof. By $(\widetilde{\text{P.-S.}})$ any sequence (u_m) in K_β is relatively compact and accumulates at some point $u \in K_\beta$. To prove the second assertion, suppose by contradiction that for some neighborhood N of K_β and any $L \in \mathbb{N}$ there is a point $u_L \in N_{\beta,L} \setminus N$. Consider the sequence (u_L). Since for any L' we have

$$\|D_{L'}E(u_L)\|^*_{L'} \leq \|D_L E(u_L)\|^*_L \leq 1/L \to 0$$

as $L \to \infty$, $L \geq L'$, from $(\widetilde{\text{P.-S.}})$ we conclude that (u_L) accumulates at a point $u \in K_\beta$, contrary to assumption. The proof for $U_{\beta,\rho}$ is similar. $\square$

Denote

$$\text{Reg}_L(E) = \{u \in \text{Dom}(E) \ ; \ D_L E(u) \neq 0\}$$

the set of regular points of E with respect to variations in T_L. Then exactly as in Lemma 3.2, using continuity of the partial derivative $D_L E$, we can construct a locally Lipschitz continuous pseudo-gradient vector field $v_L \colon \text{Reg}_L(E) \to T_L$ for E satisfying the conditions

$$\|v_L(u)\| < 2\min\{1, \|D_L E(u)\|^*_L\} \ ,$$
$$\langle v_L(u), D_L E(u)\rangle > \min\{1, \|D_L E(u)\|^*_L\}\|D_L E(u)\|^*_L \ .$$

Now the deformation lemma Theorem 3.4 can be carried over easily. Given $\beta \in \mathbb{R}$, $N \supset K_\beta$ we determine $L \in \mathbb{N}$, $\rho > 0$ such that

$$N \supset U_{\beta,\rho} \supset U_{\beta,\rho/2} \supset N_{\beta,L} \ .$$

Choose a locally Lipschitz continuous function η, $0 \leq \eta \leq 1$, such that $\eta = 0$ if $\|D_L E(u)\|^*_L \leq \frac{1}{2L}$ and such that $\eta(u) = 1$ if $\|D_L E(u)\|^*_L \geq \frac{1}{L}$. Thus, in particular, we have $\eta(u) = 1$ for $u \notin N_{\beta,L}$ satisfying $|E(u) - \beta| \leq \frac{1}{L}$. Let Φ_L be the flow corresponding to the vector field $e_L(u) = -\eta(u)v_L(u)$, defined by solving the initial value problem

$$\frac{\partial}{\partial t}\Phi_L(u,t) = e_L\big(\Phi_L(u,t)\big)$$
$$\Phi_L(u,0) = u$$

for $u \in \text{Dom}(E)$, $t \geq 0$. By local Lipschitz continuity and uniform boundedness of e_L, the flow Φ_L exists globally on $\text{Dom}(E) \times [0,\infty[$, is continuous, and fixes critical points of E. Moreover, E is non-increasing along flow-lines and we have

$$\Phi_L(E_{\beta+\varepsilon},1) \subset E_{\beta-\varepsilon} \cup N$$

respectively

$$\Phi_L(E_{\beta+\varepsilon} \setminus N, 1) \subset E_{\beta-\varepsilon} \ ,$$

where $\varepsilon = \min\left\{\frac{1}{8L}, \frac{\rho}{4L^2}\right\}$; see the proof of Theorem 3.4.

Note that we do not require E to be continuous on its domain. Thus, we have to be careful with truncating the vector field e_L outside some energy range. However, with this crude tool already, many of our abstract existence results may be carried over.

Proof of Theorem 10.2. Let us now implement the above scheme with E given by (10.3) on

$$V = \{u \in D^{1,2}(\mathbb{R}^n) \; ; \; u(x) = u(|x|)\} =: D^{1,2}_{rad}(\mathbb{R}^n) \,,$$

with norm $\|\cdot\| = \|\cdot\|_{D^{1,2}}$. (We focus on radially symmetric solutions to remove translation invariance.) Also let

$$T_L = \{u \in D^{1,2}_{rad}(\mathbb{R}^n) \; ; \; u(x) = 0 \ \text{ for } |x| \geq L\} \,,$$

with norm $\|\cdot\|_L = \|\cdot\| = \|\cdot\|_{D^{1,2}}$, $L \in \mathbb{N}$. Note that in this way $T = \bigcup_{L \in \mathbb{N}} T_L$ simply consists of all functions $u \in D^{1,2}_{rad}(\mathbb{R}^n)$ with compact support. Since variations in T_L for any L only involve the evaluation of g, respectively G on a compact domain $B_L(0)$, it is clear that $E(u + \cdot)$ is Fréchet differentiable in T_L, for any $u \in \text{Dom}(E)$, any $L \in \mathbb{N}$. Moreover, the differential $D_L E: \text{Dom}(E) \to T_L^*$ is continuous in the topology of V for any $L \in \mathbb{N}$.

Note that by radial symmetry any function $u \in D^{1,2}_{rad}(\mathbb{R}^n)$ is represented by a function (indiscriminately denoted by u), which is continuous on $\mathbb{R}^n \setminus \{0\}$. More precisely, by Hölders's inequality there holds

$$
\begin{aligned}
|u(x)|^2 &\leq \int_{|x|}^\infty \frac{d}{dr}|u(r)|^2 \, dr \\
&\leq 2 \left(\int_{|x|}^\infty r^{(2-n)n-1} \, dr \right)^{\frac{1}{n}} \left(\int_{|x|}^\infty |u(r)|^{\frac{2n}{n-2}} r^{n-1} \, dr \right)^{\frac{n-2}{2n}} \\
&\qquad \cdot \left(\int_{|x|}^\infty |\frac{d}{dr}u(r)|^2 r^{n-1} \, dr \right)^{\frac{1}{2}} \\
&\leq C \, |x|^{(2-n)} \|u\|^2 \,,
\end{aligned}
$$
(10.6)

for any $u \in D^{1,2}_{rad}(\mathbb{R}^n)$.

Decompose $g = g_+ - g_-$, $G = G_+ - G_-$ with $g_\pm(u)u = \max\{\pm g(u)u, 0\}$ and $G_\pm(u) = \int_0^u g_\pm(v) \, dv$. We assert that if $u_m \rightharpoonup u$ weakly in $D^{1,2}_{rad}(\mathbb{R}^n)$ then

$$\int_{\mathbb{R}^n} G_+(u) \, dx = \lim_{m \to \infty} \int_{\mathbb{R}^n} G_+(u_m) \, dx \,.$$
(10.7)

Indeed, by (10.6) for any $\delta > 0$ there exists $R > 0$ such that $|u_m(x)| \leq \delta$ for $|x| \geq R$. Moreover, by assumption $(1°)$ for any $\varepsilon > 0$ there is $\delta > 0$ such that

$$G_+(u_m) \leq \varepsilon |u_m|^{\frac{2n}{n-2}} \,,$$

if $|u| \leq \delta$. Hence for R large we can estimate

$$\int_{\mathbb{R}^n \setminus B_R(0)} G_+(u_m)\, dx \le \varepsilon \int_{\mathbb{R}^n \setminus B_R(0)} |u_m|^{\frac{2n}{n-2}}\, dx$$

$$\le C\,\varepsilon\, \|u_m\|^{\frac{2n}{n-2}} \le C\,\varepsilon\ ,$$

uniformly in m.

On the other hand, by assumption $(2°)$, for any $\varepsilon > 0$ there is a constant $C(\varepsilon)$ such that for all $u \in \mathbb{R}$ there holds

$$G_+(u) \le \varepsilon\, |u|^{\frac{2n}{n-2}} + C(\varepsilon)\ .$$

Hence for $\Omega \subset \mathbb{R}^n$ with sufficiently small measure $\mathcal{L}^n(\Omega) < \frac{\varepsilon}{C(\varepsilon)}$, we have

$$\int_{\Omega} G_+(u_m)\, dx \le \varepsilon \int_{\Omega} |u|^{\frac{2n}{n-2}}\, dx + C(\varepsilon)\mathcal{L}^n(\Omega) \le C\,\varepsilon\ ,$$

uniformly in m, and (10.7) follows by Vitali's convergence theorem.

Since $G_- \ge 0$, by Fatou's lemma of course also

$$\int_{\mathbb{R}^n} G_-(u)\, dx \le \liminf_{m \to \infty} \int_{\mathbb{R}^n} G_-(u_m)\, dx\ ,$$

and together with (10.7) we obtain that if $u_m \rightharpoonup u$ weakly

$$(10.8) \qquad \int_{\mathbb{R}^n} G(u)\, dx \ge \limsup_{m \to \infty} \int_{\mathbb{R}^n} G(u_m)\, dx\ .$$

Moreover, in order to verify (P.-S.) we need the following estimate whose proof is inspired by Pohožaev [1]; also see Lemma III.1.4.

10.5 Lemma. *Suppose that $(1°)$ and $(2°)$ of Theorem 10.2 hold. Then any weak solution $u \in D^{1,2}_{rad}(\mathbb{R}^n)$ of equation (10.1) satisfies the estimate*

$$\int_{\mathbb{R}^n} |\nabla u|^2\, dx \ge \frac{2n}{n-2} \int_{\mathbb{R}^n} G(u)\, dx \ge -\infty\ .$$

Proof. By $(2°)$ and our local regularity result Lemma B.3 of the appendix, any weak solution to (10.2) is in L^p_{loc} and hence also in $H^{2,p}$ locally, for any $p < \infty$. Moreover, using (10.6) as above, by $(1°),(2°)$ we have $G_+(u) \in L^1(\mathbb{R}^n)$. Testing equation (10.1) with the function $x \cdot \nabla u$ (the generator of the family $u_R(x) = u(Rx)$ of dilatations of u) we may write the product as

$$(10.9) \quad \frac{n-2}{2}|\nabla u|^2 - \operatorname{div}\left(x\frac{|\nabla u|^2}{2} - \nabla u\,(x \cdot \nabla u) \right) = (x \cdot \nabla u)\,\Delta u$$

$$= -(x \cdot \nabla u)\,g(u) = -x \cdot \nabla G(u) = -\operatorname{div}\big(xG(u)\big) + nG(u)\ .$$

Integrating over $B_R(0)$ and using the radial symmetry of u we thus obtain that

$$\int_{B_R(0)} |\nabla u|^2 \, dx = \frac{2n}{n-2} \int_{B_R(0)} G(u) \, dx$$
$$- \frac{2R}{n-2} \int_{\partial B_R(0)} \left(\frac{1}{2}|\nabla u|^2 + G(u)\right) do$$
$$\geq \frac{2n}{n-2} \int_{B_R(0)} G(u) \, dx$$
$$- \frac{2R}{n-2} \int_{\partial B_R(0)} \left(\frac{1}{2}|\nabla u|^2 + G_+(u)\right) do \ .$$

Since $\nabla u \in L^2(\mathbb{R}^n)$, $G_+(u) \in L^1(\mathbb{R}^n)$, if we let $R \to \infty$ in a suitable way the boundary integral tends to zero. Moreover,

$$\int_{B_R(0)} |\nabla u|^2 \, dx \to \int_{\mathbb{R}^n} |\nabla u|^2 \, dx \ ,$$
$$\int_{B_R(0)} G_+(u) \, dx \to \int_{\mathbb{R}^n} G_+(u) \, dx \ ,$$

while from Beppo Levi's theorem it follows that

$$\int_{\mathbb{R}^n} G_-(u) \, dx \geq \limsup_{R \to \infty} \int_{B_R(0)} G_-(u) \, dx \ .$$

The proof is complete.   □

Remark that under scaling $u \mapsto u_R(x) = u(Rx)$ the functional E behaves like

$$E(u_R) = \frac{1}{2} R^{2-n} \int_{\mathbb{R}^n} |\nabla u|^2 \, dx - R^{-n} \int_{\mathbb{R}^n} G(u) \, dx \ ;$$

that is,

$$\frac{d}{dR} E(u_R)|_{R=1} = \frac{2-n}{2} \int_{\mathbb{R}^n} |\nabla u|^2 \, dx + n \int_{\mathbb{R}^n} G(u) \, dx \ .$$

Hence we may perform a preliminary normalization of admissible functions by restricting E to the set

$$M = \{u \in \mathrm{Dom}(E) \ ; \ u \neq 0, \ \int_{\mathbb{R}^n} |\nabla u|^2 \, dx = \frac{2n}{n-2} \int_{\mathbb{R}^n} G(u) \, dx\}$$

of functions which are stationary for E with respect to dilatations. Note that for $u \in M$ we have

$$E(u) = \frac{1}{n} \int_{\mathbb{R}^n} |\nabla u|^2 \, dx \ ;$$

that is, $E|_M$ is continuous and coercive with respect to the norm in $D^{1,2}(\mathbb{R}^n)$.

Moreover, $E|_M$ satisfies the following compactness condition:

10.6 Lemma. *Suppose that for a sequence* (u_m) *in* M *as* $m \to \infty$ *we have* $E(u_m) \to \beta$ *while* $D_L E(u_m) \to 0 \in T_L^*$ *for any* L. *Then* (u_m) *accumulates at a critical point* $u \in M$ *of* E *and* $E(u) = \beta$. *That is,* E *satisfies* $(\widetilde{P.\text{-}S.})$ *on* M *(while it seems unlikely that one can even show boundedness of a* $(\widetilde{P.\text{-}S.})$ *sequence in general).*

Proof. Let (u_m) be a $(\widetilde{P.\text{-}S.})$ sequence for E in M. By coerciveness, (u_m) is bounded and we may assume that $u_m \to u$ weakly in $D_{rad}^{1,2}(\mathbb{R}^n)$ – which implies strong convergence $g(u_m) \to g(u)$ in $L^1(\Omega)$ for any $\Omega \subset\subset \mathbb{R}^n$. Thus for any $\varphi \in C_0^\infty(\mathbb{R}^n)$ we have

$$\langle \varphi, D_L E(u_m) \rangle = \int_{\mathbb{R}^n} \left(\nabla u_m \nabla \varphi - g(u_m)\varphi \right) dx$$

$$\to \int_{\mathbb{R}^n} \left(\nabla u \nabla \varphi - g(u)\varphi \right) dx = \langle \varphi, D_L E(u) \rangle = 0 \ ;$$

that is, $u \in D_{rad}^{1,2}(\mathbb{R}^n)$ is a critical point of E and hence weakly solves (10.1). By Lemma 10.5 and (10.8) therefore

$$\int_{\mathbb{R}^n} |\nabla u|^2 \, dx \geq \frac{2n}{n-2} \int_{\mathbb{R}^n} G(u) \, dx \geq \frac{2n}{n-2} \limsup_{m \to \infty} \int_{\mathbb{R}^n} G(u_m) \, dx$$

$$= \limsup_{m \to \infty} \int_{\mathbb{R}^n} |\nabla u_m|^2 \, dx \geq \int_{\mathbb{R}^n} |\nabla u|^2 \, dx \ .$$

Here we have also used the normalization condition for $u_m \in M$. Hence $u_m \to u$ strongly in $D_{rad}^{1,2}(\mathbb{R}^n)$, and also $\int_{\mathbb{R}^n} G(u_m) \, dx \to \int_{\mathbb{R}^n} G(u) \, dx$; in particular, $u \in M$ and $E(u_m) \to E(u) = \beta$, as claimed. $\square$

Now we investigate the set M more closely.

Denote $k: \mathbb{R}_+ \times \text{Dom}(E) \to \mathbb{R}$ the mapping

$$k(R, u) = \frac{2R^{n+1}}{2-n} \frac{d}{d\rho} E(u_\rho)|_{\rho=R}$$
(10.10)
$$= R^2 \int_{\mathbb{R}^n} |\nabla u|^2 \, dx - \frac{2n}{n-2} \int_{\mathbb{R}^n} G(u) \, dx \ .$$

Then for any $L \in \mathbb{N}$ and any $u \in \text{Dom}(E)$ we have $k(\cdot, u + \cdot) \in C^1(\mathbb{R}_+ \times T_L)$. Moreover, by Hölder's inequality the partial derivatives $\partial_L k(R, u) \in T_L^*$, $\frac{\partial}{\partial R} k(R, u)$, and $\frac{\partial^2}{\partial R^2} k(R, u)$ are continuous and uniformly bounded on bounded sets $\{(R, u) \in \mathbb{R}_+ \times M + T_L \ ; \ R + \|u\| \leq C\}$. Finally, for $u \in M$, at $R = 1$ we have

$$\frac{\partial}{\partial R} k(R, u)|_{R=1} = 2 \int_{\mathbb{R}^n} |\nabla u|^2 \, dx = 2n \, E(u) > 0 \ .$$

Thus, by the implicit function theorem, for any $\beta > 0$ and any $L \in \mathbb{N}$ there exists $\rho = \rho_{\beta,L} > 0$ and a continuous map $R = R_{\beta,L}$ on a neighborhood

$$V_{\beta,L} = \{u \in M \ ; \ \beta/2 \leq E(u) \leq 2\beta\} + B_{2\rho}(0; T_L)$$

such that $R(v+\cdot) \in C^1(T_L)$ and $k\big(R(v), v\big) \equiv 0$ for $v \in V_{\beta,L}$; that is, $v_{R(v)}(x) = v\big(R(v)x\big) \in M$. Denote

$$\pi_{\beta,L} \colon V_{\beta,L} \to M$$

the map $v \mapsto \pi_{\beta,L}(v) = v_{R(v)}$. Remark that $\pi_{\beta,L}$ is continuous. For the next lemma let

$$K_\beta = \{u \in M \ ; \ E(u) = \beta, \ D_L E(u) = 0 \text{ for all } L\} \ ,$$
$$N_{\beta,L} = \{u \in M \ ; \ |E(u) - \beta| < 1/L, \ \|D_L E(u)\| < 1/L\} \ .$$

Note that, by Lemma 10.6, Lemma 10.2 continues to hold for K_β, $N_{\beta,L}$ as above. We now construct a pseudo-gradient flow for E on M.

10.7 Lemma. *For any $\beta > 0$, any $\bar\varepsilon > 0$ and any neighborhood N of K_β there exist $\varepsilon \in]0, \bar\varepsilon[$ and a continuous family $\Phi \colon M \times [0,1] \to M$ of odd continuous maps $\Phi(\cdot, t) \colon M \to M$ such that*
(1°) $\Phi(u,t) = u$ if $D_L E(u) = 0$ for all $L \in \mathbb{N}$, or $t = 0$, or if $|E(u) - \beta| \geq \bar\varepsilon$,
(2°) $E\big(\Phi(u,t)\big)$ is non-increasing in t for any $u \in M$,
(3°) $\Phi(E_{\beta+\varepsilon} \setminus N, 1) \subset E_{\beta-\varepsilon}$.

Proof. Choose integers $L' < L$ such that

$$N \supset N_{\beta,L'} \supset N_{\beta,L}$$

and let $v_L \colon \{u \in M \ ; \ D_L E(u) \neq 0\} \to T_L$ be an odd, continuous pseudo-gradient vector field for E, satisfying

$$\|v_L(u)\|_L < 2 \min\{1, \|D_L E(u)\|\} \ ,$$
$$\langle v_L(u), D_L E(u)\rangle > \min\{1, \|D_L E(u)\|\} \|D_L E(u)\| \ ,$$

for all $u \in M$ such that $D_L E(u) \neq 0$. Let η, φ be continuous cut-off functions $0 \leq \eta, \varphi \leq 1$, $\eta(u) = \eta(-u)$, $\eta \equiv 0$ on $N_{\beta,L}$, $\eta \equiv 1$ off $N_{\beta,L'}$, $\varphi(s) = 0$ for $|s - \beta| \geq 2\varepsilon$, $\varphi(s) = 1$ for $|s - \beta| < \varepsilon$, where $\varepsilon \leq \min\{\bar\varepsilon/2, (4L)^{-1}\}$. We truncate v_L as usual and let

$$e_L(u) = -\varphi\big(E(u)\big)\eta(u)v_L(u) \ .$$

Note that $e_L \colon M \to T_L$ is odd and continuous. Let $\rho_{\beta,L}$, and $\pi_{\beta,L}$ be defined as above. For $\varepsilon \leq \frac{\beta}{4}$ and $t \leq \rho_{\beta,L}$ then let

$$\Phi(u,t) = \pi_{\beta,L}\big(u + te_L(u)\big) \ .$$

Φ is continuous, odd, and satisfies (1°). Moreover, for fixed $u \in M$ the term $E\big(\Phi(u,t)\big)$ is differentiable. Indeed, letting

$$R(t) = R\big(u + te_L(u)\big), \quad u_R(x) = u\big(Rx\big)$$

for brevity, we have

$$\frac{d}{dt} E\big(\Phi(u,t)\big)\big|_{t=t_0} = \frac{d}{dt} E\big((u + t e_L(u)) R(t)\big)\big|_{t=t_0} =$$

$$= \left(\frac{d}{dR} E\big((u + t_0 e_L(u))_R\big)\big|_{R=R(t_0)} \right) \frac{d}{dt} R(t)\big|_{t=t_0} +$$

$$+ \big\langle (e_L(u))_{R(t_0)}, D_L E\big(\Phi(u,t_0)\big) \big\rangle .$$

Since $\big(u + t_0 e_L(u)\big)_{R=R(t_0)} = \Phi(u,t_0) \in M$ the first term vanishes. Moreover, the second up to a factor $-\varphi(E(u))\eta(u)$ equals

$$\big\langle (v_L(u))_{R(t_0)}, D_L E\big(\Phi(u,t_0)\big) \big\rangle$$

$$= R(t_0)^{2-n} \int_{\mathbb{R}^n} \nabla\big(u + t_0 e_L(u)\big) \nabla v_L(u)\, dx$$

$$- R(t_0)^{-n} \int_{\mathbb{R}^n} g\big(u + t_0 e_L(u)\big) v_L(u)\, dx$$

$$=: \big\langle v_L(u), D_L E(u) \big\rangle - \gamma(t_0) ,$$

where the last line defines the "error function" γ.

Note that $t \to R(t)$ is differentiable with $\frac{d}{dt} R(t)$ uniformly bounded on bounded sets and that by condition (5°) above and Vitali's convergence theorem also

$$\int_{B_L(0)} \big|g\big(u + t e_L(u)\big) - g(u)\big|^{\frac{2n}{n+2}}\, dx \to 0 \text{ as } t \to 0 ,$$

uniformly on bounded sets of functions $u \in M$. Hence also the error

$$\gamma(t) \le c\big(|R(t) - 1| + |t|\big) + \left| \int_{\mathbb{R}^n} \big(g\big(u + t e_L(u)\big) - g(u)\big) e_L(u)\, dx \right|$$

$$\to 0 \text{ as } t \to 0 .$$

In particular, we can achieve that uniformly in $u \in E_{\beta + \bar\varepsilon}$ we have $\gamma(t) \le \frac{1}{2L}$ for $0 \le t \le \bar t$ sufficiently small. By choice of φ and η this implies that

$$\frac{d}{dt} E\big(\Phi(u,t)\big) \le -\frac{\varphi(E(u))\eta(u)}{2L} \qquad \text{for } u \in M,\ 0 \le t \le \bar t .$$

Hence, with $\varepsilon \le \frac{\bar t^2}{4L}$, (2°) and (3°) follow. Rescaling time we may assume $\bar t = 1$.   □

Finally, we can conclude the proof of Theorem 10.2. For $l \in \mathbb{N}$ let

$$\sum_l = \{A \subset M \ ;\ A \text{ closed}, \ A = -A, \ \gamma(A) \ge l\} ,$$

where γ denotes the Krasnoselskii genus introduced in Section 5.1, and define

$$\beta_l = \inf_{A \in \sum_l} \sup_{u \in A} E(u) .$$

10.8 Lemma. *(1°) For any $l \in \mathbb{N}$ the class $\sum_l$ is nonvoid, in particular, $M \neq \emptyset$.*
(2°) The numbers β_l are critical values of E for any $l \in \mathbb{N}$. Moreover, $\beta_l \to \infty$
as $l \to \infty$.

Proof. (1°) Fix $l \in \mathbb{N}$. By condition (6°) on G we can find an l-dimensional
subspace $W \subset C_{0,rad}^\infty(\mathbb{R}^n)$ and a constant $\alpha_1 > 0$ such that for $w \in S = \{w \in W \; ; \; \|w\| = \alpha_1\}$ we have

$$\int_{\mathbb{R}^n} G(w) \, dx > 0 .$$

By (10.10) we can find $\tau > 0$ such that

$$(10.11) \qquad\qquad k(\tau, w) < 0$$

for all $w \in S$. With no loss of generality we may assume that $\tau = 1$. Since
$W \subset C_0^\infty$ there exists another constant α_2 such that $\|w\|_{L^\infty} \leq \alpha_2$ for $w \in S$.
Consider the truncation mapping $\delta \colon W \to D_{rad}^{1,2}(\mathbb{R}^n)$ given by

$$\delta(w)(x) = \begin{cases} \alpha_2, & w(x) > \alpha_2, \\ w(x), & |w(x)| \leq \alpha_2, \\ -\alpha_2, & w(x) < -\alpha_2. \end{cases}$$

Note that δ is continuous and odd. Since the functions $\delta(w)$ are uniformly
bounded and have uniform compact support, clearly

$$(10.12) \qquad\qquad \int_{\mathbb{R}^n} G\big(\delta(w)\big) \, dx \leq c$$

uniformly in $w \in W$. On the other hand, for any $w \in S$ there holds

$$(10.13) \qquad\qquad \int_{\mathbb{R}^n} |\nabla \delta(\mu w)|^2 \, dx \to \infty \qquad (\mu \to \infty) ,$$

as is easily verified.

For $w \in W$, $\|w\| \geq \alpha_1$, let a mapping J be defined by the ralation

$$J(w) = k\big(1, \delta(w)\big)$$

and let J be continuously extended as an even map onto all of W in such a way
that $J(w) < 0$ for $\|w\| < \alpha_1$. (Note that by (10.11) we have $J(w) = k(1, w) < 0$
for $w \in S$; that is, for $\|w\| = \alpha_1$.) By (10.12), (10.13) the set $\Omega = \{w \in W \; ; \; J(w) < 0\}$ then is an open, bounded, symmetric neighborhood of $0 \in W$.
Hence, from Proposition 5.2 we deduce that the boundary A of Ω relative
to W has genus $\gamma(A) \geq l$. Since the mapping δ is odd and continuous, by
supervariance of the genus, Proposition 5.4.(4°), also $\gamma\big(\delta(A)\big) \geq l$. Moreover,
since $J(A) = \{0\}$ and $\delta(A) \not\ni 0$, we clearly have $\delta(A) \subset M$, concluding the
proof of (1°).

($2°$) By part ($1°$) and Lemma 10.7 the numbers β_l are well-defined and critical; see the proof of Theorem 4.2. To show that $\beta_l \to \infty$ ($l \to \infty$) assume by contradiction that $\beta_l \leq \beta$ uniformly in l. Thus, we can find a sequence of sets $A_l \in \Sigma_l$ such that $E(u) \leq 2\beta$ for $u \in A_l$, $l \in \mathbb{N}$. Letting $A = \bigcup_l A_l$ by Proposition 5.4.($2°$) and Proposition 5.3 there exists an infinite sequence of mutually orthogonal vectors $u_m \in A$. By coerciveness and uniform boundedness of E on A, there holds $\|u_m\| \leq C$ for all m, and hence we may extract a weakly convergent subsequence (u_m) (relabelled). By mutual orthogonality, $u_m \rightharpoonup 0$ weakly ($m \to \infty$). Decomposing $g = g_+ - g_-$ as above, with $u \cdot g_\pm(u) = \max\{0, \pm u \cdot g(u)\}$, by (10.7) therefore

$$(10.14) \qquad \int_{\mathbb{R}^n} G_+(u_m)\, dx \to 0 \qquad (m \to \infty) .$$

But for any $u \in M$

$$(10.15) \qquad \begin{aligned} \int_{\mathbb{R}^n} |\nabla u|^2\, dx &\leq \int_{\mathbb{R}^n} |\nabla u|^2\, dx + \frac{2n}{n-2}\int_{\mathbb{R}^n} G_-(u)\, dx \\ &= \frac{2n}{n-2}\int_{\mathbb{R}^n} G_+(u)\, dx . \end{aligned}$$

Thus, $u_m \to 0$ strongly in $D^{1,2}_{rad}(\mathbb{R}^n)$ as $m \to \infty$. On the other hand, by assumptions ($1°$), ($2°$) there exists a constant $c > 0$ such that for all u there holds

$$g(u)u \leq c|u|^{\frac{2n}{n-2}} ,$$

and consequently

$$G(u) \leq c|u|^{\frac{2n}{n-2}} .$$

Hence for $u \in M$ by the Sobolev embedding

$$\|u\|^2 = \int_{\mathbb{R}^n} |\nabla u|^2\, dx = \frac{2n}{n-2}\int_{\mathbb{R}^n} G(u)\, dx \leq c\|u\|^{\frac{2n}{n-2}}_{L^{\frac{2n}{n-2}}} \leq c\|u\|^{\frac{2n}{n-2}} .$$

Dividing by $\|u\|^2$ we conclude that

$$\|u\| \geq c > 0$$

is uniformly bounded away from 0 for $u \in M$, and a contradiction to (10.14), (10.15) results. This concludes the proof. □

10.9 Notes. In a recent paper, Duc [1] has developed a variational approach to singular elliptic boundary value problems which is similar to the method outlined above. Duc only requires continuity of directional derivatives of E. However, in exchange, only a weaker form of the deformation lemma can be established; see Duc [1; Lemma 2.5].

Lack of differentiability is encountered in a different way when dealing with functionals involving a combination of a differentiable and a convex term, as in the case of free boundary problems. For such functionals E, using the notion of sub-differential introduced in Section I.6, a differential DE may be defined as a set-valued map. Suitable extensions of minimax techniques to such problems have been obtained by Chang [2] and Szulkin [1]. More generally, using the concept of generalized gradients introduced by Clarke [1], [5], Chang [2] develops a complete variational theory also for Lipschitz maps satisfying a (P.-S.)-type compactness condition. In Ambrosetti-Struwe [2] these results are used to establish the existence of steady vortex rings in an ideal fluid for a prescribed, positive, non-decreasing vorticity function. Previously, this problem had been studied by Fraenkel-Berger [1] by a constrained minimization technique; however, their approach gave rise to a Lagrange multiplier that could not be controlled.

11. Ljusternik-Schnirelman Theory on Convex Sets

In applications we also frequently encounter functionals on closed and convex subsets of Banach spaces. In fact, this is the natural setting for variational inequalities where the class of admissible functions is restricted by inequality constraints; see Sections I.2.3, I.2.4. Functionals on closed convex sets also arise in certain geometric problems, as we have seen in our discussion of the classical Plateau problem in Sections I.2.7–I.2.10. In fact it was precisely for the latter problem, with the aim of re-deriving the mountain pass lemma for minimal surfaces due to Morse-Tompkins [1], [2] and Shiffman [2], [3], that variational methods for functionals on closed convex sets were first systematically developed; see Struwe [17].

Suppose M is a closed, convex subset of a Banach space V, and suppose that $E\colon M \to \mathbb{R}$ possesses an extension $E \in C^1(V;\mathbb{R})$ to V. For $u \in M$ define

$$g(u) = \sup_{\substack{v \in M \\ \|u-v\|<1}} \langle u - v, DE(u)\rangle$$

as a measure for the slope of E in M. Clearly, if $M = V$ we have $g(u) = \|DE(u)\|$. More generally, we obtain

11.1 Lemma. *If $E \in C^1(V)$, the function g is continuous in M.*

Proof. Suppose $u_m \to u$ $(m \to \infty)$, where $u_m, u \in M$. Then for any $v \in M$ such that $\|u - v\| < 1$ and sufficiently large m there also holds $\|u_m - v\| < 1$. Hence for any such $v \in M$ we may estimate

$$\langle u - v, DE(u)\rangle = \lim_{m\to\infty} \langle u_m - v, DE(u_m)\rangle$$
$$\leq \limsup_{m\to\infty} g(u_m) \ .$$

Passing to the supremum with respect to v in this inequality, we infer that

$$g(u) \leq \limsup_{m\to\infty} g(u_m) \ .$$

On the other hand, if for $\varepsilon_m \searrow 0$ we choose $v_m \in M$ such that $\|u_m - v_m\| < 1$ and

$$\langle u_m - v_m, DE(u_m) \rangle \geq g(u_m) - \varepsilon_m ,$$

by convexity of M also the vectors

$$w_m = \|u_m - u\| u_m + \left(1 - \|u_m - u\|\right) v_m$$
$$= u_m + \left(1 - \|u_m - u\|\right)(v_m - u_m) \in M$$

and satisfy

$$\|w_m - u\| \leq \|u_m - u\| + \left(1 - \|u_m - u\|\right)\|v_m - u_m\|$$
$$< \|u_m - u\| + \left(1 - \|u_m - u\|\right) = 1 ,$$

while

$$\|v_m - w_m\| = \|u_m - u\| \, \|v_m - u_m\| \leq \|u_m - u\| \to 0 .$$

Thus

$$g(u) \geq \limsup_{m \to \infty} \langle u - w_m, DE(u) \rangle$$
$$= \limsup_{m \to \infty} \langle u_m - v_m, DE(u_m) \rangle = \limsup_{m \to \infty} g(u_m) ,$$

and the proof is complete.   □

11.2 Definition. *A point $u \in M$ is critical if $g(u) = 0$, otherwise u is regular. If $E(u) = \beta$ for some critical point $u \in M$ of E, the value β is critical; otherwise β is regular.*

This definition coincides with the definition of regular or critical points (values) of a functional given earlier, if $M = V$.

Moreover, as usual we let

$$M_\beta = \{u \in M \; ; \; E(u) < \beta\} ,$$
$$K_\beta = \{u \in M \; ; \; E(u) = \beta, \; g(u) = 0\} ,$$
$$N_{\beta,\delta} = \{u \in M \; ; \; |E(u) - \beta| < \delta, \; g(u) < \delta\} ,$$
$$U_{\beta,\rho} = \{u \in M \; ; \; \exists v \in K_\beta : \|u - v\| < \rho\} ,$$

denote the sub-level sets, critical sets and families of neighborhoods of K_β, for any $\beta \in \mathbb{R}$.

11.3 Definition. *E satisfies the Palais-Smale condition on M if the following is true:*

$(P.\text{-}S.)_{\mathrm{M}}$ *Any sequence (u_m) in M such that $|E(u_m)| \leq c$ uniformly while $g(u_m) \to 0$ $(m \to \infty)$, is relatively compact.*

11.4 Lemma. *Suppose E satisfies $(P.-S.)_M$. Then for any $\beta \in \mathbb{R}$ the set K_β is compact. Moreover, the families $\{N_{\beta,\delta} \; ; \; \delta > 0\}$, respectively $\{U_{\beta,\rho} \; ; \; \rho > 0\}$ constitute fundamental systems of neighborhoods of K_β.*

The proof is identical with that of Lemma 2.3.

    Denote

$$\tilde{M} = \{u \in M \; ; \; g(u) \neq 0\}$$

the set of regular points of E, and let

$$K = \{u \in M \; ; \; g(u) = 0\} = M \setminus \tilde{M}$$

be the set of critical points .

11.5 Definition. *A locally Lipschitz vector field $v \colon \tilde{M} \to V$ is a pseudo-gradient vector field for E on M if there exists $c > 0$ such that*
$(1°)$ $u + v(u) \in M$,
$(2°)$ $\|v(u)\| < \min\{1, g(u)\}$,
$(3°)$ $\langle v(u), DE(u)\rangle < -c \min\{1, g(u)\}g(u)$,
for all $u \in \tilde{M}$.

Arguing as in the proof of Lemma 3.2 we establish:

11.6 Lemma. *There exists a pseudo-gradient vector field $v \colon \tilde{M} \to V$, satisfying $(3°)$ of Definition 11.5 with $c = \frac{1}{2}$. Moreover, v extends to a locally Lipschitz continuous vector field on $V \setminus K$.*

Proof. For $u \in \tilde{M}$ choose $w = w(u) \in M$ such that

$$(11.1) \qquad\qquad \|u - w\| < \min\{1, g(u)\} \ ,$$

$$(11.2) \qquad\qquad \langle u - w, DE(u)\rangle > \frac{1}{2} \min\{1, g(u)\}g(u) \ .$$

Now, as in the proof of Lemma 3.2, let $\{U_\iota \; ; \; \iota \in I\}$ be a locally finite open cover of $\tilde{M}$ such that for any $\iota \in I$, any $u \in U_\iota \cap \tilde{M}$ conditions (11.1–2) hold with $w = w(u_\iota)$, for some $u_\iota \in U_\iota$. Then let $\{\varphi_\iota \; ; \; \iota \in I\}$ be a locally Lipschitz partition of unity subordinate to $\{U_\iota\}$ and define

$$v(u) = \sum_{\iota \in I} \varphi_\iota(u)\big(w(u_\iota) - u\big) \ ,$$

for $u \in V$. The resulting v is a pseudo-gradient vector field on $\tilde{M}$, as claimed. Moreover, v is locally Lipschitz continuous on $V \setminus K$. $\qquad\square$

11.7 Theorem (Deformation Lemma). *Suppose $M \subset V$ is closed and convex, $E \in C^1(V)$ satisfies (P.-S.)$_M$ on M, and let $\beta \in \mathbb{R}$, $\bar{\varepsilon} > 0$ be given. Then for any neighborhoosd N of K_β there exist $\varepsilon \in]0, \bar{\varepsilon}[$ and a continuous deformation $\Phi \colon M \times [0,1] \to M$ such that*
($1°$) $\Phi(u,t) = u$ if $g(u) = 0$, of if $t = 0$, or if $|E(u) - \beta| \geq \bar{\varepsilon}$;
($2°$) $E\big(\Phi(u,t)\big)$ is non-increasing in t, for any $u \in M$;
($3°$) $\Phi(M_{\beta+\varepsilon}, 1) \subset M_{\beta-\varepsilon} \cup N$, respectively $\Phi(M_{\beta+\varepsilon} \setminus N, 1) \subset M_{\beta-\varepsilon}$.

Proof. For $\varepsilon < \min\{\bar{\varepsilon}/2, \delta/4\}$, where

$$N \supset U_{\beta,\rho} \supset U_{\beta,\rho/2} \supset N_{\beta,\delta} ,$$

as in the proof of Theorem 3.4 let η, φ satisfy $0 \leq \eta, \ \varphi \leq 1$, $\eta(u) = 0$ in $N_{\beta,\delta/2}$, $\eta(u) = 1$ on $N_{\beta,\delta}$, $\varphi(s) = 1$ for $|s - \beta| \leq \varepsilon$, $\varphi(s) = 0$ for $|s - \beta| \geq 2\varepsilon$, and define

$$e(u) = \begin{cases} \eta(u)\varphi\big(E(u)\big)v(u), & u \in V \setminus K \\ 0, & u \in K . \end{cases}$$

The vector field e is Lipschitz continuous, uniformly bounded, and induces a global flow $\Phi \colon V \times [0,1] \to V$ such that

$$\frac{\partial}{\partial t}\Phi(u,t) = e\big(\Phi(u,t)\big)$$

$$\Phi(u,0) = 0 .$$

Note that since v (and therefore e) satisfies the condition

$$u + v(u) \in M , \quad \text{for all } u \in \tilde{M} ,$$

and since M is convex, the region M is forwardly invariant under Φ. Hence the restricted flow $\Phi|_{M \times [0,1]} \colon M \times [0,1] \to M$.

Assertions ($1°$), ($2°$) are trivially satisfied by definition of e, ($3°$) is proved exactly as in Theorem 3.4. □

The results from the preceding sections now may be conveyed to functionals on closed, convex sets. In particular, we recall Theorem 9.3:

11.8 Theorem. *Suppose M is a closed, convex subset of a Banach space V, $E \in C^1(V)$ satisfies (P.-S.)$_M$ on M, and admits two distinct relative minima u_1, u_2 in M. Then either $E(u_1) = E(u_2) = \beta$ and u_1, u_2 can be connected in any neighborhood of the set of relative minima $u \in M$ of E with $E(u) = \beta$, or there exists a critical point $\bar{u}$ of E in M which is not a relative minimizer of E.*

Applications to Semilinear Elliptic Boundary Value Problems

Here we will not enter into a detailed discussion of applications of these methods to the Plateau problem for minimal surfaces and for surfaces of constant mean curvature for which they were developed. The reader will find this material in the lecture notes of Struwe [17], devoted exclusively to this topic, and in Chang-Eells [1], Jost-Struwe [1], Struwe [13].

Nor will we touch upon applications to variational inequalities. In this context, variational methods first seem to have been applied by Miersemann [1] to eigenvalue problems in a cone; see also Quittner [1]. Using the methods outlined above, a unified approach to equations and inequalities can be achieved.

Instead, we re-derive Amann's [2], [3] famous "three solution theorem" on the existence of "unstable" solutions of semilinear elliptic boundary value problems, confined in an order interval between sub- and supersolutions, in the variational case.

11.9 Theorem. *Suppose Ω is a bounded domain in $\mathbb{R}^n$ and $g \colon \mathbb{R} \to \mathbb{R}$ is of class C^1 satisfying the growth condition*

$(1°)$ $$|g_u(u)| \le c\left(1 + |u|^{p-2}\right), \qquad \text{for some } p \le \frac{2n}{n-2} \ .$$

Also suppose that the problem

$(2°)$ $$-\Delta u = g(u) \qquad \text{in } \Omega \ ,$$
$(3°)$ $$u = 0 \qquad \text{on } \partial\Omega \ ,$$

admits two pairs of sub- and supersolutions $\underline{u}_1 \le \overline{u}_1 \le \underline{u}_2 \le \overline{u}_2 \in C^2 \cap H_0^{1,2}(\Omega)$. Then either $\overline{u}_1$ or $\underline{u}_2$ weakly solves $(2°)$, $(3°)$, or $(2°)$, $(3°)$ admits at least 3 distinct solutions u_i, $\underline{u}_1 \le u_i \le \overline{u}_2$, $i = 1, 2, 3$.

Proof. By Theorem I.2.4 the functional $E \in C^1\left(H_0^{1,2}(\Omega)\right)$ related to $(2°)$, $(3°)$, given by

$$E(u) = \frac{1}{2} \int_\Omega |\nabla u|^2 \, dx - \int_\Omega G(u) \, dx \ ,$$

admits critical points u_i which are relative minima of E in the order intervals $\underline{u}_i \le u_i \le \overline{u}_i$, $i = 1, 2$. Let $\underline{u} = \underline{u}_1$, $\overline{u} = \overline{u}_2$, and define

$$M = \{u \in H_0^{1,2}(\Omega) \ ; \ \underline{u} \le u \le \overline{u} \text{ a.e.}\} \ ;$$

Observe that – unless $\overline{u}_1$ respectively $\underline{u}_2$ solves $(2°),(3°)$ – u_1 and u_2 are also relative minima of E in M. To see this, for $i = 1, 2$ consider

$$M_i = \{u \in H_0^{1,2}(\Omega) \ ; \ \underline{u}_i \le u \le \overline{u}_i \text{ almost everywhere}\}$$

and for $\rho > 0$, as in the proof of Theorem 9.5, let

$$M_i^\rho = \{u \in M \ ; \ \exists v \in M_i : \|u - v\|_{H_0^{1,2}} \le \rho\} \ .$$

Note that M_i^ρ is closed and convex, hence weakly closed, and E is coercive and weakly lower semi-continuous on M_i^ρ with respect to $H_0^{1,2}(\Omega)$. By Theorem I.1.2, therefore, E admits relative minimizers $u_i^\rho \in M_i^\rho, i = 1, 2$, for any $\rho > 0$, and there holds

$$\big\langle (u_1^\rho - \overline{u}_1)_+, DE(u_1^\rho) \big\rangle \leq 0 \leq \big\langle (\underline{u}_2 - u_2^\rho)_+, DE(u_2^\rho) \big\rangle \ ,$$

with $(s)_+ = \max\{0, s\}$, as usual. Subtracting the relations

$$\big\langle (u_1^\rho - \overline{u}_1)_+, DE(\overline{u}_1) \big\rangle \geq 0 \geq \big\langle (\underline{u}_2 - u_2^\rho)_+, DE(\underline{u}_2) \big\rangle$$

we obtain that

$$
\begin{aligned}
\|(u_1^\rho - \overline{u}_1)_+\|_{H_0^{1,2}}^2 &= \int_\Omega |\nabla(u_1^\rho - \overline{u}_1)_+|^2 \, dx \leq \int_\Omega \big((g(u_1^\rho) - g(\overline{u}_1))(u_1^\rho - \overline{u}_1)_+ \, dx \\
&\leq \int_\Omega \left(\int_0^1 g_u(\overline{u}_1 + s(u_1^\rho - \overline{u}_1)) \, dx \right) (u_1^\rho - \overline{u}_1)_+^2 \, dx \\
&\leq \sup_{v \in M_1^\rho} \|g_u(v)\|_{L^{\frac{p}{p-2}}} \|(u_1^\rho - \overline{u}_1)_+\|_{L^p}^2 \\
&\leq C \big(1 + \sup_{v \in M_1^\rho} \|v\|_{L^p}^{p-2}\big) \cdot \\
&\quad \mathcal{L}^n\big(\{x \; ; \; u_1^\rho(x) > \overline{u}_1(x)\}\big)^{2\gamma} \cdot \|(u_1^\rho - \overline{u}_1)_+\|_{L^{\frac{2n}{n-2}}}^2
\end{aligned}
$$

where $\gamma = 1 - \frac{n-2}{2n}p > 0$, and an analogous estimate for $(\underline{u}_2 - u_2^\rho)_+$. By Sobolev's embedding theorem

$$
\begin{aligned}
\|v\|_{L^p} &\leq c\|v\|_{L^{\frac{2n}{n-2}}} \leq c \inf_{w \in M_1} \big(\|v - w\|_{L^{\frac{2n}{n-2}}} + \|w\|_{L^{\frac{2n}{n-2}}}\big) \\
&\leq c \inf_{w \in M_1} \|v - w\|_{H_0^{1,2}} + C \leq C < \infty
\end{aligned}
$$

for $v \in M_1^\rho, \rho \leq 1$. Similarly, $\|(u_1^\rho - \overline{u}_1)_+\|_{L^{\frac{2n}{n-2}}} \leq c\|(u_1^\rho - \overline{u}_1)_+\|_{H^{1,2}}$, whence with a uniform constant C for all $\rho > 0$ there holds

$$(11.5) \quad \|(u_1^\rho - \overline{u}_1)_+\|_{H_0^{1,2}}^2 \leq C\mathcal{L}^n\big(\{x \; ; \; u_1^\rho(x) > \overline{u}_1(x)\}\big)^{2\gamma} \|(u_1^\rho - \overline{u}_1)_+\|_{H_0^{1,2}}^2 \ ,$$

and an analogous estimate for $(\underline{u}_2 - u_2^\rho)_+$.

As $\rho \to 0$ the functions u_i^ρ accumulate at minimizers $\tilde{u}_i$ of E in $M_i, i = 1, 2$. Arguing as in the proof of Theorem I.2.4, and using the regularity result Lemma B.3 of the appendix, these functions $\tilde{u}_i$ are classical solutions of $(2°),(3°)$. Now, if $\overline{u}_1, \underline{u}_2$ do not solve problem $(2°),(3°)$, then in particular we have $\tilde{u}_1 \neq \overline{u}_1, \tilde{u}_2 \neq \underline{u}_2$. From the strong maximum principle we then infer that $\tilde{u}_1 < \overline{u}_1, \underline{u}_2 < \tilde{u}_2$. Hence

$$\mathcal{L}^n\big(\{x; u_1^\rho(x) > \overline{u}_1(x)\}\big) \to 0 \qquad (\rho \to 0) \ ,$$

and from (11.5) it follows that $u_1^\rho \in M_1$ for sufficiently small $\rho > 0$, showing that u_1 is relatively minimal for E in M. Similarly for u_2. Thus, u_1 and u_2 are relative minima of E in M.

By Theorem 11.8 the functional E either admits infinitely many relative minima in M or possesses at least one critical point $u_3 \in M$ which is not a relative minimizer of E, hence distinct from u_1, u_2.

Finally, observe that any critical point u of E in M weakly solves (2°), (3°): Indeed for any $\varphi \in H_0^{1,2}(\Omega)$, $\varepsilon > 0$, let $v_\varepsilon = \min\{\overline{u}, \max\{\underline{u}, u+\varepsilon\varphi\}\} = u - \varepsilon\varphi - \varphi^\varepsilon + \varphi_\varepsilon \in M$ with $\varphi^\varepsilon = \max\{0, u+\varepsilon\varphi - \overline{u}\} \geq 0$, $\varphi_\varepsilon = -\min\{0, u+\varepsilon\varphi - \underline{u}\} \geq 0$, as in the proof of Theorem I.2.4. Then if $u \in M$ is critical, as $\varepsilon \to 0$ we have

$$\langle u - v_\varepsilon, DE(u)\rangle \leq \|u - v_\varepsilon\|_{H_0^{1,2}} \sup_{\substack{v \in M \\ \|u-v\|_{H_0^{1,2}} \leq 1}} \langle u - v, DE(u)\rangle$$

$$\leq O(\varepsilon)g(u) = 0 \ .$$

From this inequality the desired conclusion follows exactly as in the proof of Theorem I.2.4. $\qquad\square$

11.10 Notes. A different variational approach to this result was suggested by Chang [5], who used the regularizing properties of the heat flow to reduce the problem to a variational problem on an open subset of a suitable Banach space to which the standard methods could be applied.

A combination of topological degree and variational methods was used by Hofer [1] to obtain even higher multiplicity results for certain problems.

Chapter III

Limit Cases of the Palais-Smale Condition

Condition (P.-S.) may seem rather restrictive. Actually, as Hildebrandt [4; p. 324] records, for quite a while many mathematicians felt convinced that in-spite of its success in dealing with one-dimensional variational problems like geodesics (see Birkhoff's Theorem I.4.4 for example, or Palais' [3] work on closed geodesics), the Palais-Smale condition could never play a role in the solution of "interesting" variational problems in higher dimensions.

Recent advances in the Calculus of Variations have changed this view and it has become apparent that the methods of Palais and Smale apply to many problems of physical and/or geometric interest and – in particular – that the Palais-Smale condition will in general hold true for such problems in a broad range of energies. Moreover, the failure of (P.-S.) at certain levels reflects highly interesting phenomena related to internal symmetries of the systems under study, which geometrically can be described as "separation of spheres", or mathematically as "singularities", respectively as "change in topology". Again speaking in physical terms, we might observe "phase transitions" or "particle creation" at the energy levels where (P.-S.) fails.

Such phenomena seem to have first been observed by Sacks-Uhlenbeck [1] and – independently – by Wente [5] in the context of harmonic maps of surfaces, respectively in the context of surfaces of prescribed constant mean curvature. (See Sections 4 and 5 below.) In these cases the term "separation of spheres" has a clear geometric meaning. More recently, Sedlacek [1] has uncovered similar results also for Yang-Mills connections. If interpreted appropriately, very early indications of such phenomena already may be found in the work of Douglas [2], Morse-Tompkins [2] and Shiffman [2] on minimal surfaces of higher genus and/or connectivity. In this case, a "change in topology" in fact sometimes may be observed even physically as one tries to realize a multiply connected or higher genus minimal surface in a soap film experiment. See Jost-Struwe [1] for a modern approach to these results.

Mathematically, it seems that non-compact group actions give rise to these effects. In physics and geometry, of course, such group actions arise naturally as "symmetries" from the requirements of scale or gauge invariance; in particular, in the examples of Sacks-Uhlenbeck and Wente cited above, from conformal invariance.

Moreover, a symmetry may be "manifest" or "broken", that is, perturbed by interaction terms; surprisingly, existence results for problems with non-compact internal symmetries seem to depend on the extent to which the symmetry is broken or perturbed. As in the case of non-compact minimization problems studied in Section I.4, sometimes the perturbation from symmetry can be measured by comparing with a suitable (family of) *limiting problem(s)* where the symmetry is acting. Existence results – for example in the line of Theorem II.6.1 – therefore will strongly depend on energy estimates for critical values.

We start with a simple example.

1. Pohožaev's Non-Existence Result

Let Ω be a domain in $\mathbb{R}^n$, $n > 2$. Consider the limit case $p = 2^* = \frac{2n}{n-2}$ in Theorem I.2.1. Given $\lambda \in \mathbb{R}$ we would like to solve the problem

(1.1) $$-\Delta u = \lambda u + u|u|^{2^*-2} \qquad \text{in } \Omega \ ,$$

(1.2) $$u > 0 \qquad \text{in } \Omega \ ,$$

(1.3) $$u = 0 \qquad \text{on } \partial\Omega \ .$$

(Note that in order to be consistent with the literature, in this section we reverse the sign of λ as compared with Section I.2.1 or Section II.5.8.) As in Theorem I.2.1 we can approach this problem by a direct method and attempt to obtain non-trivial solutions of (1.1), (1.3) as relative minima of the functional

$$I_\lambda(u) = \frac{1}{2} \int_\Omega \left(|\nabla u|^2 - \lambda|u|^2\right) \, dx \ ,$$

on the unit sphere in $L^{2^*}(\Omega)$,

$$M = \{u \in H_0^{1,2}(\Omega) \ ; \ \|u\|_{L^{2^*}}^{2^*} = 1\} \ .$$

Equivalently, we may seek to minimize the Sobolev quotient

$$S_\lambda(u; \Omega) = \frac{\int_\Omega \left(|\nabla u|^2 - \lambda|u|^2\right) \, dx}{\left(\int_\Omega |u|^{2^*} \, dx\right)^{2/2^*}} \ , \qquad u \neq 0 \ .$$

Note that for $\lambda = 0$, as in Section I.4.4,

$$S(\Omega) = \inf_{\substack{u \in H_0^{1,2} \\ u \neq 0}} S_0(u; \Omega) = \inf_{\substack{u \in H_0^{1,2} \\ u \neq 0}} \frac{\int_\Omega |\nabla u|^2 \, dx}{\left(\int_\Omega |u|^{2^*} \, dx\right)^{2/2^*}}$$

is related to the (best) Lipschitz constant for the Sobolev embedding $H_0^{1,2}(\Omega) \to L^{2^*}(\Omega)$.

Recall that for any $u \in H_0^{1,2}(\Omega) \subset D^{1,2}(\mathbb{R}^n)$ the ratio $S_0(u; \mathbb{R}^n)$ is invariant under scaling $u \mapsto u_R(x) = u(x/R)$; that is, we have

(1.4) $$S_0(u; \mathbb{R}^n) = S_0(u_R; \mathbb{R}^n), \qquad \text{for all } R > 0 \ .$$

Hence, in particular, we have (see Remark I.4.5):

1.1 Lemma. $S(\Omega) = S$ is independent of Ω.

Moreover, this implies (see Remark I.4.7):

1.2 Theorem. S is never attained on a domain $\Omega \subseteq \mathbb{R}^n$, $\Omega \neq \mathbb{R}^n$.

Hence, for $\lambda = 0$, the proof of Theorem I.2.1 necessarily fails in the limit case $p = 2^*$. More generally, we have the following uniqueness result, due to Pohožaev [1]:

1.3 Theorem. Suppose $\Omega \neq \mathbb{R}^n$ is a smooth (possibly unbounded) domain in $\mathbb{R}^n$, $n \geq 3$, which is strictly star-shaped with respect to the origin in $\mathbb{R}^n$, and let $\lambda \leq 0$. Then any solution $u \in H_0^{1,2}(\Omega)$ of the boundary value problem (1.1), (1.3) vanishes identically.

The proof is based on the following "Pohožaev identity":

1.4 Lemma. Let $g: \mathbb{R} \to \mathbb{R}$ be continuous with primitive $G(u) = \int_0^u g(v)\, dv$ and let $u \in C^2(\Omega) \cap C^1(\overline{\Omega})$ be a solution of the equation

$$(1.5) \qquad\qquad -\Delta u = g(u) \qquad in\ \Omega$$
$$(1.6) \qquad\qquad u = 0 \qquad on\ \partial\Omega$$

in a domain $\Omega \subset\subset \mathbb{R}^n$. Then there holds

$$\frac{n-2}{2} \int_\Omega |\nabla u|^2\, dx - n \int_\Omega G(u)\, dx + \frac{1}{2} \int_{\partial\Omega} \left|\frac{\partial u}{\partial \nu}\right|^2 x \cdot \nu\, do = 0\ ,$$

where ν denotes the exterior unit normal.

Proof of Theorem 1.3. Let $g(u) = \lambda u + u|u|^{2^*-2}$ with primitive

$$G(u) = \frac{\lambda}{2}|u|^2 + \frac{1}{2^*}|u|^{2^*}\ .$$

By Theorem I.2.2 and Lemma B.3 of the appendix, any solution of (1.1), (1.3) is smooth on $\overline{\Omega}$. Hence from Pohožaev's identity we infer that

$$\int_\Omega |\nabla u|^2\, dx - 2^* \int_\Omega G(u)\, dx + \frac{1}{n-2} \int_{\partial\Omega} \left|\frac{\partial u}{\partial \nu}\right|^2 x \cdot \nu\, do$$
$$= \int_\Omega \left(|\nabla u|^2 - |u|^{2^*}\right) dx + \frac{n|\lambda|}{n-2} \int_\Omega |u|^2\, dx$$
$$+ \frac{1}{n-2} \int_{\partial\Omega} \left|\frac{\partial u}{\partial \nu}\right|^2 x \cdot \nu\, do = 0\ .$$

However, testing the equation (1.1) with u, we infer that

$$\int_\Omega \left(|\nabla u|^2 - \lambda |u|^2 - |u|^{2^*} \right) dx = 0 \ ,$$

whence

$$2|\lambda| \int_\Omega |u|^2 \, dx + \int_{\partial\Omega} \left| \frac{\partial u}{\partial\nu} \right|^2 x \cdot \nu \, do = 0 \ .$$

Moreover, since Ω is strictly star-shaped with respect to $0 \in \mathbb{R}^n$, we have $x \cdot \nu > 0$ for all $x \in \partial\Omega$. Thus $\frac{\partial u}{\partial\nu} = 0$ on $\partial\Omega$, and hence $u \equiv 0$ by the principle of unique continuation. $\qquad\square$

Proof of Lemma 1.4. Multiply (1.5) by $x \cdot \nabla u$ and compute

$$0 = \left(\Delta u + g(u) \right)(x \cdot \nabla u) =$$

$$= \mathrm{div}\left(\nabla u(x \cdot \nabla u) \right) - |\nabla u|^2 - x \cdot \nabla\left(\frac{|\nabla u|^2}{2} \right) + x \cdot \nabla G(u)$$

$$= \mathrm{div}\left(\nabla u(x \cdot \nabla u) - x\frac{|\nabla u|^2}{2} + xG(u) \right) + \frac{n-2}{2}|\nabla u|^2 - nG(u) \ .$$

Upon integrating this identity over Ω and taking account of the fact that by (1.6) we have

$$x \cdot \nabla u = x \cdot \nu \frac{\partial u}{\partial\nu} \qquad \text{on } \partial\Omega \ ,$$

the lemma follows. $\qquad\square$

1.5 Interpretation. Theorem 1.3 goes beyond Theorem 1.2, as the former applies to any solution, whereas the latter is limited to minima of $S_0(\cdot\,; \Omega)$. However, Theorem 1.2 applies to any domain.

The connection between the scale invariance of $S = S_0$ and Theorem 1.3 is given by the fact that the function $x \cdot \nabla u = \frac{d}{dR} u_R$ used in the proof of Lemma 1.4 is the generator of the family of scaled maps $\{u_R \ ; \ 0 < R < \infty\}$. We interpret Theorem 1.3 as reflecting the non-compactness of the multiplicative group $\mathbb{R}_+ = \{R \ ; \ 0 < R < \infty\}$ acting on S via scaling. Note that this group action is manifest for $S_\lambda(\cdot\,; \Omega)$ only if $\lambda = 0$ and $\Omega = \mathbb{R}^n$. In case of a bounded domain Ω not all scalings $u \to u_R$ will map $H_0^{1,2}(\Omega)$ into itself. For instance, if Ω is an annular region $\Omega = \{x \ ; \ a < |x| < b\}$, in fact, $H_0^{1,2}(\Omega)$ does not admit any of these scalings as symmetries. (In Section 3 we will see that in this case (1.1)–(1.3) does have nontrivial solutions.) However, if Ω is star-shaped with respect to the origin, all scalings $u \to u_R$, $R \leq 1$ will be symmetries of $H_0^{1,2}(\Omega)$, and compactness is lost as $R \to 0$. The effect is shown in Theorem 1.3.

Remark that it is also possible to characterize solutions $u \in H_0^{1,2}(\Omega)$ of equation (1.1) as critical points of a functional E_λ on $H_0^{1,2}(\Omega)$ given by

$$(1.7) \qquad E_\lambda(u) = \frac{1}{2} \int_\Omega \left(|\nabla u|^2 - \lambda |u|^2 \right) dx - \frac{1}{2^*} \int_\Omega |u|^{2^*} \, dx \ .$$

By continuity of the embedding $H_0^{1,2}(\Omega) \hookrightarrow L^{2^*}(\Omega) \hookrightarrow L^2(\Omega)$, the functional E_λ is Fréchet differentiable on $H_0^{1,2}(\Omega)$. Moreover, for $\lambda < \lambda_1$, the first Dirichlet eigenvalue of the operator $-\Delta$, E_λ satisfies the conditions (1°)–(3°) of the mountain pass lemma Theorem II.6.1; compare the proof of Theorem I.2.1. In view of Theorem II.6.1, the absence of a critical point $u > 0$ of E_λ for any $\lambda \leq 0$ proves that E_λ for such λ cannot satisfy the Palais-Smale condition (P.-S.) on a star-shaped domain. Again the non-compact action $R \mapsto u_R(x) = u(Rx)$ can be held responsible.

2. The Brezis-Nirenberg Result

In contrast to Theorem 1.3, for $\lambda > 0$ problem (1.1)–(1.3) may admit non-trivial solutions. However, a subtle dependence on the dimension n is observed.

The first result in this direction is due to Brezis and Nirenberg [2]; their approach is related to ideas of Trudinger [1] and Aubin [2].

2.1 Theorem. *Suppose Ω is a domain in $\mathbb{R}^n$, $n \geq 3$, and let $\lambda_1 > 0$ denote the first eigenvalue of the operator $-\Delta$ with homogeneous Dirichlet boundary conditions.*
(1°) If $n \geq 4$, then for any $\lambda \in]0, \lambda_1[$ there exists a (positive) solution of (1.1)–(1.3).
(2°) If $n = 3$, there exists $\lambda_* \in [0, \lambda_1[$ such that for any $\lambda \in]\lambda_*, \lambda_1[$ problem (1.1)–(1.3) admits a solution.
(3°) If $n = 3$ and $\Omega = B_1(0) \subset \mathbb{R}^3$, then $\lambda_* = \frac{\lambda_1}{4}$ and for $\lambda \leq \frac{\lambda_1}{4}$ there is no solution to (1.1)–(1.3).

As we have seen in Section 1, there are (at least) two different approaches to this theorem. The first, which is the one primarily chosen by Brezis and Nirenberg [2], involves the quotient

$$S_\lambda(u; \Omega) = \frac{\int_\Omega \left(|\nabla u| - \lambda|u|^2\right) dx}{\left(\int_\Omega |u|^{2^*} dx\right)^{2/2^*}} \; .$$

A second proof can be given along the lines of Theorem II.6.1, applied to the "free" functional E_λ

$$E_\lambda(u) = \frac{1}{2} \int_\Omega \left(|\nabla u|^2 - \lambda|u|^2\right) dx - \frac{1}{2^*} \int_\Omega |u|^{2^*} dx$$

defined earlier. Recall that $E_\lambda \in C^1\left(H_0^{1,2}(\Omega)\right)$. As we shall see, while it is not true that E_λ satisfies the Palais-Smale condition "globally", some compactness will hold in an energy range determined by the best Sobolev constant S; see Lemma 2.3 below. A similar compactness property holds for the functional S_λ. We will first pursue the approach involving S_λ.

Constrained Minimization

Denote

$$S_\lambda(\Omega) = \inf_{u \in H_0^{1,2}(\Omega) \setminus \{0\}} S_\lambda(u; \Omega) .$$

Note that $S_\lambda(\Omega) \leq S$ for all $\lambda \geq 0$ (in fact, for all $\lambda \in \mathbb{R}$), and $S_\lambda(\Omega)$ in general is not attained. Similar to Theorem I.4.2 now there holds:

2.2 Lemma. *If Ω is a bounded domain in $\mathbb{R}^n$, $n \geq 3$, and if*

$$S_\lambda(\Omega) < S ,$$

then there exists $u \in H_0^{1,2}(\Omega)$, $u > 0$, such that $S_\lambda(\Omega) = S_\lambda(u; \Omega)$.

Proof. Consider a minimizing sequence (u_m) for S_λ in $H_0^{1,2}(\Omega)$; normalize $\|u_m\|_{L^{2^*}} = 1$. Replacing u_m by $|u_m|$, if necessary, we may assume that $u_m \geq 0$. Since by Hölder's inequality

$$S_\lambda(u_m; \Omega) = \int_\Omega \left(|\nabla u_m|^2 - \lambda |u_m|^2\right) dx \geq \int_\Omega |\nabla u_m|^2 \, dx - c ,$$

we also may assume that $u_m \rightharpoonup u$ weakly in $H_0^{1,2}(\Omega)$ and strongly in $L^2(\Omega)$ as $m \to \infty$.

To proceed, observe that like (I.4.4) by Vitali's convergence theorem we have

$$\int_\Omega \left(|u_m|^{2^*} - |u_m - u|^{2^*}\right) dx =$$

$$= \int_\Omega \int_0^1 \frac{d}{dt} \left|u_m + (t-1)u\right|^{2^*} dt \, dx$$

(2.1)

$$= 2^* \int_0^1 \int_\Omega \left(u_m + (t-1)u\right)\left|u_m + (t-1)u\right|^{2^*-2} u \, dx \, dt$$

$$\to 2^* \int_0^1 \int_\Omega t \, u|t \, u|^{2^*-2} u \, dx \, dt = \int_\Omega |u|^{2^*} \, dx \quad \text{as } m \to \infty .$$

Also note that

$$(2.2) \qquad \int_\Omega |\nabla u_m|^2 \, dx = \int_\Omega |\nabla(u_m - u)|^2 \, dx + \int_\Omega |\nabla u|^2 \, dx + o(1) ,$$

where $o(1) \to 0$ as $m \to \infty$. Hence we obtain:

$$S_\lambda(\Omega) = S_\lambda(u_m; \Omega) + o(1) = \int_\Omega |\nabla(u_m - u)|^2 \, dx + \int_\Omega \left(|\nabla u|^2 - \lambda |u|^2\right) dx + o(1)$$

$$\geq S\|u_m - u\|_{L^{2^*}}^2 + S_\lambda(\Omega)\|u\|_{L^{2^*}}^2 + o(1)$$

$$\geq S\|u_m - u\|_{L^{2^*}}^{2^*} + S_\lambda(\Omega)\|u\|_{L^{2^*}}^{2^*} + o(1)$$

$$\geq \left(S - S_\lambda(\Omega)\right)\|u_m - u\|_{L^{2^*}}^{2^*} + S_\lambda(\Omega) + o(1) .$$

Since $S > S_\lambda(\Omega)$ by assumption, this implies that $u_m \to u$ in $L^{2^*}(\Omega)$; that is $u \in M$, and by weak lower semi-continuity of the norm in $H_0^{1,2}(\Omega)$ it follows that

$$S_\lambda(u; \Omega) \leq \lim_{m \to \infty} S_\lambda(u_m; \Omega) = S_\lambda(\Omega) ,$$

as desired.

Computing the first variation of $S_\lambda(u; \Omega)$, as in the proof of Theorem I.2.1 we see that a positive multiple of u satisfies (1.1), (1.3). Since $u \geq 0$, $u \neq 0$, from the strong maximum principle (Theorem B.4 of the appendix) we infer that $u > 0$ in Ω. The proof is complete. □

The Unconstrained Case: Local Compactness

Postponing the complete proof of Theorem 2.1 for a moment, we now also indicate the second approach, based on a careful study of the compactness properties of the free functional E_λ. Note that in the case of Theorem 2.1 both approaches are completely equivalent – and the final step in the proof of Theorem 2.1 actually is identical in both cases. However, for more general non-linearities with critical growth it is not always possible to reduce a boundary value problem like (II.6.1),(II.6.2) to a constrained minimization problem and we will have to use the free functional instead. Moreover, this second approach will bring out the peculiarities of the limiting case more clearly. Our presentation follows Cerami-Fortunato-Struwe [1]. An indication of Lemma 2.3 below is also given by Brezis-Nirenberg [2; p.463].

2.3 Lemma. *Let Ω be a bounded domain in $\mathbb{R}^n$, $n \geq 3$. Then for any $\lambda \in \mathbb{R}$, any sequence (u_m) in $H_0^{1,2}(\Omega)$ such that*

$$E_\lambda(u_m) \to \beta < \frac{1}{n}S^{n/2}, \qquad DE_\lambda(u_m) \to 0 ,$$

as $m \to \infty$, is relatively compact.

Proof. To show boundedness of (u_m), compute

$$o(1)\left(1 + \|u_m\|_{H_0^{1,2}}\right) + \frac{2}{n}S^n \geq 2E_\lambda(u_m) - \langle u_m, DE_\lambda(u_m)\rangle$$

$$= \left(1 - \frac{2}{2^*}\right) \int_\Omega |u_m|^{2^*} dx \geq c\left(\int_\Omega |u_m|^2 dx\right)^{2^*/2} ,$$

where $c > 0$ and $o(1) \to 0$ as $m \to \infty$. Hence

$$\|u_m\|_{H_0^{1,2}}^2 = 2E_\lambda(u_m) + \lambda\int_\Omega |u_m|^2 dx + \frac{2}{2^*}\int_\Omega |u_m|^{2^*} dx$$

$$\leq C + o(1)\|u_m\|_{H_0^{1,2}} ,$$

and it follows that (u_m) is bounded.

Hence we may assume that $u_m \rightharpoonup u$ weakly in $H_0^{1,2}(\Omega)$, and therefore also strongly in $L^p(\Omega)$ for all $p < 2^*$ by the Rellich-Kondrakov theorem; see Theorem A.5 of the appendix.

In particular, for any $\varphi \in C_0^\infty(\Omega)$ we obtain that

$$\langle \varphi, DE_\lambda(u_m) \rangle = \int_\Omega \left(\nabla u_m \nabla \varphi - \lambda u_m \varphi - u_m |u_m|^{2^*-2} \varphi \right) dx$$

$$\rightarrow \int_\Omega \left(\nabla u \nabla \varphi - \lambda u \varphi - u |u|^{2^*-2} \varphi \right) dx = \langle \varphi, DE_\lambda(u) \rangle = 0 ,$$

as $m \to \infty$. Hence, $u \in H_0^{1,2}(\Omega)$ weakly solves (1.1). Moreover, choosing $\varphi = u$, we have

$$0 = \langle u, DE_\lambda(u) \rangle = \int_\Omega \left(|\nabla u|^2 - \lambda |u|^2 - |u|^{2^*} \right) dx ,$$

and hence

$$E_\lambda(u) = \left(\frac{1}{2} - \frac{1}{2^*} \right) \int_\Omega |u|^{2^*} dx = \frac{1}{n} \int_\Omega |u|^{2^*} dx \geq 0 .$$

To proceed, note that by (2.1) and (2.2) we have

$$\int_\Omega |\nabla u_m|^2 dx = \int_\Omega |\nabla(u_m - u)|^2 dx + \int_\Omega |\nabla u|^2 dx + o(1) ,$$

$$\int_\Omega |u_m|^{2^*} dx = \int_\Omega |(u_m - u)|^{2^*} dx + \int_\Omega |u|^{2^*} dx + o(1) ,$$

and similarly, again using (2.1),

$$\int_\Omega \left(u_m |u_m|^{2^*-2} - u|u|^{2^*-2} \right)(u_m - u) \, dx$$

$$= \int_\Omega \left(|u_m|^{2^*} - u_m |u_m|^{2^*-2} u \right) dx + o(1)$$

$$= \int_\Omega \left(|u_m|^{2^*} - |u|^{2^*} \right) dx + o(1) = \int_\Omega |u_m - u|^{2^*} dx + o(1) ,$$

where $o(1) \to 0$ $(m \to \infty)$. Hence

$$E_\lambda(u_m) = E_\lambda(u) + E_0(u_m - u) + o(1) ,$$

$$o(1) = \langle u_m - u, DE_\lambda(u_m) \rangle = \langle u_m - u, DE_\lambda(u_m) - DE_\lambda(u) \rangle$$

$$= \int_\Omega \left(|\nabla(u_m - u)|^2 - |u_m - u|^{2^*} \right) dx + o(1) .$$

In particular, from the last equation

$$E_0(u_m - u) = \frac{1}{n} \int_\Omega |\nabla(u_m - u)|^2 \, dx + o(1) ,$$

while

$$E_0(u_m - u) = E_\lambda(u_m) - E_\lambda(u) + o(1)$$

$$\leq E_\lambda(u_m) + o(1) \leq c < \frac{1}{n} S^{n/2} \qquad \text{for } m \geq m_0 .$$

Therefore

$$\|u_m - u\|_{H_0^{1,2}}^2 \leq c < S^{n/2} \qquad \text{for } m \geq m_0 .$$

But then Sobolev's inequality

$$\|u_m - u\|_{H_0^{1,2}}^2 \left(1 - S^{-2^*/2}\|u_m - u\|_{H_0^{1,2}}^{2^*-2}\right) \leq$$

$$\leq \int_\Omega \left(|\nabla(u_m - u)|^2 - |u_m - u|^{2^*}\right) dx = o(1)$$

shows that $u_m \to u$ strongly in $H_0^{1,2}(\Omega)$, as desired. □

Lemma 2.3 motivates to introduce the following variant of (P.-S.), which seems to appear first in Brezis-Coron-Nirenberg [1].

2.4 Definition. *Let V be a Banach space, $E \in C^1(V)$, be $\in \mathbb{R}$. E satisfies condition (P.-S.)$_\beta$, if any sequence (u_m) in V such that $E(u_m) \to \beta$ while $DE(u_m) \to 0$ as $m \to \infty$ is relatively compact. (Such sequences in the sequel for brevity will be referred to as (P.-S.)$_\beta$-sequences.)*

Now recall that E_λ for $\lambda < \lambda_1$ satisfies conditions (1°)–(3°) of Theorem II.6.1.

By Lemma 2.3, therefore, the proof of the first two parts of Theorem 2.1 will be complete if we can show that for $\lambda > 0$ (respectively $\lambda > \lambda_*$) there holds

$$(2.3) \qquad \beta = \inf_{p \in P} \sup_{u \in p} E_\lambda(u) < \frac{1}{n} S^{n/2} ,$$

where, for a suitable function u_1 satisfying $E(u_1) \leq 0$, we let

$$P = \{p \in C^0([0,1] ; H_0^{1,2}(\Omega)) ; p(0) = 0, p(1) = u_1\} ,$$

as in Theorem 6.1.

Of course, (2.3) and the condition $S_\lambda(\Omega) < S$ of Lemma 2.2 are related. Given $u \in H_0^{1,2}(\Omega)$, $\|u\|_{L^{2^*}} = 1$, we may let $p(t) = t\,u$, $u_1 = t_1 u$ for sufficiently large t_1 to obtain

$$\beta \leq \sup_{0 \leq t < \infty} E_\lambda(t\,u) = \sup_{0 \leq t < \infty} \left(\frac{t^2}{2} S_\lambda(u; \Omega) - \frac{t^{2^*}}{2^*}\right) = \frac{1}{n} S_\lambda^{n/2}(u; \Omega) .$$

Likewise, for $p \in P$ there exists $u \in p$ such that $u \neq 0$ and

$$\langle u, DE_\lambda(u) \rangle = \int_\Omega \left(|\nabla u|^2 - \lambda |u|^2 - |u|^{2^*}\right) dx = 0 .$$

Indeed, since $\lambda < \lambda_1$, for $u = p(t)$ with t close to 0 we have $\langle u, DE_\lambda(u) \rangle > 0$, while for $u = p(1) = u_1$ we have

$$\langle u_1, DE_\lambda(u_1) \rangle < 2E_\lambda(u_1) \le 0 ,$$

and by the intermediate value theorem there exists u, as claimed. But for such u we easily compute

$$S_\lambda(u; \Omega) = \left(\int_\Omega |\nabla u|^2 - \lambda |u|^2 \, dx \right)^{1-2/2^*}$$

$$= (n \, E_\lambda(u))^{2/n} \le \left(n \sup_{u \in p} E_\lambda(u) \right)^{2/n} .$$

That is,

(2.4) $$\beta = \inf_{p \in P} \sup_{u \in p} E_\lambda(u) = \frac{1}{n} S_\lambda^{n/2}(\Omega) ,$$

and (2.3) and the condition $S_\lambda < S$ are in fact equivalent.

Proof of Theorem 2.1(1°). It suffices to show that $S_\lambda < S$. Consider the family

(2.5) $$u_\varepsilon^*(x) = \frac{[n(n-2)\varepsilon^2]^{\frac{n-2}{4}}}{[\varepsilon^2 + |x|^2]^{\frac{n-2}{2}}}, \quad \varepsilon > 0 ,$$

of functions $u_\varepsilon^* \in D^{1,2}(\mathbb{R}^n)$. Note that $u_\varepsilon^*(u) = \varepsilon^{\frac{2-n}{2}} u_1^* \left(\frac{x}{\varepsilon} \right)$, and u_ε^* satisfies the equation

(2.6) $$-\Delta u_\varepsilon^* = u_\varepsilon^* |u_\varepsilon^*|^{2^*-2} \quad \text{in } \mathbb{R}^n ,$$

as is easily verified by a direct computation. We claim that $S_0(u_\varepsilon^*; \mathbb{R}^n) = S$; that is, the best Sobolev constant is achieved by the family u_ε^*, $\varepsilon > 0$. Indeed, let $u \in D^{1,2}(\mathbb{R}^n)$ satisfy $S_0(u; \mathbb{R}^n) = S$. (The existence of such a function u can be deduced for instance from Theorem I.4.9.) Using Schwarz-symmetrization we may assume that u is radially symmetric; that is, $u(x) = u(|x|)$. Moreover, u solves (2.6). Choose $\varepsilon > 0$ such that $u_\varepsilon^*(0) = u(0)$. Then u and u_ε^* both are solutions of the ordinary differential equation of second order in $r = |x|$,

$$r^{1-n} \frac{\partial}{\partial r} \left(r^{n-1} \frac{\partial}{\partial r} \right) u = u|u|^{2^*-2} \quad \text{for } r > 0 ,$$

sharing the initial data $u(0) = u_\varepsilon^*(0)$, $\partial_r u(0) = \partial_r u_\varepsilon^*(0) = 0$. It is not hard to prove that this initial value problem admits a unique solution, and thus $u = u_\varepsilon^*$, which implies that $S_0(u_\varepsilon^*; \mathbb{R}^n) = S_0(u; \mathbb{R}^n) = S$.

In particular,

$$\|u_\varepsilon^*\|_{H_0^{1,2}}^2 = \|u_\varepsilon^*\|_{L^{2^*}}^{2^*} = S^{n/2}, \quad \text{for all } \varepsilon > 0 .$$

We may suppose that $0 \in \Omega$. Let $\eta \in C_0^\infty(\Omega)$ be a fixed cut-off function, $\eta \equiv 1$ in a neighborhood $B_\rho(0)$ of 0. Let $u_\varepsilon = \eta\, u_\varepsilon^*$ and compute

$$
\int_\Omega |\nabla u_\varepsilon|^2 \, dx = \int_\Omega |\nabla u_\varepsilon^*|^2 \eta^2 \, dx + O\big(\varepsilon^{n-2}\big)
$$

$$
= \int_{\mathbb{R}^n} |\nabla u_\varepsilon^*|^2 \, dx + O\big(\varepsilon^{n-2}\big) = S^{n/2} + O\big(\varepsilon^{n-2}\big) \; .
$$

$$
\int_\Omega |u_\varepsilon|^{2^*} \, dx = \int_{\mathbb{R}^n} |u_\varepsilon^*|^{2^*} \, dx + O\big(\varepsilon^n\big) = S^{n/2} + O\big(\varepsilon^n\big)
$$

$$
\int_\Omega |u_\varepsilon|^2 \, dx = \int_{B_\rho(0)} |u_\varepsilon^*|^2 \, dx + O\big(\varepsilon^{n-2}\big)
$$

(2.7)
$$
\geq \int_{B_\varepsilon(0)} \frac{[n(n-2)\varepsilon^2]^{\frac{n-2}{2}}}{[2\varepsilon^2]^{n-2}} \, dx
$$

$$
+ \int_{B_\rho(0) \backslash B_\varepsilon(0)} \frac{[n(n-2)\varepsilon^2]^{\frac{n-2}{2}}}{[2|x|^2]^{n-2}} \, dx + O\big(\varepsilon^{n-2}\big)
$$

$$
= c_1 \cdot \varepsilon^2 + c_2 \varepsilon^{n-2} \int_\varepsilon^\rho r^{3-n} \, dr + O(\varepsilon^{n-2})
$$

$$
= \begin{cases} c\varepsilon^2 + O\big(\varepsilon^{n-2}\big), & \text{if } n > 4 \\ c\varepsilon^2 |\ln \varepsilon| + O\big(\varepsilon^2\big), & \text{if } n = 4 \\ c\varepsilon + O\big(\varepsilon^2\big), & \text{if } n = 3 \end{cases}
$$

with positive constants $c, c_1, c_2 > 0$. Thus, if $n \geq 5$

$$
S_\lambda(u_\varepsilon) \leq \frac{(S^{n/2} - c\lambda\varepsilon^2 + O(\varepsilon^{n-2}))}{(S^{n/2} + O(\varepsilon^n))^{2/2^*}}
$$

$$
= S - c\lambda\varepsilon^2 + O\big(\varepsilon^{n-2}\big) < S \; ,
$$

if $\varepsilon > 0$ is sufficiently small. Similarly, if $n = 4$, we have

$$
S_\lambda(u_\varepsilon) \leq S - c\lambda\varepsilon^2 |\ln \varepsilon| + O\big(\varepsilon^2\big) < S
$$

for $\varepsilon > 0$ sufficiently small.

Remark on Theorem 2.1.($2°$),($3°$). If $n = 3$, estimate (2.7) shows that the "gain" due to the presence of λ and the "loss" due to truncation of u_ε^* may be of the same order in ε; hence S_λ can only be expected to be smaller than S for "large" λ. To see that $\lambda_* < \lambda_1$, choose the first eigenfunction $u = \varphi_1$ of $(-\Delta)$ as comparison function. The non-existence result for $\Omega = B_1(0)$, $\lambda \leq \frac{\lambda_1}{4}$ follows from a weighted estimate similar to Lemma 1.4; see Brezis-Nirenberg [2; Lemma 1.4]. We omit the details. □

Theorem 2.1 should be viewed together with the global bifurcation result of Rabinowitz [1; p. 195 f.]. Intuitively, Theorem 2.1 indicates that the branch of positive solutions found by Rabinowitz in dimension $n \geq 4$ on a star-shaped domain bends back to $\lambda = 0$ and becomes asymptotic to this axis.

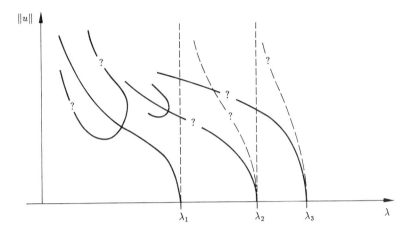

Fig. 2.1. Solution "branches" for (1.1), (1.3) depending on λ

However, equation (1.1) may have many positive solutions. In Section 3 we shall see that if Ω is an annulus, positive radial solutions exist for any value of $\lambda < \lambda_1$. But note that, by Theorem 1.2, for $\lambda = 0$ these cannot minimize S_λ. Hence we may have many different branches of (or secondary bifurcations of branches of) positive solution, in general.

Multiple Solutions

2.5 Bifurcation from higher eigenvalue. The local Palais-Smale condition Lemma 2.3 permits to obtain bifurcation of non-trivial solutions of (1.1), (1.3) from higher eigenvalues, as well. Recall that by a result of Böhme [1] and Marino [1] it is known that any eigenvalue of $-\Delta$ on $H_0^{1,2}(\Omega)$ is a point of bifurcation from the trivial solution of (1.1), (1.3). However, the variational method may give better estimates for the λ-interval of existence for such solutions.

The following result is due to Cerami-Fortunato-Struwe [1]:

2.6 Theorem. *Let Ω be a bounded domain in $\mathbb{R}^n$, $n \geq 3$, and let $0 < \lambda_1 < \lambda_2 \leq \ldots$ denote the eigenvalues of $-\Delta$ in $H_0^{1,2}(\Omega)$. Also let*

$$\nu = S\left(\mathcal{L}^n(\Omega)\right)^{-2/n} > 0 .$$

Then if

$$m = m(\lambda) = \sharp\{j \; ; \; \lambda < \lambda_j < \lambda + \nu\} ,$$

problem (1.1), (1.3) admits at least m distinct pairs of non-trivial solutions.

Remark. From the Weyl formula $\lambda_j \sim C(\Omega)j^{2/n}$ for the asymptotic behavior of the eigenvalues λ_j we conclude that $m(\lambda) \to \infty$ as $\lambda \to \infty$.

Proof of Theorem 2.6. Consider the functional

$$S_\lambda(u; \Omega) = \frac{\int_\Omega (|\nabla u|^2 - \lambda|u|^2) \, dx}{\left(\int_\Omega |u|^{2^*} \, dx\right)^{2/2^*}}$$

on the unit sphere

$$M = \{u \in H_0^{1,2}(\Omega) \; ; \; \|u\|_{L^{2^*}} = 1\} \; .$$

$M \subset H_0^{1,2}(\Omega)$ is a complete Hilbert manifold, invariant under the involution $u \to -u$. Recall that S_λ is differentiable on M and that if $u \in M$ is a critical point of S_λ with $S_\lambda(u) = \beta > 0$, then $\tilde{u} = (S_\lambda(u))^{\frac{1}{2^*-2}} u$ solves (1.1), (1.3) with

$$E_\lambda(\tilde{u}) = \frac{1}{n} S_\lambda(u)^{n/2} \; ;$$

see the proof of Theorem I.2.1.

Moreover, (u_m) is a (P.-S.)$_\beta$-sequence for S_λ if and only if $\left(\tilde{u}_m = \beta^{\frac{1}{2^*-2}} u\right)$ is a (P.-S.)$_{\tilde\beta}$ sequence for E_λ with $\tilde\beta = \frac{1}{n}\beta^{n/2}$. In particular, by Lemma 2.3 we have that S_λ satisfies (P.-S.)$_\beta$ for any $\beta \in]0, S[$.

Now let γ denote the Krasnoselskii genus, and for $j \in \mathbb{N}$ such that $\lambda_j \in]\lambda, \lambda + \nu[$ let

$$\beta_j = \inf_{\substack{A \subset M \\ \gamma(A) \geq j}} \sup_{u \in A} S_\lambda(u) \; .$$

Note that by Proposition II.5.3 for any $A \subset M$ such that $\gamma(A) \geq j$ there exist at least j mutually orthogonal vectors in A, whence $\beta_j > 0$ for all j as above. Moreover, if we denote by $\varphi_k \in H_0^{1,2}(\Omega)$ the k-th eigenfunction of $-\Delta$ and let $A_j = \text{span}\{\varphi_1, \ldots, \varphi_j\} \cap M$, we obtain that

$$\beta_j \leq (\lambda_j - \lambda) \int_\Omega |u|^2 \, dx \Big/ \left(\int_\Omega |u|^{2^*} \, dx\right)^{2/2^*}$$
$$< \nu \left(\mathcal{L}^n(\Omega)\right)^{\frac{2}{n}} = S \; .$$

Hence the theorem follows from Theorem II.4.2 and Lemma II.5.6.   □

We do not know how far the solution branches bifurcating off the trivial solution $u \equiv 0$ at $\lambda = \lambda_j$ extend. However, the following result has been obtained by Capozzi-Fortunato-Palmieri [1]:

2.7 Theorem. *Let Ω be a bounded domain in $\mathbb{R}^n$, $n \geq 4$. Then for any $\lambda > 0$ problem (1.1), (1.3) admits a non-trivial solution.*

The proof of Capozzi-Fortunato-Palmieri is based on a linking-argument and the local Palais-Smale condition Lemma 2.3. A simpler proof for λ not belonging to the spectrum of $-\Delta$, using the duality method, was worked out by Ambrosetti and Struwe [1].

Theorem 2.7 leaves open the question of multiplicity. For $n \geq 4$, $\lambda > 0$ Fortunato-Jannelli [1] present results in this direction using symmetries of the domain to restrict the space of admissible functions to certain symmetric subspaces, where the best Sobolev constant and hence the compactness threshold of Lemma 2.3 is increased. For instance, if $\Omega = B_1(0; \mathbb{R}^2) \times \Omega' \subset \mathbb{R}^n$, $n \geq 4$, $\lambda > 0$, essentially Fortunato-Jannelli solve (1.1), (1.3) by constructing a positive solution u to (1.1), (1.3) on a "slice"

$$\Omega_m = \{x = re^{i\varphi} \in B_1(0; \mathbb{R}^2) \; ; \; 0 \leq \varphi \leq \frac{\pi}{m}\} \times \Omega'$$

and reflecting u in the "vertical" edges of Ω_m a total of $2m$ times. Since the first eigenvalue $\lambda_1^{(m)}$ of $(-\Delta)$, acting on $H_0^{1,2}(\Omega_m)$, tends to ∞ as $m \to \infty$, we can achieve that $\lambda_1^{(m)} > \lambda$ for $m \geq m_0$.

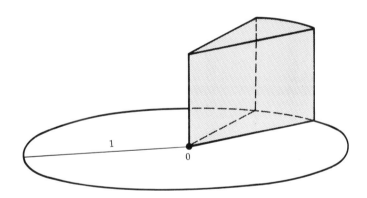

Fig. 2.2. A slice Ω_m of the pie Ω

Hence the existence of a positive solution u to (1.1), (1.3) on Ω_m is guaranteed by Theorem 2.1.

Somewhat surprisingly, radially symmetric solutions on balls $\Omega = B_R(0) \subset \mathbb{R}^n$ are harder to obtain. The reason is that the best Sobolev constant S is attained in the class of radially symmetric functions (on $\mathbb{R}^n$) and the construction of solutions must proceed at energies above the compactness threshold given by Lemma 2.3. The following result is due to Cerami-Solimini-Struwe [1] and Solimini [1] :

2.8 Theorem. *Suppose $\Omega = B_R(0)$ is a ball in $\mathbb{R}^n, n \geq 7$. Then for any $\lambda > 0$ problem (1.1), (1.3) admits infinitely many radially symmetric solutions.*

The proof of Theorem 2.8 uses a characterization of different solutions by the nodal properties they possess, as in the locally compact case; see Remark II.7.3.

2.9 Notes. (1°)By recent results of Atkinson-Brezis-Peletier [1], Castro-Kurepa [1] the restriction on the dimension n in Theorem 2.8 appears to be sharp.
(2°) Egnell [1], Guedda-Veron [1], and Lions-Pacella-Tricarico [1] have studied problems of the type (1.1), (1.3) involving the (degenerate) pseudo Laplace operator and partially free boundary conditions.
(3°) The linear term λu in equation (1.1) may be replaced by other compact perturbations; see Brezis-Nirenberg [2]. In this regard we also mention the results by Mancini-Musina [1] concerning obstacle problems of type (1.1), (1.3). Still a different kind of "perturbation" will be considered in the next section.

3. The Effect of Topology

Instead of a star-shaped domain, as in Theorem 1.3, consider an annulus

$$\Omega = \{x \in \mathbb{R}^n \; ; \; r_1 < |x| < r_2\} \, ,$$

and for $\lambda \in \mathbb{R}$ let E_λ be given by (1.7). Note that if we restrict our attention to radial functions

$$H^{1,2}_{0,rad}(\Omega) = \{u \in H^{1,2}_0(\Omega) \; ; \; u(x) = u(|x|)\} \, ,$$

by estimate (II.10.6) the embedding

$$H^{1,2}_{0,rad}(\Omega) \hookrightarrow L^p(\Omega)$$

is compact for any $p < \infty$. Hence $DE_\lambda \colon H^{1,2}_{0,rad}(\Omega) \to H^{-1}(\Omega)$ is of the form $id + compact$. Since any (P.-S.)-sequence for E_λ by the proof of Lemma 2.3 is bounded, from Proposition II.2.2 we thus infer that E_λ satisfies (P.-S.) globally on $H^{1,2}_{0,rad}(\Omega)$ and hence from Theorem II.5.7 or Theorem II.6.5 that problem (1.1), (1.3) possesses infinitely many radially symmetric solutions on Ω, for any $n \geq 2$, and any $\lambda \in \mathbb{R}$, in particular, for $\lambda = 0$. This result stands in striking contrast with Theorem 1.3 (or Theorem 2.8, as regards the restriction of the dimension).

In the following we shall investigate whether the solvability of (1.1), (1.3) on an annulus is a singular phenomenon, observable only in a highly symmetric case, or is stable and survives perturbations of the domain.

A Global Compactness Result

Remark that by Theorem 1.2, for $\lambda = 0$ no non-trivial solution $u \in H_0^{1,2}(\Omega)$ of (1.1) can satisfy $S_\lambda(u; \Omega) \leq S$. Hence the local compactness of Lemma 2.3 will not suffice to produce such solutions and we must study the compactness properties of E_λ, respectively S_λ, at higher energy levels as well. The next result can be viewed as an extension of P.-L. Lions' concentration-compactness method for minimization problems (see Section I.4) to problems of minimax type. The idea of analyzing the behavior of a (P.-S.)-sequence near points of concentration by "blowing up" the singularities seems to appear first in papers by Sacks and Uhlenbeck [1] and Wente [5] where variants of the local compactness condition Lemma 2.3 are obtained (see Sacks-Uhlenbeck [1; Lemma 4.2]). In the next result, due to Struwe [8], we systematically employ the blow-up technique to characterize *all* energy values β of a variational problem where (P.-S.)$_\beta$ may fail in terms of "critical points at infinity".

3.1 Theorem. *Suppose Ω is a bounded domain in $\mathbb{R}^n$, $n \geq 3$, and for $\lambda \in \mathbb{R}$ let (u_m) be a (P.-S.)-sequence for E_λ in $H_0^{1,2}(\Omega) \subset D^{1,2}(\mathbb{R}^n)$. Then there exist an index $k \in \mathbb{N}_0$, sequences (R_m^j), (x_m^j), $1 \leq j \leq k$, of radii $R_m^j \to \infty$ $(m \to \infty)$ and points $x_m^j \in \Omega$, a solution $u^0 \in H_0^{1,2}(\Omega) \subset D^{1,2}(\mathbb{R}^n)$ to (1.1), (1.3) and non-trivial solutions $u^j \in D^{1,2}(\mathbb{R}^n)$, $1 \leq j \leq k$, to the "limiting problem" associated with (1.1) and (1.3),*

$$(3.1) \qquad\qquad -\Delta u = u|u|^{2^*-2} \qquad in \ \mathbb{R}^n \ ,$$

such that a subsequence (u_m) satisfies

$$\left\| u_m - u^0 - \sum_{j=1}^{k} u_m^j \right\|_{D^{1,2}(\mathbb{R}^n)} \to 0 \ .$$

Here u_m^j denotes the rescaled function

$$u_m^j(x) = (R_m^j)^{\frac{n-2}{2}} u^j \left(R_m^j(x - x_m^j) \right), \ 1 \leq j \leq k, \ m \in \mathbb{N} \ .$$

Moreover,

$$E_\lambda(u_m) \to E_\lambda(u^0) + \sum_{j=1}^{k} E_0(u^j) \ .$$

3.2 Remark. In particular, if Ω is a ball $\Omega = B_R(0)$, $u_m \in H_{0,rad}^{1,2}(\Omega)$, from the uniqueness of the family $(u_\varepsilon^*)_{\varepsilon > 0}$ of radial solutions to (3.1) – see the proof of Theorem 2.1.(1°) – it follows that each u^j is of the form (2.5) with $E_0(u^j) = \frac{1}{n} S^{n/2} =: \beta^*$. Hence in this case (P.-S.)$_\beta$ holds for E_λ for all levels β which cannot be decomposed

$$\beta = \beta_0 + k\beta^* \ ,$$

where $k \geq 1$ and $\beta_0 = E_\lambda(u^0)$ is the energy of some radial solution of (1.1), (1.3). Similarly, if Ω is an arbitrary bounded domain and $u_m \geq 0$ for all m, then also $u^j \geq 0$ for all j, and by a result of Gidas-Ni-Nirenberg [1; p. 210 f.] and Obata [1] again each function u^j will be radially symmetric about some point x^j. Therefore also in this case each u^j is of the form $u^j = u_\varepsilon^*(\cdot - x^j)$ for some $\varepsilon > 0$, and (P.-S.)$_\beta$ holds for all β which are not of the form

$$\beta = \beta_0 + k\beta^* \ ,$$

where $k \geq 1$ and $\beta_0 = E_\lambda(u_0)$ is the energy of some non-negative solution u^0 of (1.1), (1.3).

For some time it was believed that the family (2.5) gives all non-trivial solutions of (3.1). Surprisingly, Ding [1] was able to establish that (3.1) also admits infinitely many solutions of changing sign which are distinct modulo scaling.

In general, decomposing a solution v of (3.1) into positive and negative parts $v = v_+ + v_-$, where $v_\pm = \pm \max\{\pm v, 0\}$, upon testing (3.1) with $v_\pm$ from Sobolev's inequality we infer that

$$0 = \int_{\mathbb{R}^n} \left(-\Delta v - v|v|^{2^*-2}\right) v_\pm \ dx$$

$$= \int_{\mathbb{R}^n} \left(|\nabla v_\pm|^2 - |v_\pm|^{2^*}\right) dx \geq \left(1 - S^{-2^*/2} \|v_\pm\|_{D^{1,2}}^{2^*-2}\right) \|v_\pm\|_{D^{1,2}}^2 \ .$$

Hence $v_\pm \equiv 0$ or

$$E_0(v_\pm) = \frac{1}{n} \|v_\pm\|_{D^{1,2}}^2 \geq \frac{1}{n} S^{n/2} = \beta^* \ ,$$

and therefore any solution v of (3.1) that changes sign satisfies

$$E_0(v) = E_0(v_+) + E_0(v_-) \geq 2\beta^* \ .$$

In fact, $E_0(v) > 2\beta^*$; otherwise S would be achieved at v_+ and v_-, which would contradict Theorem 1.2. Thus, in Theorem 3.1 we can assert that $E_0(u^j) \in \{\beta^*\} \cup \ ]2\beta^*, \infty[$.

In particular, if (1.1), (1.3) does not admit any solution but the trivial solution $u \equiv 0$, the local Palais-Smale condition (P.-S.)$_\beta$ will hold for all $\beta < 2\beta^*$, except for $\beta = \beta^*$.

Proof of Theorem 3.1. First recall that as in the proof of Lemma 2.3 any (P.-S.)-sequence for E_λ is bounded. Hence we may assume that $u_m \rightharpoonup u^0$ weakly in $H_0^{1,2}(\Omega)$, and u^0 solves (1.1), (1.3). Moreover, if we let $v_m = u_m - u^0$ we have $v_m \to 0$ strongly in $L^2(\Omega)$, and by (2.1), (2.2) also that

$$\int_\Omega |v_m|^{2^*} \ dx = \int_\Omega |u_m|^{2^*} \ dx - \int_\Omega |u^0|^{2^*} \ dx + o(1) \ ,$$

$$\int_\Omega |\nabla v_m|^2 \ dx = \int_\Omega |\nabla u_m|^2 \ dx - \int_\Omega |\nabla u^0|^2 \ dx + o(1) \ ,$$

where $o(1) \to 0$ $(m \to \infty)$. Hence, in particular, we obtain that

$$E_\lambda(u_m) = E_\lambda(u^0) + E_0(v_m) + o(1) .$$

Also note that

$$DE_\lambda(u_m) = DE_\lambda(u^0) + DE_0(v_m) + o(1) = DE_0(v_m) + o(1) ,$$

where $o(1) \to 0$ in $H^{-1}(\Omega)$ $(m \to \infty)$. Using the following lemma, we can now proceed by induction:

3.3 Lemma. *Suppose (v_m) is a (P.-S.)-sequence for $E = E_0$ in $H_0^{1,2}(\Omega)$ such that $v_m \to 0$ weakly. Then there exists a sequence (x_m) of points $x_m \in \Omega$, a sequence (R_m) of radii $R_m \to \infty$ $(m \to \infty)$, a non-trivial solution v^0 to the limiting problem (3.1) and a (P.-S.)-sequence (w_m) for E_0 in $H_0^{1,2}(\Omega)$ such that for a subsequence (v_m) there holds*

$$w_m = v_m - R_m^{\frac{n-2}{2}} v^0 \left(R_m (\cdot - x_m) \right) + o(1) ,$$

where $o(1) \to 0$ in $D^{1,2}(\mathbb{R}^n)$ as $m \to \infty$. In particular, $w_m \to 0$ weakly. Furthermore,

$$E_0(w_m) = E_0(v_m) - E_0(v^0) + o(1) .$$

Moreover,

$$R_m \, dist(x_m, \partial\Omega) \to \infty .$$

Finally, if $E_0(v_m) \to \beta < \beta^*$, the sequence (v_m) is relatively compact and hence $v_m \to 0$, $E_0(v_m) \to \beta = 0$.

Proof of Theorem 3.1 (completed). Apply Lemma 3.3 to the sequences $v_m^1 = u_m - u^0$, $v_m^j = u_m - u^0 - \sum_{i=1}^{j-1} u_m^i = v_m^{j-1} - u_m^{j-1}$, $j > 1$, where

$$u_m^i(x) = (R_m^i)^{\frac{n-2}{2}} u^i \left(R_m^i(x - x_m^i) \right) .$$

By induction

$$E_0(v_m^j) = E_\lambda(u_m) - E_\lambda(u^0) - \sum_{i=1}^{j-1} E_0(u^i)$$

$$\leq E_\lambda(u_m) - (j-1)\beta^* .$$

Since the latter will be negative for large j, by Lemma 3.3 the induction will terminate after some index $k \geq 0$. Moreover, for this index we have

$$v_m^{k+1} = u_m - u^0 - \sum_{j=1}^{k} u_m^i \to 0$$

strongly in $D^{1,2}(\mathbb{R}^n)$, and

$$E_\lambda(u_m) - E_\lambda(u^0) - \sum_{j=1}^{k} E_0(u^j) \to 0 ,$$

as desired.   □

Proof of Lemma 3.3. If $E_0(v_m) \to \beta < \beta^*$, by Lemma 2.3 the sequence (v_m) is strongly relatively compact and hence $v_m \to 0$, $\beta = 0$. Therefore, we may assume that $E_0(v_m) \to \beta \geq \beta^* = \frac{1}{n} S^{n/2}$. Moreover, since $DE_0(v_m) \to 0$ we also have

$$\frac{1}{n} \int_\Omega |\nabla v_m|^2 \, dx = E_0(v_m) - \frac{1}{2^*} \langle v_m, DE_0(v_m) \rangle \to \beta \geq \frac{1}{n} S^{n/2}$$

and hence that

$$(3.2) \qquad \liminf_{m \to \infty} \int_\Omega |\nabla v_m|^2 \, dx = n\beta \geq S^{n/2} .$$

Denote

$$Q_m(r) = \sup_{x \in \Omega} \int_{B_r(x)} |\nabla v_m|^2 \, dx$$

the concentration function of v_m, introduced in Section I.4.3. Choose $x_m \in \overline{\Omega}$ and scale

$$v_m \mapsto \tilde{v}_m(x) = R_m^{\frac{2-n}{2}} v_m\big(x/R_m + x_m\big)$$

such that

$$\tilde{Q}_m(1) = \sup_{\substack{x \in \mathbb{R}^n \\ x/R_m + x_m \in \Omega}} \int_{B_1(x)} |\nabla \tilde{v}_m|^2 \, dx = \int_{B_1(0)} |\nabla \tilde{v}_m|^2 \, dx = \frac{1}{2L} S^{n/2} ,$$

where L is a number such that $B_2(0)$ is covered by L balls of radius 1. Clearly, by (3.2) we have $R_m \geq R_0 > 0$, uniformly in m.

Considering $\tilde{\Omega}_m = \{x \in \mathbb{R}^n \; ; \; x/R_m + x_m \in \Omega\}$, we may regard $\tilde{v}_m \in H_0^{1,2}(\tilde{\Omega}_m) \subset D^{1,2}(\mathbb{R}^n)$.

Moreover,

$$\|\tilde{v}_m\|_{D^{1,2}}^2 = \|v_m\|_{D^{1,2}}^2 \to n\beta < \infty$$

and we may assume that $\tilde{v}_m \to v^0$ weakly in $D^{1,2}(\mathbb{R}^n)$. We claim that $\tilde{v}_m \to v^0$ strongly in $H^{1,2}(\Omega')$, for any $\Omega' \subset\subset \mathbb{R}^n$. It suffices to consider $\Omega' = B_1(x_0)$ for any $x_0 \in \mathbb{R}^n$. (For brevity $B_r(x_0) =: B_r$.) Indeed, by Fubini's theorem and since

$$\int_1^2 \left(\int_{\partial B_r} |\nabla \tilde{v}_m|^2 \, do \right) dr \leq \int_{B_2} |\nabla \tilde{v}_m|^2 \, dx \leq n\beta + o(1) ,$$

where $o(1) \to 0$ $(m \to \infty)$, there is a radius $\rho \in [1, 2]$ such that

$$\int_{\partial B_\rho} |\nabla \tilde{v}_m|^2 \, do \leq 2n\beta$$

for infinitely many $m \in \mathbb{N}$. (Relabelling, we may assume that this estimate holds for all $m \in \mathbb{N}$.) By compactness of the embedding $H^{1,2}(\partial B_\rho) \hookrightarrow H^{1/2,2}(\partial B_\rho)$, we deduce that a subsequence $\tilde{v}_m \to \tilde{v}^0$ strongly in $H^{1/2,2}(\partial B_\rho)$; see Theorem A.8 of the appendix. Moreover, since also the trace operator $H^{1,2}(B_2) \to L^2(\partial B_\rho)$ is compact, we conclude that $\tilde{v}^0 = v^0$. Now let

$$\varphi_m = \begin{cases} \tilde{v}_m - v^0 & \text{in } B_\rho \\ \tilde{w}_m & \text{in } B_3 \setminus B_\rho , \end{cases}$$

where $\tilde{w}_m$ denotes the solution to the Dirichlet problem $\Delta \tilde{w}_m = 0$ in $B_3 \setminus B_\rho$, $\tilde{w}_m = \tilde{v}_m - v^0$ on ∂B_ρ, $\tilde{w}_m = 0$ on ∂B_3. By continuity of the solution operator to the Dirichlet problem on the annulus $B_3 \setminus B_\rho$ in the $H^{1/2,2}$-norm (see for instance Lions-Magenes [1; Theorem 8.2]), we have

$$\|\tilde{w}_m\|_{H^{1,2}(B_3 \setminus B_\rho)} \leq c\|\tilde{v}_m - v^0\|_{H^{1/2,2}(\partial B_\rho)} \to 0 .$$

Hence $\varphi_m = \tilde{\varphi}_m + o(1) \in H_0^{1,2}(\tilde{\Omega}_m) + D^{1,2}(\mathbb{R}^n)$, where $\tilde{\varphi}_m \in H_0^{1,2}(\tilde{\Omega}_m)$ and $o(1) \to 0$ in $D^{1,2}(\mathbb{R}^n)$ as $m \to \infty$. Thus

$$\langle \varphi_m, DE_0(\tilde{v}_m; \mathbb{R}^n) \rangle = \langle \tilde{\varphi}_m, DE_0(\tilde{v}_m; \tilde{\Omega}_m) \rangle + o(1) \to 0 .$$

On the other hand, using convergence arguments familiar by now and Sobolev's inequality, we obtain

$$o(1) = \langle \varphi_m, DE_0(\tilde{v}_m; \mathbb{R}^n) \rangle =$$

$$= \int_{\mathbb{R}^n} \left(\nabla \tilde{v}_m \nabla \varphi_m - \tilde{v}_m |\tilde{v}_m|^{2^*-2} \varphi_m \right) dx$$

(3.3)
$$= \int_{B_\rho} \left(|\nabla(\tilde{v}_m - v^0)|^2 - |\tilde{v}_m - v^0|^{2^*} \right) dx + o(1)$$

$$= \int_{\mathbb{R}^n} \left(|\nabla \varphi_m|^2 - |\varphi_m|^{2^*} \right) dx + o(1)$$

$$\geq \|\varphi_m\|_{D^{1,2}(\mathbb{R}^n)}^2 \left(1 - S^{-2^*/2} \|\varphi_m\|_{D^{1,2}(\mathbb{R}^n)}^{2^*-2} \right) ,$$

where $o(1) \to 0$ as $m \to \infty$. But now we note that

$$\int_{\mathbb{R}^n} |\nabla \varphi_m|^2 \, dx = \int_{B_\rho} |\nabla(\tilde{v}_m - v^0)|^2 \, dx + o(1) \leq \int_{B_2} |\nabla \tilde{v}_m|^2 \, dx + o(1)$$

$$\leq L \, \tilde{Q}_m(1) = \frac{1}{2} S^{n/2} ,$$

from which via (3.3) we deduce that $\varphi_m \to 0$ in $D^{1,2}(\mathbb{R}^n)$; that is, $\tilde{v}_m \to v^0$ locally in $H^{1,2}$, as desired.

In particular,

$$\int_{B_1(0)} |\nabla v^0|^2 \, dx = \frac{1}{2L} S^{n/2} > 0 ,$$

and $v^0 \not\equiv 0$. Since the original sequence $v_m \rightharpoonup 0$ weakly, thus it also follows that $R_m \to \infty$ as $m \to \infty$. Now we distinguish two cases:

(1°) $R_m \, dist(x_m, \partial\Omega) \le c < \infty$, uniformly, in which case (after rotation of coordinates) we may assume that the sequence $\tilde{\Omega}_m$ exhausts the half-space

$$\tilde{\Omega}_\infty = \mathbb{R}^n_+ = \{x = (x_1, \ldots, x_n) \; ; \; x_1 > 0\} \, ,$$

or

(2°) $R_m \, dist(x_m, \partial\Omega) \to \infty$, in which case $\tilde{\Omega}_m \to \tilde{\Omega}_\infty = \mathbb{R}^n$.

Since in each case for any $\varphi \in C_0^\infty(\tilde{\Omega}_\infty)$ we have that $\varphi \in C_0^\infty(\tilde{\Omega}_m)$ for large m, there holds

$$\langle \varphi, DE_0(v^0; \tilde{\Omega}_\infty) \rangle = \lim_{m \to \infty} \langle \varphi, DE_0(\tilde{v}_m; \tilde{\Omega}_m) \rangle = 0 \, ,$$

for all such φ, and $v^0 \in H_0^{1,2}(\tilde{\Omega}_\infty)$ is a weak solution of (3.1) on $\tilde{\Omega}_\infty$. But if $\tilde{\Omega}_\infty = \mathbb{R}^n_+$, by Theorem 1.3 then v^0 must vanish identically. Thus (1°) is impossible, and we are left with (2°).

To conclude the proof, let $\varphi \in C_0^\infty(\mathbb{R}^n)$ satisfy $0 \le \varphi \le 1$, $\varphi \equiv 1$ in $B_1(0)$, $\varphi \equiv 0$ outside $B_2(0)$, and let

$$w_m(x) = v_m(x) - R_m^{\frac{n-2}{2}} v^0 \big(R_m(x - x_m)\big) \cdot \varphi\big(\overline{R}_m(x - x_m)\big) \in H_0^{1,2}(\Omega) \, ,$$

where the sequence $(\overline{R}_m)$ is chosen such that $\tilde{R}_m := R_m(\overline{R}_m)^{-1} \to \infty$ while $\overline{R}_m \, dist(x_m, \partial\Omega) \to \infty$ as $m \to \infty$; that is,

$$\tilde{w}_m(x) = R_m^{\frac{2-n}{2}} w_m(x/R_m + x_m) = \tilde{v}_m(x) - v^0(x)\varphi(x/\tilde{R}_m) \, .$$

Set $\varphi_m(x) = \varphi\big(x/\tilde{R}_m\big)$. Note that

$$\int_{\mathbb{R}^n} \big|\nabla\big(v^0(\varphi_m - 1)\big)\big|^2 \, dx \le$$

$$\le C \int_{\mathbb{R}^n} |\nabla v^0|^2 (\varphi_m - 1)^2 \, dx + C \int_{\mathbb{R}^n} |v^0|^2 \big|\nabla(\varphi_m - 1)\big|^2 \, dx$$

$$\le C \int_{\mathbb{R}^n \setminus B_{\tilde{R}_m}(0)} |\nabla v^0|^2 \, dx + C\tilde{R}_m^{-2} \int_{B_{2\tilde{R}_m}(0) \setminus B_{\tilde{R}_m}(0)} |v^0|^2 \, dx \, .$$

But $\nabla v^0 \in L^2(\mathbb{R}^n)$. Therefore the first term tends to 0 as $m \to \infty$, while by Hölder's inequality also the second term

$$\tilde{R}_m^{-2} \int_{B_{2\tilde{R}_m}(0) \setminus B_{\tilde{R}_m}(0)} |v^0|^2 \, dx \le C \left(\int_{B_{2\tilde{R}_m}(0) \setminus B_{\tilde{R}_m}(0)} |v^0|^{2^*} \, dx \right)^{2/2^*} \to 0$$

as $m \to \infty$. Thus we have $\tilde{w}_m = \tilde{v}_m - v^0 + o(1)$, where $o(1) \to 0$ in $D^{1,2}(\mathbb{R}^n)$. Hence, as in the proof of Lemma 2.3, also

$$E_0(w_m) = E_0(\tilde{w}_m) = E_0(\tilde{v}_m) - E_0(v^0) + o(1) ,$$
$$\|DE_0(w_m; \Omega)\| = \|DE_0(\tilde{w}_m; \tilde{\Omega}_m)\|$$
$$\leq \|DE_0(\tilde{v}_m; \tilde{\Omega}_m)\| + \|DE_0(v^0; \mathbb{R}^n)\| + o(1)$$
$$= \|DE_0(v_m; \Omega)\| + o(1) \to 0 \qquad (m \to \infty) .$$

This concludes the proof. □

Positive Solutions on Annular-Shaped Regions

With the aid of Theorem 3.2 we can now show the existence of solutions to (1.1),(1.3) on perturbed annular domains for $\lambda = 0$.

The following result is due to Coron [2]:

3.4 Theorem. *Suppose Ω is a bounded domain in $\mathbb{R}^n$ satisfying the following condition: There exist constants $0 < R_1 < R_2 < \infty$ such that*

(1°) $\qquad\qquad \Omega \supset \{x \in \mathbb{R}^n \; ; \; R_1 < |x| < R_2\} ,$
(2°) $\qquad\qquad \overline{\Omega} \not\supset \{x \in \mathbb{R}^n \; ; \; |x| < R_1\} .$

Then, if R_2/R_1 is sufficiently large, problem (1.1), (1.3) for $\lambda = 0$ admits a positive solution to $u \in H_0^{1,2}(\Omega)$.

Again remark that the solution u must have an energy above the compactness threshold given by Lemma 2.3.

The idea of the proof is to argue by contradiction and to use a minimax method for $S = S_0(\,\cdot\,; \Omega)$ based on a set A of non-negative functions which is homeomorphic to a sphere Σ around 0 in Ω. Note that A is contractible in the positive cone in $H_0^{1,2}(\Omega)$. Moreover, if (1.1), (1.3) does not admit a positive solution, then under certain conditions such a contraction of A in $H_0^{1,2}(\Omega)$ will induce a contraction of Σ in Ω, and the desired contradiction will result.

Proof. We may assume $R_1 = (4R)^{-1} < 1 < 4R = R_2$. Consider the unit sphere

$$\Sigma = \{x \in \mathbb{R}^n \; ; \; |x| = 1\} .$$

For $\sigma \in \Sigma$, $x \in \mathbb{R}^n$, $0 \leq t < 1$ let

$$u_t^\sigma(x) = \left[\frac{1-t}{(1-t)^2 + |x - t\sigma|^2} \right]^{\frac{n-2}{2}} \in D^{1,2}(\mathbb{R}^n) .$$

Note that S is attained on any such function u_t^σ, and u_t^σ "concentrates" at σ as $t \to 1$. Moreover, letting $t \to 0$ we have

$$u_t^\sigma \to u_0 = \left[\frac{1}{1 + |x|^2} \right]^{\frac{n-2}{2}} ,$$

for any $\sigma \in \Sigma$. Choose a radially symmetric function $\varphi \in C_0^\infty(\Omega)$ such that $0 \leq \varphi \leq 1$ on Ω, $\varphi \equiv 1$ on the annulus $\{x \; ; \; \frac{1}{2} < |x| < 2\}$ and $\varphi \equiv 0$ outside the annulus $\{x \; ; \; \frac{1}{4} < |x| < 4\}$. For $R \geq 1$ scale

$$\varphi_R(x) = \begin{cases} \varphi(Rx), & 0 \leq |x| < R^{-1} \\ 1, & R^{-1} \leq |x| < R \\ \varphi(x/R), & R \leq |x| \end{cases}$$

and let

$$w_t^\sigma = u_t^\sigma \cdot \varphi_R, \ w_0 = u_0 \cdot \varphi_R \in H_0^{1,2}(\Omega) \ .$$

Remark that

$$\int_{\mathbb{R}^n} |\nabla(w_t^\sigma - u_t^\sigma)|^2 \, dx \leq$$

$$\leq c \int_{(\mathbb{R}^n \setminus B_{2R}) \cup B_{(2R)^{-1}}} |\nabla u_t^\sigma|^2 \, dx + c \cdot R^{-2} \int_{B_{4R} \setminus B_{2R}} |u_t^\sigma|^2 \, dx$$

$$+ cR^2 \int_{B_{(2R)^{-1}}} |u_t^\sigma|^2 \, dx$$

$$\to 0 \qquad (R \to \infty) \ ,$$

uniformly in $\sigma \in \Sigma$, $t \in [0,1[$, and the same holds for the normalized functions

$$v_t^\sigma = w_t^\sigma / \|w_t^\sigma\|_{H_0^{1,2}}, \ v_0 = w_0 / \|w_0\|_{H_0^{1,2}} \ ;$$

that is, $\|v_t^\sigma - u_t^\sigma\| \to 0$ as $R \to \infty$. Hence

$$S(v_t^\sigma; \Omega) \to S(u_t^\sigma; \mathbb{R}^n) = S$$

as $R \to \infty$, uniformly in $\sigma \in \Sigma$, $0 \leq t < 1$. In particular, if $R \geq 1$ is sufficiently large, we can achieve that

$$\sup_{\sigma, t} S(v_t^\sigma; \Omega) < S_1 < 2^{2/n} S$$

for some constant $S_1 \in \mathbb{R}$.

Fix such an R and suppose that Ω satisfies $(1^\circ), (2^\circ)$ of the Theorem. Let M be the set

$$M = \{u \in H_0^{1,2}(\Omega) \; ; \; \int_\Omega |u|^{2^*} \, dx = 1\} \ ,$$

and for $u \in M$ let

$$F(u) = \int_\Omega x |\nabla u|^2 \, dx$$

denote its center of mass. Suppose (1.1), (1.3) does not admit a positive solution. This is equivalent to the fact that

$$E(u) = \frac{1}{2} \int_\Omega |\nabla u|^2 \, dx - \frac{1}{2^*} \int_\Omega |u|^{2^*} \, dx$$

does not admit a critical point $u > 0$. By Remark 3.2, therefore, E satisfies (P.-S.)$_\beta$ on $H_0^{1,2}(\Omega)$ for $\frac{1}{n} S^{n/2} < \beta < \frac{2}{n} S^{n/2}$. Equivalently, $S_0(\,\cdot\,;\Omega)$ satisfies (P.-S.)$_\beta$ on M for $S < \beta < 2^{2/n} S$. Moreover, $S_0(\,\cdot\,;\Omega)$ does not admit a critical value in this range.

By the deformation lemma Theorem II.3.11 and Remark II.3.12, therefore, for any β in this range there exists $\varepsilon > 0$ and a flow $\Phi \colon M \times [0,1] \to M$ such that

$$\Phi(M_{\beta+\varepsilon}, 1) \subset M_{\beta-\varepsilon} \ ,$$

where

$$M_\beta = \{u \in M \ ; \ S(u;\Omega) < \beta\} \ .$$

Given $\delta > 0$, we may cover the interval $[S + \delta, S_1]$ by finitely many such ε-intervals and compose the corresponding deformations to obtain a flow $\Phi \colon M \times [0,1] \to M$ such that

$$\Phi(M_{S_1}, 1) \subset M_{S+\delta} \ .$$

Moreover, we may assume that $\Phi(u,t) = u$ for all u with $S(u;\Omega) \le S + \delta/2$.

On the other hand, it easily follows either from Theorem I.4.7 or Theorem 3.1 that, given any neighborhood U of $\overline{\Omega}$, there exists $\delta > 0$ such that $F(M_{S+\delta}) \subset U$. Since Ω is smooth we may choose a neighborhood U of $\overline{\Omega}$ such that any point $p \in U$ has a unique nearest neighbor $q = \pi(p) \in \Omega$ and such that the projection π is continuous. Let $\delta > 0$ be determined for such a neighborhood U, and let $\Phi \colon M \times [0,1] \to M$ be the corresponding flow constructed above. The map $h \colon \Sigma \times [0,1] \to \Omega$, given by

$$h(\sigma, t) = \pi \left(F\big(\Phi(v_t^\sigma, 1)\big) \right) ,$$

then is well-defined, continuous, and satisfies

$$h(\sigma, 0) = \pi \left(F\big(\Phi(v_0, 1)\big) \right) =: x_0 \in \Omega \ , \qquad \text{for all } \sigma \in \Sigma$$

$$h(\sigma, 1) = \sigma, \qquad \text{for all } \sigma \in \Sigma \ .$$

Hence h is a contraction of Σ in Ω, contradicting (2°). $\qquad\square$

Actually, the effect of topology is much stronger than indicated by Theorem 3.4. In a penetrating analysis, Bahri and Coron [1] have obtained the following result; see also Bahri [2]:

3.5 Theorem. *Suppose Ω is a domain in $\mathbb{R}^n$ such that*

$$H_d(\Omega, \mathbb{Z}_2) \ne 0$$

for some $d > 0$. Then (1.1), (1.3) admits a positive solution for $\lambda = 0$.

Remark that if $\Omega \subset \mathbb{R}^3$ is non-contractible then either $H_1(\Omega, \mathbb{Z}_2)$ of $H_2(\Omega, \mathbb{Z}_2)$ $\ne 0$ and the conclusion of Theorem 3.5 holds. It is conjectured that a similar result will also hold for $n \ge 4$.

Remarks on the Yamabe Problem

Equation (1.1) arises in a geometric context in the problem whether a given metric g on a manifold $\mathcal{M}$ with scalar curvature k can be conformally deformed to a metric g_0 of constant scalar curvature. If we let

$$g_0 = u^{\frac{4}{n-2}} g \, ,$$

where $u > 0$ gives the conformal factor, the requirement of g_0 to have constant scalar curvature k_0 is equivalent to the equation

(3.4)
$$-\frac{4(n-1)}{n-2} \Delta_{\mathcal{M}} u + ku = k_0 u |u|^{2^*-2},$$

where $\Delta_{\mathcal{M}}$ is the Laplace-Beltrami operator on the manifold $\mathcal{M}$ with respect to the original metric g; see Yamabe [1], T. Aubin [2], [3]. Observe that equation (3.4) is conformally invariant; that is, if u solves (3.4) on $(\mathcal{M}, g)$ and if

$$g = v^{\frac{4}{n-2}} \tilde{g}, \qquad v > 0$$

is conformal to a metric $\tilde{g}$ on $\mathcal{M}$, then $\tilde{u} = uv$ satisfies (3.4) on $(\mathcal{M}, \tilde{g})$ with k replaced by the scalar curvature $\tilde{k}$ of the metric $\tilde{g}$; see for instance Aubin [3; Proposition, p. 126].

   Of particular interest is the case of the sphere $\mathcal{M} = S^n$ with the standard metric induced by the embedding $S^n \hookrightarrow \mathbb{R}^{n+1}$. We center S^n at $(0,1) \in \mathbb{R}^n \times \mathbb{R}$. By conformal projection

$$\mathbb{R}^n \ni x \mapsto \frac{2}{1+|x|^2}(x,1) \in S^n \subset \mathbb{R}^{n+1} \, ,$$

S^n induces a metric $g_0 = \frac{4}{[1+|x|^2]^2} g$ of constant scalar curvature $k_0 = n(n-1)$ on $\mathbb{R}^n$, conformal to the Euclidean metric g on $\mathbb{R}^n$. Therefore the classical solution

$$u(x) = \frac{[n(n-2)]^{\frac{n-2}{4}}}{[1+|x|^2]^{\frac{n-2}{2}}}$$

of (3.1) on $\mathbb{R}^n$ reappears in the conformal factor that changes the flat Euclidean metric to a metric of constant scalar curvature k_0, and the invariance of (3.4) under scaling

$$u \mapsto u_R(x) = R^{\frac{n-2}{2}} u\big(R(x - x_0)\big)$$

reflects the group of conformal diffeomorphism of S^n which act as translations and dilatations in the chart $\mathbb{R}^n$.

   We can give a variational formulation of (3.4) analogous to the one given for (1.1)–(1.3). Let

$$H^{1,2}(\mathcal{M}; g)$$

be the set of Sobolev functions $u \colon \mathcal{M} \to \mathbb{R}$ such that (in local coordinates on $\mathcal{M}$, with $g^{ij} = (g_{ij})^{-1}$ denoting the coefficients of the inverse of the metric tensor, and with $g = \det(g_{ij})$) the norm

$$\|u\|^2 = \int_{\mathcal{M}} \big(g^{ij}(x)\partial_i u \partial_j u + |u|^2\big) \sqrt{g}\, dx$$

is finite. Also let

$$S(u;\mathcal{M}) = \int_{\mathcal{M}} \Big(\frac{4(n-1)}{n-2} g^{ij}(x)\partial_i u \partial_j u + k|u|^2\Big) \sqrt{g}\, dx\ .$$

Then $S \in C^1\big(H^{1,2}(\mathcal{M},g)\big)$ and critical points of S on the unit sphere in $L^{2^*}(\mathcal{M},g)$ correspond to solutions of (3.4) for some Lagrange multiplier k_0.

In particular, if the best Sobolev constant on $\mathcal{M}$,

$$S(\mathcal{M}) = \inf \left\{ S(u;\mathcal{M})\ ;\ u \in H^{1,2}(\mathcal{M};g),\ \int_{\mathcal{M}} |u|^{2^*} \sqrt{g}\, dx = 1 \right\},$$

is attained at a function u, (3.4) will admit a scalar multiple of u as a positive solution. Depending on the sign of $S(\mathcal{M})$, the latter question can also be approached as a minimax problem for the free functional

$$E(u;\mathcal{M}) = \frac{1}{2} \int_{\mathcal{M}} \Big(\frac{4(n-1)}{n-2} g^{ij}\partial_i u \partial_j u + k|u|^2\Big) \sqrt{g}\, dx \pm$$

$$\pm \frac{1}{2^*} \int_{\mathcal{M}} |u|^{2^*} \sqrt{g}\, dx\ ,$$

where the "+"-sign ("−"-sign) is valid if $S(\mathcal{M})$ is negative (positive). If $S(\mathcal{M}) = 0$, $k_0 = 0$ and (3.4) reduces to a linear equation.

If $S(\mathcal{M}) < 0$, the functional E is coercive and weakly lower semi-continuous on $H^{1,2}(\mathcal{M};g)$, as may be easily verified by the reader. Hence in this case the existence of a non-negative critical point follows from the direct methods; see Chapter I. In particular, we recall the approach via sub- and supersolutions to (3.4) used by Kazdan-Warner [1]; see also Sections I.2.3–I.2.6.

The difficult case is $S(\mathcal{M}) > 0$. Years after Yamabe's [1] first – unsuccessful – attempt to solve (3.4) for general manifolds $\mathcal{M}$, Trudinger [1; Theorem 2, p. 269] obtained a rigorous existence result for small positive $S(\mathcal{M})$. His approach then was refined by Aubin [2]. By using optimal Sobolev estimates, Aubin was able to show the following result on which our Lemma 2.2 above was modelled.

3.6 Lemma. *If $S(\mathcal{M}) < S$, then $S(\mathcal{M})$ is attained at a positive solution to (3.3).*

More generally, in our terms we can say that $E(\,\cdot\,;\mathcal{M})$ satisfies (P.-S.)$_\beta$ on $\mathcal{M}$ for $\beta < \frac{1}{n} S^{n/2}$ (compare Lemma 2.3). Lemma 3.6 now almost led to a complete existence proof for (3.4), as Aubin was able to show that the assumption $S(\mathcal{M}) < S$ is always satisfied if $\mathcal{M}$ is not locally conformally flat or has dimension ≥ 6. Finally, Schoen [1] recently was able to show that $S(\mathcal{M}) < S$ also in the remaining cases. His proof uses the "positive mass theorem", another deep result in differential topology.

For further material and references on the Yamabe problem we refer the reader to the survey by Lee-Parker [1].

A related problem is the Kazdan-Warner problem; see Kazdan-Warner [1], Aubin [3], Bahri-Coron [1], Chang-Yang [1], Schoen [2].

4. The Dirichlet Problem for the Equation of Constant Mean Curvature

Another border-line case of a variational problem is the following: Let Ω be a bounded domain in $\mathbb{R}^2$ with generic point $z = (x, y)$ and let $u_0 \in C^0(\overline{\Omega}; \mathbb{R}^3)$, $H \in \mathbb{R}$ be given. Find a solution $u \in C^2(\Omega; \mathbb{R}^3) \cap C^0(\overline{\Omega}; \mathbb{R}^3)$ to the problem

$$(4.1) \qquad\qquad \Delta u = 2 H u_x \wedge u_y \qquad \text{in } \Omega \ ,$$

$$(4.2) \qquad\qquad u = u_0 \qquad \text{on } \partial\Omega \ .$$

Here, for $a = (a_1, a_2, a_3)$, $b = (b_1, b_2, b_3) \in \mathbb{R}^3$, $a \wedge b$ denotes the wedge product $a \wedge b = (a_2 b_3 - b_2 a_3, a_3 b_1 - b_3 a_1, a_1 b_2 - b_1 a_2)$ and, for instance, $u_x = \frac{\partial}{\partial x} u$. (4.1) is the equation satisfied by surfaces of mean curvature H in conformal representation.

Surprisingly, (4.1) is of variational type. In fact, solutions of (4.1) may arise as "soap bubbles", that is, surfaces of least area enclosing a given volume. Also for prescribed Dirichlet data, where a geometric interpretation of (4.1) is impossible, we may recognize (4.1) as the Euler-Lagrange equations associated with the variational integral

$$E_H(u) = \frac{1}{2} \int_\Omega |\nabla u|^2 \, dz + \frac{2H}{3} \int_\Omega u \cdot u_x \wedge u_y \, dz \ .$$

For smooth "surfaces" u, the term

$$V(u) := \frac{1}{3} \int_\Omega u \cdot u_x \wedge u_y \, dz$$

may be interpreted as the algebraic volume enclosed between the "surface" parametrized by u and a fixed reference surface spanning the "curve" defined by the Dirichlet data u_0; see Figure 4.1.

Indeed, computing the variation of the volume V at a point $u \in C^2(\Omega; \mathbb{R}^3)$ in direction of a vector $\varphi \in C_0^\infty(\Omega; \mathbb{R}^3)$, we obtain

$$3 \frac{d}{d\varepsilon} V(u + \varepsilon\varphi)|_{\varepsilon=0} =$$

$$= \int_\Omega \left(\varphi \cdot u_x \wedge u_y + u \cdot \varphi_x \wedge u_y + u \cdot u_x \wedge \varphi_y \right) dz$$

$$= 3 \int_\Omega \varphi \cdot u_x \wedge u_y \, dz + \int_\Omega \varphi \cdot (u \wedge u_{yx} + u_{xy} \wedge u) \, dz \ ,$$

and the second integral vanishes by anti-symmetry of the wedge product. Hence critical points $u \in C^2(\Omega; \mathbb{R}^3)$ of E solve (4.1).

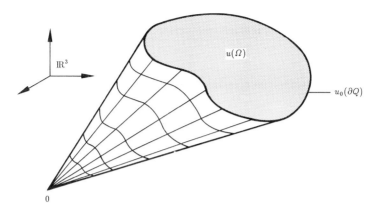

$u(\Omega)$

$\mathbb{R}^3$

$u_0(\partial Q)$

0

Fig. 4.1. On the volume functional

Small Solutions

Since V is cubic, the Dirichlet integral dominates if u is "small" and we can expect that for "small data" and "small" H a solution of (4.1–2) can be obtained by minimizing E_H in a suitable convex set. Generalizing earlier results by Heinz [1] and Werner [1], Hildebrandt [2] has obtained the following result which is conjectured to give the best possible bounds for the type of constraint considered:

4.1 Theorem. *Suppose $u_0 \in H^{1,2} \cap L^\infty(\Omega; \mathbb{R}^3)$. Then for any $H \in \mathbb{R}$ such that*

$$\|u_0\|_{L^\infty} \cdot |H| < 1 ,$$

there exists a solution $u \in u_0 + H_0^{1,2}(\Omega; \mathbb{R}^3)$ of (4.1), (4.2) such that

$$\|u\|_{L^\infty} \leq \|u_0\|_{L^\infty} .$$

The solution u is characterized by the condition

$$E_H(u) = \min\{E_H(v) ; \ v \in u_0 + H_0^{1,2} \cap L^\infty(\Omega, \mathbb{R}^3), \ \|v\|_{L^\infty}|H| \leq 1\}$$

In particular, u is a relative minimizer of E_H in $u_0 + H_0^{1,2} \cap L^\infty(\Omega; \mathbb{R}^3)$.

4.2 Remark. Working with a different geometric constraint Wente [1; Theorem 6.1] and Steffen [1; Theorem 2.2] prove the existence of a relative minimizer provided

$$E_0(u_0)H^2 < \frac{2}{3}\pi ,$$

where

$$E_0(u) = \frac{1}{2} \int_\Omega |\nabla u|^2 \, dz$$

is the Dirichlet integral of u. The bound is not optimal, see Struwe [17; Remark IV.4.14]; it is conjectured that $E_0(u_0)H^2 < \pi$ suffices.

Proof of Theorem 4.1. Let

$$M = \{v \in u_0 + H_0^{1,2}(\Omega; \mathbb{R}^3) \; ; \; v \in L^\infty(\Omega; \mathbb{R}^3), \; \|v\|_{L^\infty}|H| \le 1\} \; .$$

M is closed and convex, hence weakly closed in $H^{1,2}(\Omega)$. Note that for $u \in M$ we can estimate

(4.3) $$|H \, u \cdot u_x \wedge u_y| \le \frac{1}{2}|H| \, \|u\|_{L^\infty} |\nabla u|^2 \le \frac{1}{2} |\nabla u|^2 \; ,$$

almost everywhere in Ω. Hence

$$E_H(u) \ge \frac{1}{3} E_0(u) \; ;$$

that is, E_H is coercive on M with respect to the $H^{1,2}$-norm. Moreover, by (4.3) on M the functional E_H may be represented by an integral

$$E_H(u) = \int_\Omega F(u, \nabla u) \, dx$$

where

$$F(u, p) = \frac{1}{2}|p|^2 + \frac{2}{3} H u \cdot p_1 \wedge p_2$$

is non-negative, continuous in $u \in \mathbb{R}^3$, and convex in $p = (p_1, p_2) \in \mathbb{R}^3 \times \mathbb{R}^3$. Hence from Theorem 1.6 we infer that E_H is weakly lower semi-continuous on M.

By Theorem I.1.2, therefore, E_H attains its infimum on M at a point $u \in M$. Moreover, E_H is analytic in $H^{1,2} \cap L^\infty(\Omega; \mathbb{R}^3)$. Therefore we may compute the directional derivative of E_H in direction of any vector pointing from u into M. Let $\varphi \in C_0^\infty(\Omega)$, $0 \le \varphi \le 1$ and choose $v = u(1 - \varphi) \in M$ as comparison function. Then by (4.3) we have

$$0 \ge \langle u - v, DE_H(u) \rangle = \langle u\varphi, DE_H(u) \rangle$$

$$= \int_\Omega \left(\nabla u \nabla(u\varphi) + 2H \, u \cdot u_x \wedge u_y \varphi \right) dz$$

$$= \int_\Omega \left(|\nabla u|^2 + 2H u \cdot u_x \wedge u_y \right) \varphi \, dz + \frac{1}{2} \int_\Omega \nabla(|u|^2) \nabla \varphi \, dz$$

$$\ge \frac{1}{2} \int_\Omega \nabla(|u|^2) \nabla \varphi \, dz \; .$$

Hence $|u|^2$ is weakly sub-harmonic on Ω. By the weak maximum principle, Theorem B.6 of the appendix, $|u|^2$ attains its supremum on $\partial\Omega$. That is, there holds

$$\|u\|_{L^\infty} \le \|u_0\|_{L^\infty} \, ,$$

as desired. But then u lies interior to M relative to $H^{1,2} \cap L^\infty$ and $DE_H(u) = 0$; which means that u weakly solves (4.1), (4.2). By a result of Wente [1; Theorem 5.5], finally, any weak solution of (4.1) is also regular, in fact, analytic in Ω. $\square$

In view of the cubic character of V, having established the existence of a relatively minimal solution to (4.1),(4.2), we are now led to expect the existence of a second solution for $H \neq 0$. This is also supported by geometrical evidence; see Figure 4.2.

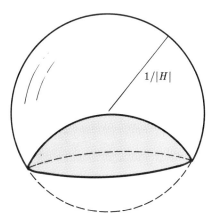

Fig. 4.2. A small and a large spherical cap of radius $1/|H|$ for $0 < |H| < 1$ give rise to distinct solutions of (4.1) with boundary data $u_0(z) = z$ on $\partial B_1(0; \mathbb{R}^2)$.

We start with an analysis of V.

The Volume Functional

In the preceding theorem we have used the obvious fact that V is smooth on $H^{1,2} \cap L^\infty(\Omega; \mathbb{R}^3)$ – but much more is true. Without proof we state the following result due to Wente [1; Section III]:

4.3 Lemma. *For any $u_0 \in H^{1,2} \cap L^\infty(\Omega; \mathbb{R}^3)$ the volume functional V extends to an analytic functional on the affine space $u_0 + H_0^{1,2}(\Omega; \mathbb{R}^3)$ and the following expansion holds:*

$$(4.4) \qquad V(u_0 + \varphi) = V(u_0) + \langle \varphi, DV(u_0) \rangle + \frac{1}{2} D^2 V(u_0)(\varphi, \varphi) + V(\varphi) \, .$$

Moreover, the derivatives

$$DV : H^{1,2}(\Omega; \mathbb{R}^3) \to H^{-1}(\Omega; \mathbb{R}^3) = H_0^{1,2}(\Omega; \mathbb{R}^3)^* ,$$

$$D^2 V : H^{1,2}(\Omega; \mathbb{R}^3) \to \left(\left[H_0^{1,2}(\Omega; \mathbb{R}^3) \right]^2 \right)^*$$

are continuous and bounded in terms of the Dirichlet integral

$$\langle \varphi, DV(u) \rangle \le c E_0(u) E_0(\varphi)^{1/2} ,$$

$$D^2 V(u)(\varphi, \psi) \le c \big(E_0(u) E_0(\varphi) E_0(\psi) \big)^{1/2} .$$

Furthermore, DV and D^2V are weakly continuous in the sense that, if $u_m \rightharpoonup u$ weakly in $H_0^{1,2}(\Omega; \mathbb{R}^3)$, then

$$\langle \varphi, DV(u_m) \rangle \to \langle \varphi, DV(u) \rangle, \quad \text{for all } \varphi \in H_0^{1,2}(\Omega; \mathbb{R}^3) ,$$

$$D^2 V(u_m)(\varphi, \psi) \to D^2 V(u)(\varphi, \psi), \quad \text{for all } \varphi, \psi \in H_0^{1,2}(\Omega; \mathbb{R}^3) .$$

Finally, for any $u \in H^{1,2}(\Omega; \mathbb{R}^3)$ the bilinear form $D^2 V(u)$ is compact; that is, if $\varphi_m \rightharpoonup \varphi$, $\psi_m \rightharpoonup \psi$ weakly in $H_0^{1,2}(\Omega; \mathbb{R}^3)$, then

$$D^2 V(u)(\varphi_m, \psi_m) \to D^2 V(u)(\varphi, \psi) .$$

The remaining term in our functional E_H is simply Dirichlet's integral $E_0(u) = \frac{1}{2} \int_\Omega |\nabla u|^2 \, dz$, well familiar from I.2.7–I.2.10. Both E_0 and V are conformally invariant, in particular invariant under scaling $u \to u(Rx)$.

The fundamental estimate for dealing with the functional E is the isoperimetric inequality for closed surfaces in $\mathbb{R}^3$, see for instance Radó [2]. This inequality for (4.1), (4.2) plays the same role as the Sobolev inequality $S\|u\|_{L^{2^*}}^2 \le \|u\|_{H^{1,2}}^2$ played for problem (1.1), (1.3).

4.4 Theorem. *For any "closed surface" $\varphi \in H_0^{1,2}(\Omega; \mathbb{R}^3)$ there holds*

$$36\pi \big| V(\varphi) \big|^2 \le E_0(\varphi)^3 .$$

The constant 36π is best possible.

Remark. The best constant 36π is attained for instance on the function $\varphi \in D^{1,2}(\mathbb{R}^2)$,

$$\varphi(x, y) = \frac{2}{1 + x^2 + y^2}(x, y, 1) ,$$

corresponding to stereographic projection of a sphere of radius 1 above $(0,0) \in \mathbb{R}^2$ onto $\mathbb{R}^2$, and its rescalings

(4.5) $$\varphi_\varepsilon(x, y) = \varphi(x/\varepsilon, y/\varepsilon) = \frac{2\varepsilon}{\varepsilon^2 + x^2 + y^2}(x, y, \varepsilon) .$$

φ and φ_ε solve equation (4.1) on $\mathbb{R}^2$ with $H = 1$: the mean curvature of the unit sphere in $\mathbb{R}^3$.

Wente's Uniqueness Result

Using the unique continuation property for the analytic equation (4.1), analogous to Theorem 1.2 we can show that the best constant in the isoperimetric inequality is never achieved on a domain $\Omega \subseteq \mathbb{R}^2$, $\Omega \neq \mathbb{R}^2$. Moreover, similar to Theorem 1.3, a sharper result holds, due to Wente [4]:

4.5 Theorem. *If $\Omega \subset \mathbb{R}^2$ is smoothly bounded and simply connected then any solution $u \in H_0^{1,2}(\Omega; \mathbb{R}^3)$ to (4.1) vanishes identically.*

Proof. By conformal invariance of (4.1) we may assume that Ω is a ball $B_1(0; \mathbb{R}^2)$. Reflecting $u(z) = -u\left(\dfrac{z}{|z|^2}\right)$ we extend u as a (weak) solution of (4.1) on $\mathbb{R}^2$. From Wente's regularity result (Wente [1; Theorem 5.5]) we infer that u is smooth and solves (4.1) classically. Now by direct computation we see that the function

$$\Phi(x + iy) = \left(|u_x|^2 - |u_y|^2\right) - 2iu_x \cdot u_y$$

is holomorphic on $\mathbb{C}$. Since

$$\int_{\mathbb{R}^2} |\nabla u|^2 \, dz = 2 \int_{B_1(0)} |\nabla u|^2 \, dz < \infty \,,$$

it follows that $\Phi \in L^1(\mathbb{R}^2)$, and hence that $\Phi \equiv 0$ by the mean-value property of holomorphic functions. That is, u is conformal. But then, since $u \equiv 0$ on $\partial B_1(0; \mathbb{R}^2)$, it follows that also $\nabla u \equiv 0$ on $\partial B_1(0; \mathbb{R}^2)$ and hence, by unique continuation, that $u \equiv 0$; see Hartmann-Wintner [1; Corollary 1]. $\qquad \square$

Theorem 4.5 – like Theorem 1.3 in the context of problem (1.1), (1.3) – proves that E_H cannot satisfy (P.-S.) globally on $H_0^{1,2}(\Omega)$, for any $H \neq 0$. Indeed, note that $E_H(0) = 0$. Moreover, by Theorem 4.4 for $u \in H_0^{1,2}(\Omega; \mathbb{R}^3)$ with $E_0(u) = \frac{4\pi}{H^2}$ there holds

$$E_H(u) = E_0(u) + 2HV(u) \geq E_0(u)\left(1 - \sqrt{\frac{4H^2}{36\pi}E_0(u)}\right)$$

$$\geq \alpha := \frac{4\pi}{3H^2} > 0 \,,$$

while for any comparison surface u with $V(u) \neq 0$, if $HV(u) < 0$, we have

$$E_H(\rho u) = \rho^2 E_0(u) + 2H\rho^3 V(u) \to -\infty \qquad \text{as } \rho \to \infty \,.$$

Hence, if E_H satisfied (P.-S.) globally on $H_0^{1,2}(\Omega; \mathbb{R}^3)$, from Theorem II.6.1 we would obtain a contradiction to Wente's uniqueness result Theorem 4.5.

Local Compactness

However, the following analogue of Lemma 2.3 holds:

4.6 Lemma. *Suppose $u_0 \in H^{1,2} \cap L^\infty(\Omega; \mathbb{R}^3)$ is a relative minimizer of E_H in the space $u_0 + H^{1,2} \cap L^\infty(\Omega; \mathbb{R}^3)$. Then for any $\beta < E_H(u_0) + \frac{4\pi}{3H^2}$ condition $(P.\text{-}S.)_\beta$ holds on the affine space $\{u_0\} + H_0^{1,2}(\Omega; \mathbb{R}^3)$.*

For the proof of Lemma 4.6 we need $D^2 E_H(u_0)$ to be positive definite on $H_0^{1,2}(\Omega; \mathbb{R}^3)$. By (4.3) this is clear for the relative minimizers constructed in Theorem 4.1, if $\|u_0\|_{L^\infty} |H| < \frac{1}{2}$. In the general case some care is needed. Also note the subtle difference in the topology of $H^{1,2} \cap L^\infty$ considered in Theorem 4.1 and $H^{1,2}$ considered here.

Postponing the proof of Lemma 4.6 for a moment we establish the following result by Brezis and Coron [2; Lemma 3]:

4.7 Lemma. *Suppose $u_0 \in H^{1,2} \cap L^\infty(\Omega; \mathbb{R}^3)$ is a relative minimizer of E_H in $u_0 + H_0^{1,2} \cap L^\infty(\Omega; \mathbb{R}^3)$. Then u_0 is a relative minimizer of E_H in $u_0 + H_0^{1,2}(\Omega; \mathbb{R}^3)$, and there exists a constant $\delta > 0$ such that*

$$D^2 E_H(u_0)(\varphi, \varphi) \geq \delta E_0(\varphi), \quad \text{for all } \varphi \in H_0^{1,2}(\Omega; \mathbb{R}^3) .$$

Proof. By density of $C_0^\infty(\Omega; \mathbb{R}^3)$ in $H_0^{1,2}(\Omega; \mathbb{R}^3)$ clearly

$$\delta = \inf\{D^2 E_H(u_0)(\varphi, \varphi) ; E_0(\varphi) = 1\} \geq 0 .$$

Note that

$$D^2 E_H(u_0)(\varphi, \varphi) = 2E_0(\varphi) + 2H D^2 V(u_0)(\varphi, \varphi) ,$$

and $D^2 V(u_0)$ is compact by Lemma 4.3. Hence, if

$$\delta = 2 + 2H \inf\{D^2 V(u_0)(\varphi, \varphi) ; E_0(\varphi) = 1\} < 2 ,$$

then a minimizing sequence for δ is relatively compact in $H_0^{1,2}(\Omega; \mathbb{R}^n)$, and δ is attained. In particular, if $\delta = 0$ there exists φ such that $E_0(\varphi) = 1$ and

$$D^2 E_H(u_0)(\varphi, \varphi) = 0 = \inf\{D^2 E_H(u_0)(\psi, \psi) ; E_0(\psi) = 1\} .$$

Necessarily, φ satisfies the Euler equation for $D^2 E_H(u_0)$:

$$\Delta \varphi = 2H(u_{0x} \wedge \varphi_y + \varphi_x \wedge u_{0y}) .$$

It follows from a result of Wente [5; Lemma 3.1] that $\varphi \in L^\infty(\Omega; \mathbb{R}^3)$; see also Brezis-Coron [2; Lemma A.1] or Struwe [17; Theorem III.5.1]. Hence by minimality of u_0, for small $|t|$ we have from Lemma 4.3 that

$$E_H(u_0) \leq E_H(u_o + t\varphi) = E_H(u_0) + t\langle \varphi, DE_H(u_0)\rangle$$
$$+ \frac{t^2}{2} D^2 E_H(u_0)(\varphi, \varphi) + 2H t^3 V(\varphi)$$
$$= E_H(u_0) + 2H t^3 V(\varphi) ,$$

and it follows that $V(\varphi) = 0$; that is, $E_H(u_0 + t\varphi) = E_H(u_0)$ for all $t \in \mathbb{R}$. But then for small $|t|$ also $u_0 + t\varphi$ is a relative minimizer of E_H and satisfies (4.1), (4.2):

$$\Delta(u_0 + t\varphi) = 2H(u_0 + t\varphi)_x \wedge (u_0 + t\varphi)_y .$$

Differentiating twice with respect to t there results

$$0 = \varphi_x \wedge \varphi_y .$$

But then $D^2 V(u_0)(\varphi, \varphi) = 2 \int_\Omega u_0 \cdot \varphi_x \wedge \varphi_y \, dz = 0$, and we obtain

$$\delta = d^2 E_H(u_0)(\varphi, \varphi) = 2D(\varphi) = 2 ,$$

contrary to assumption about δ. □

Proof of Lemma 4.6. Let (u_m) be a (P.-S.)$_\beta$-sequence in $u_0 + H_0^{1,2}(\Omega; \mathbb{R}^3)$. Consider $\varphi_m = u_m - u_0$ and note that by Lemma 4.3

$$\begin{aligned} E_H(u_m) = E_H(u_0 + \varphi_m) &= E_H(u_0) + \langle \varphi_m, DE_H(u_0) \rangle \\ &\quad + \frac{1}{2} D^2 E_H(u_0)(\varphi_m, \varphi_m) + 2HV(\varphi_m) \\ &= E_H(u_0) + \frac{1}{2} D^2 E_H(u_0)(\varphi_m, \varphi_m) + 2HV(\varphi_m) , \end{aligned}$$

while

$$\begin{aligned} o(1)\big(E_0(\varphi_m)\big)^{\frac{1}{2}} &= \langle \varphi_m, DE_H(u_m) \rangle \\ &= \langle \varphi_m, DE_H(u_0) \rangle + D^2 E_H(u_0)(\varphi_m, \varphi_m) + 6HV(\varphi_m) \\ &= D^2 E_H(u_0)(\varphi_m, \varphi_m) + 6HV(\varphi_m) , \end{aligned}$$

where $o(1) \to 0$ $(m \to \infty)$. Subtracting, we obtain

$$D^2 E_H(u_0)(\varphi_m, \varphi_m) = 6\big(E_H(u_m) - E_H(u_0)\big) + o(1)\big(E_0(\varphi_m)\big)^{\frac{1}{2}} ,$$

and by Lemma 4.7 it follows that (φ_m) and hence (u_m) is bounded. We may assume that $u_m \rightharpoonup u$ weakly. Then for any $\varphi \in H_0^{1,2}(\Omega; \mathbb{R}^3)$ by Lemma 4.3 also

$$o(1) = \langle \varphi, DE_H(u_m) \rangle \to \langle \varphi, DE_H(u) \rangle ,$$

and it follows that u solves (4.1), (4.2). Thus $t = 1$ is a critical point for the cubic function

$$t \mapsto E_H(u_0 + t(u - u_0))$$

which also attains a relative minimum at $t = 0$, and we conclude that

$$E_H(u) \geq E_H(u_0) .$$

Now let $\psi_m = u_m - u$ and note that by Lemma 4.3 we have

$$E_H(u_m) = E_H(u) + E_0(\psi_m) + HD^2V(u)(\psi_m, \psi_m) + 2HV(\psi_m)$$
$$= E_H(u) + E_H(\psi_m) + o(1) ,$$

$$o(1) = \langle \psi_m, DE_H(u_m) \rangle = \langle \psi_m, DE_H(u) \rangle +$$
$$+ \langle \psi_m, DE_H(\psi_m) \rangle + 2HD^2V(u)(\psi_m, \psi_m)$$
$$= \langle \psi_m, DE_H(\psi_m) \rangle + o(1) ,$$

with error $o(1) \to 0$ as $m \to \infty$. That is, for m sufficiently large

$$E_H(\psi_m) = E_0(\psi_m) + 2HV(\psi_m) \le E_H(u_m) - E_H(u) + o(1) \le$$
$$\le E_H(u_m) - E_H(u_0) + o(1) \le c < \frac{4\pi}{3H^2} ,$$

while

$$\frac{1}{2} \langle \psi_m, DE_H(\psi_m) \rangle = E_0(\psi_m) + 3HV(\psi_m) = o(1) .$$

It follows that for $m \ge m_0$ there holds

$$E_0(\psi_m) \le c < \frac{4\pi}{H^2} ,$$

whence by the isoperimetric inequality Theorem 4.4 we have

$$o(1) = E_0(\psi_m) + 3HV(\psi_m)$$
$$\ge E_0(\psi_m) \left(1 - \sqrt{\frac{H^2 E_0(\psi_m)}{4\pi}} \right) \ge cE_0(\psi_m)$$

for some $c > 0$, and $\psi_m \to 0$ strongly, as desired. $\square$

Large Solutions

We can now state the main result of this section, the existence of "large" solutions to the Dirichlet problem (4.1), (4.2), due to Brezis-Coron [2] and Struwe [12], [11] with a contribution by Steffen [2].

4.8 Theorem. *Suppose $u_0 \not\equiv const.$, $H \neq 0$ and assume that E_H admits a relative minimum $\underline{u}$ on $u_0 + H_0^{1,2}(\Omega; \mathbb{R}^3)$. Then there exists a further solution $\overline{u}$, distinct from $\underline{u}$, of (4.1), (4.2).*

Proof. We may assume that $\underline{u} = u_0$. Then by Lemma 4.6 the theorem follows from Theorem II.6.1 once we establish that
(1°) there exists u_1 such that $E_H(u_1) < E_H(u_0)$;
(2°) letting

$$P = \left\{ p \in C^0 \big([0,1] ; \; u_0 + H_0^{1,2}(\Omega; \mathbb{R}^3) \big) ; \; p(0) = u_0, \; p(1) = u_1 \right\} .$$

we have

$$\beta = \inf_{p \in P} \sup_{u \in p} E_H(u) < E_H(u_0) + \frac{4\pi}{3H^2} \ .$$

$(1°)$ and $(2°)$ will be established by using the sphere-attaching mechanism of Wente [2; p. 285 f.], [3], refined by Brezis and Coron [2; Lemma 5]: Since $\underline{u} = u_0 \neq const.$, there exists some point $z_0 = (x_0, y_0) \in \Omega$ such that $\nabla \underline{u}(z_0) \neq 0$. By conformal invariance of (4.1) we may assume that $z_0 = 0$, $\underline{u}(z_0) = 0$, and that with

$$\underline{u}_x(0) = (a_1, a_2, a_3), \ \underline{u}_y = (b_1, b_2, b_3)$$

there holds

$$H(a^1 + b^2) < 0 \ .$$

For $\varepsilon > 0$ now let $\varphi_\varepsilon(x, y) = \frac{2\varepsilon}{\varepsilon^2 + x^2 + y^2}(x, y, \varepsilon)$ be the stereographic projection of $\mathbb{R}^2$ onto a sphere of radius 1 centered at $(0, 0, 1)$ considered earlier, and let $\xi \in C_0^\infty(\Omega)$ be a symmetric cut-off function such that $\xi(z) = \xi(-z)$ and $\xi \equiv 1$ in a neighborhood of $z_0 = 0$. Define

$$u_t = u_t^\varepsilon := \underline{u} + t\xi\varphi_\varepsilon \in \underline{u} + H_0^{1,2}(\Omega; \mathbb{R}^3) \ .$$

For small $\varepsilon > 0$ the surface u_t "looks" like a sphere of radius t attached to $\underline{u}$ above $\underline{u}(0)$.

Now compute, using (4.4)

$$E_H(u_t) = E_H(\underline{u}) + \frac{t^2}{2} D^2 E_H(\underline{u})(\xi\varphi_\varepsilon, \xi\varphi_\varepsilon) + 2Ht^3 V(\xi\varphi_\varepsilon)$$

$$= E_H(\underline{u}) + t^2 E_0(\xi\varphi_\varepsilon) + 2Ht^3 V(\xi\varphi_\varepsilon)$$

$$+ 2Ht^2 \int_\Omega \underline{u} \cdot (\xi\varphi_\varepsilon)_x \wedge (\xi\varphi_\varepsilon)_y \ dx \ dy \ .$$

Clearly

$$E_0(\xi\varphi_\varepsilon) \le E_0(\varphi_\varepsilon; \mathbb{R}^2) + O(\varepsilon^2) = 4\pi + O(\varepsilon^2) \ ,$$

while

$$V(\xi\varphi_\varepsilon) = V(\varphi_\varepsilon; \mathbb{R}^2) + O(\varepsilon^3) = \frac{4\pi}{3} + O(\varepsilon^3) \ .$$

Expand

$$\underline{u}(x, y) = \underline{u}(0) + \underline{u}_x(0)x + \underline{u}_y(0)y + O(r^2) = ax + by + O(r^2) \ ,$$

where $r^2 = x^2 + y^2$. Upon integrating by parts and using anti-symmetry of "$\wedge$" we obtain the following expression:

$$2H \int_\Omega (ax + by) \cdot (\xi\varphi_\varepsilon)_x \wedge (\xi\varphi_\varepsilon)_y \, dx \, dy$$

$$= H \int_\Omega \big(a \wedge (\xi\varphi_\varepsilon)_y + (\xi\varphi_\varepsilon)_x \wedge b\big) \cdot (\xi\varphi_\varepsilon) \, dx \, dy$$

$$= H \int_\Omega \big(a \wedge (0,1,0) + (1,0,0) \wedge b\big) \cdot \xi^2 \frac{4\varepsilon^2}{(\varepsilon^2 + r^2)^2} (x, y, \varepsilon) \, dx \, dy$$

$$= 4H(a_1 + b_2) \int_\Omega \frac{\varepsilon^3}{(\varepsilon^2 + r^2)^2} \xi^2 \, dx \, dy$$

$$- 4H \int_\Omega (a_3 x + b_3 y) \frac{4\varepsilon^2}{(\varepsilon^2 + r^2)^2} \xi^2 \, dx \, dy \ .$$

Since ξ is symmetric the last term vanishes, while for sufficiently small $\varepsilon > 0$

$$\int_\Omega \frac{\varepsilon^3}{(\varepsilon^2 + r^2)^2} \xi^2 \, dx \, dy \geq \frac{1}{4\varepsilon} \int_{B_\varepsilon(0)} dx \, dy = \frac{\varepsilon\pi}{4} > 0 \ .$$

Finally, since $|\nabla\varphi_\varepsilon(x, y)| \leq \frac{c\varepsilon}{\varepsilon^2 + x^2 + y^2}$, we may estimate

$$\left| \int_\Omega O(r^2) \cdot (\xi\varphi_\varepsilon)_x \wedge (\xi\varphi_\varepsilon)_y \, dx \, dy \right| \leq$$

$$\leq \int_\Omega O(r^2) |\nabla\varphi_\varepsilon|^2 \, dx \, dy + O(\varepsilon^2)$$

$$\leq c \int_\Omega \frac{\varepsilon^2 r^2}{(\varepsilon^2 + r^2)^2} \, dx \, dy + O(\varepsilon^2)$$

$$\leq c \int_{B_\varepsilon(0)} dx \, dy + c \int_{\Omega \setminus B_\varepsilon(0)} \frac{\varepsilon^2}{r^2} \, dx \, dy + O(\varepsilon^2)$$

$$\leq c\varepsilon^2 + c\varepsilon^2 |ln\ \varepsilon| \ .$$

Hence for sufficiently small $\varepsilon > 0$, $t \geq 0$ we may estimate

$$E_H(u_t) \leq E_H(\underline{u}) + t^2 \big(4\pi + \pi H(a_1 + b_2)\varepsilon + c\varepsilon^2 |ln\ \varepsilon| + c\varepsilon^2\big) + 2Ht^3 \Big(\frac{4\pi}{3} + O(\varepsilon^3)\Big) \ .$$

In particular, if $H < 0$, there exists $T > 0$ such that u_T satisfies $E_H(u_T) < E_H(\underline{u})$ and, moreover,

$$\sup_{0 \leq t \leq T} E_H(u_t) < E_H(\underline{u}) + \frac{4\pi}{3H^2} \ ,$$

as desired. The case $H > 0$ follows in a similar way by considering $(u_t)_{t \leq 0}$. □

4.9 Remarks. A "global" compactness condition analogous to Theorem 3.1 also holds in the case of equation (4.1); see Brezis-Coron [3], Struwe [11]. It would be interesting to investigate the effect of topology on the existence of non-trivial solutions to (4.1). Presumably, if Ω is an annulus or of even higher topological type, solutions $u \in H_0^{1,2}(\Omega)$ to (4.1) need not necessarily vanish, as in the case of problem (1.1)–(1.3).

For further material and references on the Dirichlet and Plateau problems for surfaces of prescribed mean curvature we refer the interested reader to the lecture notes by Struwe [17]. A result analogous to Theorem 4.8 for variable mean curvature $H = H(u)$ has recently been obtained by Struwe [18].

5. Harmonic Maps of Riemannian Surfaces

As our final example we now present a border-line variational problem involving a non-differentiable functional.

Given a (smooth) compact Riemannian surface Σ with metric γ (for simplicity without boundary) and any (smooth) compact manifold N with metric g, a natural generalization of Dirichlet's integral for C^1-functions on a domain in $\mathbb{R}^n$ is the following energy functional

$$E(u) = \int_\Sigma e(u)d\Sigma \ .$$

Here in local coordinates on Σ and N the energy density $e(u)$ is given by

$$e(u) = \sum_{1 \leq \alpha,\beta \leq 2} \sum_{1 \leq i,j \leq n} \frac{1}{2} \gamma^{\alpha\beta}(x) g_{ij}(u) \frac{\partial}{\partial x_\alpha} u^i \frac{\partial}{\partial x_\beta} u^j \ ,$$

with $\gamma^{\alpha\beta} = (\gamma_{\alpha\beta})^{-1}$ denoting the coefficients of the inverse of the matrix $(\gamma_{\alpha\beta})$ representing the metric γ, (g_{ij}) representing g, and with

$$d\Sigma = \sqrt{|\gamma|}dx, \ |\gamma| = det(\gamma_{\alpha\beta}) \ .$$

Since we assume that both Σ and N are compact this expression may be simplified considerably: First, by the Nash embedding theorem, see for instance Nash [1] or Schwartz [2; pp. 43–53], any compact manifold N may be isometrically embedded into some Euclidean $\mathbb{R}^n$. A new proof – avoiding the "hard" implicit function theorem – has recently been obtained by Günther along the lines of Günther [1].

Moreover, E is invariant under conformal mappings of Σ. Thus, by the uniformization theorem, we may assume that either $\Sigma = S^2$ or $\Sigma = T^2 = \mathbb{R}^2/\mathbb{Z}^2$, or is a quotient of the upper half-space $\mathbb{H}$, endowed with the hyperbolic metric $\frac{1}{y}dx \, dy$. In particular, if $\Sigma = T^2$, the energy density is simply given by $e(u) = \frac{1}{2}|\nabla u|^2$ and E becomes the standard Dirichlet integral for mappings $u: T^2 = \mathbb{R}^2/\mathbb{Z}^2 \to N \subset \mathbb{R}^n$.

Consider the space $C^1(\Sigma; N)$ of C^1-functions $u: \Sigma \to N$. Note that $C^1(\Sigma; N)$ is a manifold with tangent space given by

$$T_u C^1(\Sigma; N) = \{\varphi \in C^1(\Sigma; TN) \; ; \; \varphi(x) \in T_{u(x)} N \text{ for } x \in \Sigma\} \, ,$$

and E is differentiable on this space.

In fact, if we consider variations $\varphi \in T_u C^1(\Sigma; N)$ such that $\operatorname{supp}(\varphi)$ is contained in a chart U on Σ whose image $u(U)$ is contained in a coordinate patch of N, then in order to compute the variation of E in direction φ it suffices to work in *one* coordinate frame – both on Σ and N – and all computations can be done as in the "flat" case. From this, the differentiability of E is immediate and the following definition is meaningful.

5.1 Definition. *A stationary point $u \in C^1(\Sigma; N)$ of E is called a harmonic map.*

The concept of harmonic map generalizes the notion of (closed) geodesic to higher dimensions; compare Section II.4. Moreover, if we choose $N = \mathbb{R}^n$, we see that harmonic functions simply appear as special cases of harmonic maps. In order to become somewhat familiar with this new notion, let us derive the Euler-Lagrange equations for harmonic maps.

The Euler-Lagrange Equations for Harmonic Maps

As a model case we first consider the case $\Sigma = T^2 = \mathbb{R}^2/\mathbb{Z}^2$, where E reduces to the standard Dirichlet integral for doubly periodic mappings $u: \mathbb{R}^2 \to N \subset \mathbb{R}^n$, restricted to a fundamental domain.

In this case, if $u: T^2 = \mathbb{R}^2/\mathbb{Z}^2 \to N \subset \mathbb{R}^n$ is harmonic of class C^2, the first variation of E gives

$$0 = \langle \varphi, DE(u) \rangle = \int_{T^2} \nabla u \nabla \varphi \, dx = -\int_{T^2} \Delta u \, \varphi \, dx \, ,$$

for all doubly periodic $\varphi \in C^1(\mathbb{R}^2; \mathbb{R}^n)$ satisfying the condition

$$\varphi(x) \in T_{u(x)} N \text{ for all } x \in \mathbb{R}^2 \, .$$

That is, $-\Delta u(x)$ is orthogonal to the tangent space at N at any point $x \in T^2$; in symbols:

$$-\Delta u(x) = \lambda(x)\nu_u(x) \perp T_{u(x)} N \qquad \text{for all } x \in T^2$$

for some normal vector field ν_u and a scalar function λ. In general, the Laplace operator will be replaced by the Laplace-Beltrami operator Δ_Σ on Σ.

For further illustration, consider the case $N = S^{n-1} \subset \mathbb{R}^n$. In this case, if $u: T^2 \to S^{n-1} \subset \mathbb{R}^n$ is of class C^2 and harmonic, it follows that

$$-\Delta u = \lambda u$$

for some continuous function $\lambda : T^2 \to \mathbb{R}$. Testing this relation with u and noting that $|u| \equiv 1$, $u \cdot \nabla u \equiv 0$, we see that

$$\lambda = -\operatorname{div}(u \cdot \nabla u) + |\nabla u|^2 = |\nabla u|^2 \; ;$$

that is, harmonic maps into spheres satisfy the relation

$$-\Delta u = u |\nabla u|^2 \; .$$

Upon differentiating this identity in direction x_i and taking its scalar product with the ith component of the gradient of u, since $\nabla u \perp T_u S^n$, we obtain the Bochner type identity for the energy density $e(u) = \frac{1}{2} |\nabla u|^2$

$$-\Delta e(u) + |\nabla^2 u|^2 = |\nabla u|^4 \; .$$

For general target manifolds N we may proceed similarly. Given $x \in \Sigma$, choose a smooth unit normal vector field ν in a neighborhood of $u(x)$ in N such that $\nu\big(u(x)\big) = \nu_u(x)$. Then, since $\nabla u \in T_u N$, we have

$$\lambda(x) = -\Delta u(x) \nu\big(u(x)\big) =$$
$$= -\operatorname{div}\big(\nabla u \nu(u)\big)(x) + \big(\nabla u \nabla \nu(u)\big)(x) \le c |\nabla u(x)|^2 \; .$$

Thus, by compactness of N there is a uniform constant C (depending only on N) such that there holds

$$|\Delta u| \le C |\nabla u|^2 \; .$$

Similarly, for general domains, in local coordinates on Σ and N the harmonic map equation can be written down explicitly as

$$-\Delta_\Sigma u - \Gamma(u)(\nabla u, \nabla u) = 0 \; ,$$

where Γ is a bilinear form with coefficients involving the Christoffel symbols of the metric g on N, and the metric γ. Differentiating this equation and multiplying by ∇u, there results the Bochner type differential inequality

$$-\Delta_\Sigma e(u) + c |\nabla^2 u|^2 \le \kappa_N |\nabla u|^4 + C |\nabla u|^2$$

for the energy density of u, where κ_N denotes an upper bound for the sectional curvature of N, and where $c > 0$ and C denote constants depending only on Σ and N.

With these remarks we hope that the reader has become somewhat familiar with the subject of harmonic maps. Moreover, to fix ideas, in the following one may always think of mappings $u : T^2 = \mathbb{R}^2 / \mathbb{Z}^2 \to S^{n-1} \subset \mathbb{R}^n$. In this special case already, all essential difficulties appear and nothing of the flavour of the results will be lost.

The Homotopy Problem and its Functional Analytic Setting

A natural generalization of Dirichlet's problem for harmonic functions now is the following:

Problem 1. Given a map $u_0: \Sigma \to N$ is there a harmonic map u homotopic to u_0?

As in the case of a scalar function $u: \Omega \to \mathbb{R}$ we may attempt to approach this problem by direct methods: Denote

$$H^{1,2}(\Sigma; N) = \left\{ u \in H^{1,2}(\Sigma; \mathbb{R}^n) \; ; \; u(\Sigma) \subset N \text{ a.e.} \right\}$$

the space of $H^{1,2}$-mappings into N. (If $\Sigma = T^2$, then $H^{1,2}(\Sigma; N)$ is the space of mappings $u \in H^{1,2}_{loc}(\mathbb{R}^2; N)$ of period 1 in both variables, restricted to a fundamental domain.) Note that E is weakly lower semi-continuous and coercive on $H^{1,2}(\Sigma; N)$ with respect to $H^{1,2}(\Sigma; \mathbb{R}^n)$.

Moreover, by a result of Schoen-Uhlenbeck [2; Section IV] we have:

5.2 Theorem. *($1°$) The space $C^\infty(\Sigma; N)$ of smooth maps $u: \Sigma \to N \subset \mathbb{R}^n$ is dense in $H^{1,2}(\Sigma; N)$.*
($2°$) The homotopy class of an $H^{1,2}(\Sigma; N)$-map is well-defined.

Proof. ($1°$) The following argument basically is due to Courant [1; p. 214 f.]. Let $\varphi \in C_0^\infty(B_1(0))$ satisfy $o \leq \varphi \leq 1$, $\int_{B_1(0)} \varphi \, dx = 1$, and for $\varepsilon > 0$ let $\varphi_\varepsilon = \varepsilon^{-2}\varphi\left(\frac{x}{\varepsilon}\right)$. Given $u \in H^{1,2}(\Sigma; \mathbb{R}^n)$ denote

$$u * \varphi_\varepsilon(x_0) = \int_{B_\varepsilon(x_0)} u(x)\varphi_\varepsilon(x_0 - x) \, dx$$

its mollification with φ_ε (in local coordinates on Σ). Note that $u * \varphi_\varepsilon \in C^\infty(\Sigma; N)$ and $u * \varphi_\varepsilon \to u$ in $H^{1,2}(\Sigma; \mathbb{R}^n)$ as $\varepsilon \to 0$. Let $\fint$ denote average and let $\text{dist}(P, N) = \inf\{|P - Q| \; ; \; Q \in N\}$ denote the distance of a point P from N.

For $x_0 \in \Sigma, \varepsilon > 0$ and $y \in B_\varepsilon(x_0)$ now estimate

$$\text{dist}^2\left(u * \varphi_\varepsilon(x_0), N\right) \leq \left| \int_{B_\varepsilon(x_0)} \left(u(x) - u(y)\right)\varphi_\varepsilon(x_0 - x) \, dx \right|^2$$

$$\leq C \fint_{B_\varepsilon(x_0)} \left|u(x) - u(y)\right|^2 \, dx$$

Taking the average with respect to $y \in B_\varepsilon(x_0)$ therefore

$$\text{dist}^2\big(u * \varphi_\varepsilon(x_0), N\big) \le C \fint_{B_\varepsilon(x_0)} \fint_{B_\varepsilon(x_0)} |u(x) - u(y)|^2 \, dx \, dy$$

$$\le C\varepsilon^2 \fint_{B_\varepsilon(x_0)} \fint_{B_\varepsilon(x_0)} \int_0^1 \big|\nabla u\big(y + t(x - y)\big)\big|^2 \, dt \, dx \, dy$$

$$\le c \int_{B_\varepsilon(x_0)} |\nabla u|^2 \, dx \ .$$

Thus, a mollified function $u \in H^{1,2}(\Sigma; N)$ stays close to N and may be projected onto N to yield a function $\tilde{u} \in C^\infty(\Sigma; N)$ close to u in $H^{1,2}$-norm.

(2°) Define the homotopy class of $u \in H^{1,2}(\Sigma; N)$ as the homotopy class of maps in $C^\infty(\Sigma; N)$ close to u in the $H^{1,2}(\Sigma; N)$-norm. To see that this is well-defined, let $v_0, v_1 \in C^\infty(\Sigma; N)$ be close to $u \in H^{1,2}(\Sigma; N)$ and let $v_t = tv_0 + (1 - t)v_1$ be a homotopy connecting v_0 and v_1 in $C^\infty(\Sigma; \mathbb{R}^n)$. Note that

$$\|v_t - u\|_{H^{1,2}} \le \|v_0 - u\|_{H^{1,2}} + \|v_1 - u\|_{H^{1,2}} \ .$$

For suitable $\varepsilon > 0$ therefore

$$\text{dist}(v_t * \varphi_\varepsilon, N) \le \text{dist}(u * \varphi_\varepsilon, N) + \|v_t * \varphi_\varepsilon - u * \varphi_\varepsilon\|_{L^\infty}$$

$$\le \text{dist}(u * \varphi_\varepsilon, N) + \varepsilon^{-1}\|v_t - u\|_{H^{1,2}}$$

is small, if v_0, v_1 are close to u, allowing us to project the maps $v_t * \varphi_\varepsilon$ to maps $\widetilde{v_t * \varphi_\varepsilon} \in C^\infty(\Sigma; N)$, $0 \le t \le 1$, connecting $\widetilde{v_0 * \varphi_\varepsilon}$ and $\widetilde{v_1 * \varphi_\varepsilon}$. Since v_0 and $\widetilde{v_0 * \varphi_\varepsilon}$ are homotopic through $\widetilde{v_0 * \varphi_\delta}$, $0 \le \delta \le \varepsilon$, and similarly for v_1, this shows that v_0 and v_1 belong to the same homotopy class.   □

However, while $H^{1,2}(\Sigma; N)$ is weakly closed in the topology of $H^{1,2}(\Sigma; \mathbb{R}^n)$, in general this will not be the case for homotopy classes of non-constant maps. Consider for example the family $(\varphi_\varepsilon)_{\varepsilon>0}$ of stereographic projections of the standard sphere, introduced in (4.5); projecting back with φ_1, we obtain a family of maps $u_\varepsilon = (\varphi_1)^{-1} \circ \varphi_\varepsilon : S^2 \to S^2$ of degree 1, converging weakly to a constant map. Therefore, the direct method fails to be applicable for solving Problem 1.

Moreover, note that the space $H^{1,2}(\Sigma; N)$ is not a manifold and also the standard deformation lemma, that is, Theorem II.3.4 or II.3.11, cannot be applied.

Existence and Non-Existence Results

Even worse, the infimum of E in a given homotopy class in general need not be attained. In fact, Problem 1 need not always have an affirmative answer. This result is due to Eells-Wood [1]:

5.3 Theorem. *Any harmonic map $u \in C^1(T^2; S^2)$ necessarily has topological degree $\neq 1$.*

In particular, there is no harmonic map homotopic to a map $u_0 : T^2 \to S^2$ of degree $+1$.

This result is analogous to the results of Pohožaev (Theorem 1.3) and Wente (Theorem 4.1) in our previous examples. It shows that we may encounter some lack of compactness in attempting to find critical points of E.

However, compactness can be restored under suitable conditions. Imposing a restriction on the sectional curvature of the target manifold, in a pioneering paper Eells-Sampson [1] have obtained the following result:

5.4 Theorem. *Suppose the sectional curvature κ_N of N is non-positive. Then for any map $u_0 : \Sigma \to N$ there exists a harmonic map homotopic to u_0.*

We sketch the main idea of the proof. In order to overcome the difficulties mentioned above, Eells and Sampson consider the evolution problem

$$(5.1) \qquad u_t - \Delta_\Sigma u = \lambda \nu_u \perp T_u N \text{ in } \Sigma \times \mathbb{R}_+$$
$$(5.2) \qquad u|_{t=0} = u_0$$

associated with E, where ν_u is a normal vector field and λ is a scalar function. Observe that, as in the stationary case, in local coordinates on Σ and N, equation (5.1) can also be written down explicitly as

$$(5.1') \qquad u_t - \Delta_\Sigma u - \Gamma(u)(\nabla u, \nabla u) = 0 ,$$

where Γ is a bilinear form with coefficients involving the Christoffel symbols of the metric g on N. Hence for (5.1), (5.2) there holds the Bochner type estimate

$$(5.3) \qquad \partial_t e(u) - \Delta_\Sigma e(u) + c|\nabla^2 u|^2 \le \kappa_N |\nu|^4 + C|\nabla u|^2 .$$

Moreover, (5.1), (5.2) is of the type of a pseudo-gradient flow, and we obtain the *energy inequality*

$$\int_0^T \int_\Sigma |\partial_t u|^2 \, d\Sigma \, dt + E\big(u(T)\big) \le E(u_0) ,$$

for all $T > 0$; see Lemma 5.8 below.

Now, if $\kappa_N \le 0$, estimate (5.3) becomes a linear differential inequality for the energy density, and we obtain the existence of a global solution $u \in C^2(\Sigma \times \mathbb{R}_+, N)$ to the evolution problem (5.1), (5.2). Moreover, by the weak Harnack inequality for sub-solutions of parabolic equations (see Moser [2; Theorem 3]) and the energy inequality, the maximum of $|\nabla u|$ may be a priori bounded in terms of the initial energy. Again by the energy inequality, we can find a sequence of numbers $t_m \to \infty$ such that $\partial_t u(t_m) \to 0$ in L^2 as $m \to \infty$ and it follows that $(u(t_m))$ converges to a harmonic map. □

Surprisingly, also a topological condition on the target may suffice to solve the homotopy problem. The following result was obtained independently by Lemaire [1] and Sacks-Uhlenbeck [1]:

5.5 Theorem. *If $\pi_2(N) = 0$, then for any $u_0 \in H^{1,2}(\Sigma; N)$ there is a smooth harmonic map homotopic to u_0.*

We will give a proof of this result based on the analysis of the "gradient flow" (5.1), (5.2). This proof very clearly shows the analogy of the harmonic map problem with the problems (1.1), (1.3), the Yamabe problem, and the Dirichlet problem for the equation of constant mean curvature (4.1).

The Evolution of Harmonic Maps

We propose to establish that (5.1), (5.2) admits a global weak solution for arbitrary initial data $u \in H^{1,2}(\Sigma; N)$, without *any* topological or geometric restrictions on the target manifold. By Theorem 5.3 we cannot expect the existence of a smooth global solution, converging asymptotically to a harmonic map, in general. Therefore, the following theorem, due to Struwe [10], seems to be best possible. Let $\exp_x \colon T_x\Sigma \to \Sigma$ denote the exponential map at a point $x \in \Sigma$. (If $\Sigma = T^2$, then $\exp_x(y) = x + y$.)

5.6 Theorem. *For any $u_0 \in H^{1,2}(\Sigma; N)$ there exists a distribution solution $u \colon \Sigma \times \mathbb{R}_+ \to N$ of (5.1) with $E\big(u(\,\cdot\,, t)\big) \leq E(u_0) = E_0$ which is smooth on $\Sigma \times \mathbb{R}_+$ away from at most finitely many points $(\overline{x}_k, \overline{t}_k)$, $1 \leq k \leq K$, $0 < \overline{t}_k \leq \infty$, and which assumes its initial data continuously in $H^{1,2}(\Sigma; N)$. u is unique in this class. At a singularity $(\overline{x}, \overline{t})$ a smooth harmonic map $\overline{u} \colon S^2 \cong \overline{\mathbb{R}^2} \to N$ separates in the sense that for sequences $x_m \to \overline{x}$, $t_m \nearrow \overline{t}$, $R_m \searrow 0$ as $m \to \infty$ the family*

$$u_m(x) \equiv u\big(\exp_{x_m}(R_m x), t_m\big) \to \tilde{u} \qquad in \ H^{2,2}_{loc}(\mathbb{R}^2; N) \ ,$$

where $\tilde{u}$ has finite energy and extends to a smooth harmonic map $\overline{u} \colon S^2 \cong \overline{\mathbb{R}^2} \to N$. Finally, for a suitable sequence $t_m \to \infty$ the sequence of maps $u(\,\cdot\,, t_m)$ converges weakly in $H^{1,2}(\Sigma; N)$ to a smooth harmonic map $u_\infty \colon \Sigma \to N$. The convergence is strong in $H^{2,2}_{loc}(\Sigma \setminus \{\overline{x}_k \ ; \ \overline{t}_k = \infty\}, N)$. Moreover,

$$E(u_\infty) \leq E(u_0) - K \, \varepsilon_0 \ ,$$

where K is defined above, and

$$\varepsilon_0 = \inf\big\{E(u) \ ; \ u \in C^1(S^2; S^2) \ \text{ is non-constant and harmonic }\big\} > 0$$

is a constant depending only on the geometry of N. In particular, the number of singularities of u is a priori bounded, $K \leq \varepsilon_0^{-1} E(u_0)$.

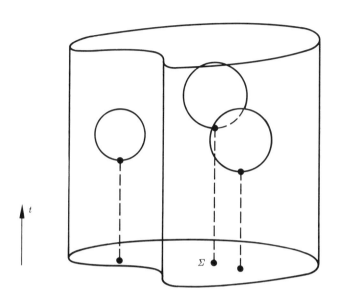

Fig. 5.1. "Separation of spheres"

Theorem 5.6 implies Theorem 5.5:

Proof of Theorem 5.5. We follow Struwe [15; p. 299 f.]. Let $[u_0]$ be a homotopy class of maps from Σ into N. We may suppose that $[u_0]$ is represented by a smooth map $u_0 \colon \Sigma \to N$ such that

$$E(u_0) \leq \inf_{u \in [u_0]} E(u) + \frac{\varepsilon_0}{4} \ .$$

Let $u \colon \Sigma \times \mathbb{R}_+ \to N$ be the solution to the evolution problem (5.1), (5.2) constructed in Theorem 5.6. Suppose u first becomes singular at a point $(\overline{x}, \overline{t})$. Then for sequences $t_m \nearrow \overline{t}$, $x_m \to \overline{x}$, $r_m \searrow 0$ we have

$$(5.4) \qquad u_m(x) := u\bigl(\exp_{x_m}(r_m x), t_m\bigr) \to \tilde{u} \qquad \text{in } H^{2,2}_{loc}(\mathbb{R}^2; N)$$

where $\tilde{u}$ may be extended to a smooth harmonic map $\overline{u} \colon S^2 \to N$.

In the image covered by u_m we now replace a large part of the "harmonic sphere" $\overline{u}$ by its "small" complement, "saving" at least $\varepsilon_0/2$ in energy. If $\pi_2(N) = 0$, this change of u_m will not affect the homotopy class $[u_m] = [u_0]$, and a contradiction will result.

More precisely, by (5.4) and since $E(\tilde{u}) < \infty$, we can find a sequence of radii $R_m \to \infty$ such that as $m \to \infty$ we have

$$r_m R_m \to 0 \ ,$$

$$\sup_{\partial B_{R_m}} |u_m - \tilde{u}| + R_m \int_{\partial B_{R_m}} |\nabla(u_m - \tilde{u})|^2 \, do \to 0 \,,$$

$$\int_{\mathbb{R}^2 \setminus B_{R_m}} |\nabla \tilde{u}|^2 \, dx \to 0 \,,$$

$$\int_{B_{R_m}} |\nabla(u_m - \tilde{u})|^2 \, dx \to 0 \,.$$

Now let $\psi \in H^{1,2}(B_{R_m}; \mathbb{R}^n)$ solve Dirichlet's problem $\Delta \psi_m = 0$ in $B_{R_m} = B_{R_m}(0) \subset \mathbb{R}^2$ with boundary data

$$\psi_m = u_m - \tilde{u} \qquad \text{on } \partial B_{R_m} \,.$$

By the maximum principle

$$\sup_{B_{R_m}} |\psi_m| = \sup_{\partial B_{R_m}} |u_m - \tilde{u}| \to 0 \,.$$

Moreover, by classical potential estimates and interpolation (see Lions-Magenes [1; Theorem 8.2]), the Dirichlet integral of ψ_m may be estimated in terms of the semi-norm of $\psi_m|_{\partial B_{R_m}} \in H^{1/2,2}(\partial B_{R_m})$:

$$\int_{B_{R_m}} |\nabla \psi_m|^2 \, dx \le c |\psi_m|^2_{H^{1/2,2}(\partial B_{R_m})}$$

$$\le c \int_{\partial B_{R_m}} \int_{\partial B_{R_m}} \frac{|\psi_m(x) - \psi_m(y)|^2}{|x - y|^2} \, dx \, dy$$

$$\le c R_m \int_{\partial B_{R_m}} |\nabla \psi_m|^2 \, do = c R_m \int_{\partial B_{R_m}} |\nabla(u_m - \tilde{u})|^2 \, do \to 0 \,,$$

as $m \to \infty$.

(The equivalent integral representation of the $H^{1/2,2}$-semi-norm may be found for instance in Adams [1; Theorem 7.48].)

Hence if we replace $u(\cdot, t_m)$ by the map

$$v_m\big(\exp_{x_m}(r_m x)\big) = \begin{cases} u_m(x), & |x| \ge R_m \\ \tilde{u}\left(R_m^2 \dfrac{x}{|x|^2} \right) + \psi_m(x), & |x| < R_m \,, \end{cases}$$

defined via the exponential map in a small coordinate chart around x_m, we obtain that, as $m \to \infty$,

$$\text{dist}\big(v_m(x), N\big) \le \sup_{B_{R_m}} |\psi_m| \to 0 \,,$$

uniformly in $x \in \Sigma$. But then we may project v_m onto N to obtain a map $w_m \in H^{1,2}(\Sigma; N)$ satisfying, with error $o(1) \to 0$ as $m \to \infty$,

$$E(w_m) = E(v_m) + o(1)$$

$$= E(u_m) - \int_{B_{R_m}} |\nabla u_m|^2 \, dx + \int_{\mathbb{R}^2 \setminus B_{R_m}} |\nabla \tilde{u}|^2 \, dx +$$

$$+ \int_{B_{R_m}} |\nabla \psi_m|^2 \, dx + o(1)$$

$$\leq E(u_0) - \int_{B_{R_m}} |\nabla \tilde{u}|^2 \, dx + o(1)$$

$$\leq \inf_{u \in [u_0]} E(u) + \varepsilon_0/4 - \varepsilon_0/2 + o(1)$$

$$< \inf_{u \in [u_0]} E(u),$$

if $m \geq m_0$.

On the other hand, since $\pi_2(N) = 0$, the maps $x \mapsto \tilde{u}(x)$, resp. $x \mapsto \tilde{u}\left(R_m^2 \dfrac{x}{|x|^2}\right)$ for any m are homotopic as maps from B_{R_m} into N with fixed boundary; thus w_m is homotopic to $u(\cdot, t_m)$, and the latter is homotopic to u_0 via the flow u, which is regular on $\Sigma \times [0, t_m]$ for all m, by assumption. Thus, if u develops a singularity in finite or infinite time, we obtain a contradiction to our choice of u_0. Hence, the flow u exists and is regular for all time and, as $t_m \to \infty$ suitably, $u(\cdot, t_m) \to u_\infty$ in $H^{1,2}(\Sigma; N)$ where u_∞ is harmonic. But by Theorem 5.2, for sufficiently large m the maps u_∞ and $u(\cdot, t_m)$ are homotopic while clearly the flow $u(\cdot, t), 0 \leq t \leq t_m$, yields a homotopy of u_0 with $u(\cdot, t_m)$. The proof is complete.   □

We now enter into the detailed proof of Theorem 5.6. It may be helpful to keep in mind the special case $\Sigma = T^2$, where (5.1) becomes the heat equation, projected into the tangent space of N at each point $u(x)$.

The proof of Theorem 5.6 is based on the following Sobolev type inequality, essentially due to Ladyzhenskaya [1]. Observe that, as in the stationary case, by (5.1) and since $\partial_t u, \ \nabla u \in T_u N$ we have

$$\lambda(x) \leq c |\nabla u(x)|^2 \, ,$$

and we can find a uniform constant C (depending only on N) such that for any solution u of (5.1) there holds

(5.5) $$|\partial_t u - \Delta_\Sigma u| \leq C |\nabla u|^2 \, .$$

Hence an L^4-bound for $|\nabla u|$ will imply an L^2-estimate for $|\nabla^2 u|$; see Lemma 5.10 below. Moreover, we will show that solutions u to (5.1), satisfying the energy inequality Lemma 5.8 and with $|\nabla^2 u| \in L^2$ are regular; see Lemma 5.11.

5.7 Lemma. *For any function $u \in H^{1,2}_{loc}(\mathbb{R}^2)$ and any function $\varphi \in C^\infty_0(B_R)$ with $0 \le \varphi \le 1$ and such that $|\nabla\varphi| \le 4/R$, there holds*

$$\int_{\mathbb{R}^2} |u|^4\varphi^2 \, dx \le c_0 \left(\int_{B_R} |u|^2 dx \right) \cdot \left(\int_{B_R} |\nabla u|^2 \varphi^2 \, dx + R^{-2} \int_{B_R} |u|^2 dx \right),$$

with c_0 independent of u and R.

Proof. The function $|u|^2\varphi$ has a compact support. Therefore, for any $x = (\xi, \eta) \in \mathbb{R}^2$ we have

$$\left| (|u|^2\varphi)(\xi, \eta) \right| = \left| \int_{-\infty}^{\xi} \frac{\partial}{\partial\xi} (|u|^2\varphi)(\xi', \eta) \, d\xi' \right|$$

$$\le 2 \left(\int_{-\infty}^{\infty} (|\nabla u| \, |u|\varphi + |u|^2 |\nabla\varphi|)(\xi', \eta) \, d\xi' \right) ,$$

and an analogous estimate with integration in η-direction.

Hence by Fubini's theorem and Hölder's inequality

$$\int_{\mathbb{R}^2} |u|^4\varphi^2(x) \, dx = \int_{-\infty}^{\infty} \int_{-\infty}^{\infty} |u|^4\varphi^2(\xi, \eta) \, d\xi \, d\eta \le$$

$$\le 4 \cdot \int_{-\infty}^{\infty} \int_{-\infty}^{\infty} \left\{ \left(\int_{-\infty}^{\infty} |\nabla u| \, |u|\varphi + |u|^2 \, |\nabla\varphi|)(\xi', \eta) \, d\xi' \right) \cdot \right.$$

$$\left. \cdot \left(\int_{-\infty}^{\infty} (|\nabla u| \, |u|\varphi + |u|^2 \, |\nabla\varphi|)(\xi, \eta') \, d\eta' \right) \right\} d\xi \, d\eta$$

$$= 4 \int_{-\infty}^{\infty} \int_{-\infty}^{\infty} (|\nabla u| \, |u|\varphi + |u|^2|\nabla\varphi|)(\xi', \eta) \, d\xi' \, d\eta \cdot$$

$$\cdot \int_{-\infty}^{\infty} \int_{-\infty}^{\infty} (|\nabla u| \, |u|\varphi + |u|^2|\nabla\varphi|)(\xi, \eta') d\xi \, d\eta'$$

$$= 4 \left(\int_{\mathbb{R}^2} (|\nabla u| \, |u|\varphi + |u|^2|\nabla\varphi|) \, dx \right)^2$$

$$\le 8 \left(\int_{\mathrm{supp}\,\varphi} |u|^2 \, dx \right) \cdot \int_{\mathbb{R}^2} (|\nabla u|^2\varphi^2 + |u|^2|\nabla\varphi|^2) \, dx$$

$$\le 8 \left(\int_{B_R} |u|^2 dx \right) \cdot \left(\int_{B_R} |\nabla u|^2\varphi^2 dx + 16R^{-2} \int_{B_R} |u|^2 dx \right),$$

which proves the lemma.   □

Moreover, we have the following "energy inequality", showing that (5.1), (5.2) is a "gradient-like" flow for E.

5.8 Lemma. *Suppose $u \in C^2(\Sigma \times [0, T[; N)$ solves (5.1), (5.2). Then*

$$\sup_{0 \leq t < T} \left(\int_0^t \int_\Sigma |\partial_t u|^2 \, d\Sigma \, dt' + E(u(t)) \right) \leq E(u_0) \ .$$

Proof. Multiply (5.1) by $\partial_t u \in T_u N$ and integrate by parts over $\Sigma \times \{t\}$ to obtain that

$$\frac{d}{dt} E(u(t)) + \int_\Sigma |\partial_t u|^2 \, d\Sigma = 0 \ ,$$

for all $t < T$. The claim follows upon integrating in t. □

Lemma 5.7 also calls for control of the energy density, locally. For this let $\imath_\Sigma$ be the injectivity radius of the exponential map on Σ. Then on any ball $B_R(x_0) \subset \Sigma$ of radius $R < \imath_\Sigma$ we may introduce Euclidean coordinates by a conformal change of variables. Recall that E is conformally invariant, and introduce the local energy

$$E(u; B_R(x_0)) = \frac{1}{2} \int_{B_R(x_0)} |\nabla u|^2 \, dx \ , \qquad R < \imath_\Sigma \ ,$$

with reference to such a conformal chart. Shift $x_0 = 0$, and let as always $B_R = B_R(0)$ for brevity.

5.9 Lemma. *There is an absolute constant c_1 such that for any $R < \frac{1}{2}\imath_\Sigma$ and any solution $u \in C^2(B_{2R} \times [0, T]; N)$ to (5.1) with $E(u(t); B_{2R}) \leq E_0$ there holds*

$$E(u(T); B_R) \leq E(u_0; B_{2R}) + c_1 \frac{T}{R^2} E_0 \ .$$

Proof. Consider $\Sigma = T^2$, for simplicity. (In the general case we may pass to a conformal chart.)
Choose $\varphi \in C_0^\infty(B_{2R})$ such that $0 \leq \varphi \leq 1$, $\varphi \equiv 1$ in B_R and $|\nabla \varphi| \leq 2/R$. Multiplying (5.1) by $\partial_t u \varphi^2 \in T_u N$ and integrating by parts, there results

$$\int_0^T \int_{B_{2R}} \left\{ |\partial_t u|^2 \varphi^2 + \frac{1}{2} \frac{d}{dt} (|\nabla u|^2 \varphi^2) \right\} dx \, dt =$$

$$= -2 \int_0^T \int_{B_{2R}} \nabla u \partial_t u \nabla \varphi \varphi \, dx \, dt$$

$$\leq \int_0^T \int_{B_{2R}} |\partial_t u|^2 \varphi^2 \, dx \, dt + \int_0^T \int_{B_{2R}} |\nabla u|^2 |\nabla \varphi|^2 \, dx \, dt \ .$$

Hence

$$E(u(T); B_R) - E(u_0, B_{2R}) \leq \frac{1}{2} \int_{B_{2R}} (|\nabla u|^2 \varphi^2) \, dx \Big|_{t=0}^T$$

$$\leq 4R^{-2} \int_0^T \int_{B_{2R}} |\nabla u|^2 \, dx \, dt \leq 8 \frac{T}{R^2} E_0 \ ,$$

as claimed. □

Given some uniform control of the local energy densities, we can control higher derivatives, as well.

5.10 Lemma. *There exists $\varepsilon_1 = \varepsilon_1(\Sigma, N) > 0$ with the following properties: If $u \in C^2\big(B_{2R} \times [0, T[; N)$ with $E\big(u(t)\big) \leq E_0$ solves (5.1) on $B_{2R} \times [0, T[$, for some $R \in]0, \frac{1}{2}\imath_\Sigma[$, and if*

$$\sup_{x \in B_{2R}, \, 0 \leq t \leq T} E\big(u(t), B_R(x)\big) < \varepsilon_1 \;,$$

then we have

$$\int_0^T \int_{B_R} |\nabla^2 u|^2 \, dx \, dt \leq c \, E_0 \left(1 + \frac{T}{R^2}\right) \;,$$

where c only depends on Σ and N.

Proof. Let $Q = B_{2R} \times [0, T[$. Choose $\varphi \in C_0^\infty(B_{2R})$ such that $0 \leq \varphi \leq 1$, $\varphi \equiv 1$ on $B_R(x)$, $|\nabla \varphi| \leq 2/R$ and multiply (5.1) by $\Delta u \varphi^2$. Then from (5.5) we obtain that

$$\int_Q \partial_t \left(\frac{|\nabla u|^2 \varphi^2}{2}\right) + |\Delta u|^2 \varphi^2 \, dx \, dt$$

$$\leq c \int_Q |\nabla u|^2 |\Delta u| \varphi^2 \, dx \, dt + 2 \int_Q \partial_t u \nabla u \nabla \varphi \varphi \, dx \, dt$$

$$\leq c \int_Q \big(|\nabla u|^4 \varphi^2 + |\nabla u|^2 |\nabla \varphi|^2\big) \, dx \, dt + \frac{1}{8} \int_Q |\Delta u|^2 \varphi^2 \, dx \, dt$$

$$\leq \left(c \, c_0 \varepsilon_1 + \frac{1}{8}\right) \int_Q |\nabla^2 u|^2 \varphi^2 \, dx \, dt + c \, R^{-2} \int_Q |\nabla u|^2 \, dx \, dt \;,$$

where we have used the fact that $\partial_t u \nabla u = \Delta u \nabla u$, by (5.1), and the binomial inequality $2|ab| \leq \delta a^2 + \delta^{-1} b^2$ for any $a, b \in \mathbb{R}, \delta > 0$, to pass from the second to the third line; in the final estimate also Lemma 5.7 was used.

Integrating by parts twice and using the binomial inequality once again

$$\int_Q |\Delta u|^2 \varphi^2 \, dx \geq \frac{1}{2} \int_Q |\nabla^2 u|^2 \varphi^2 \, dx - c \int_Q |\nabla u|^2 |\nabla \varphi|^2 \, dx \;.$$

Hence for suitable $\varepsilon_1 > 0$ we obtain

$$\int_Q |\nabla^2 u|^2 \varphi^2 \, dx \, dt \leq 4 \int_\Sigma \frac{|\nabla u_0|^2 \varphi^2}{2} \, dx + c R^{-2} \int_0^T \int_{B_{2R}} |\nabla u|^2 \, dx \, dt$$

$$\leq C \, E_0 \left(1 + \frac{T}{R^2}\right) \;.$$

By choice of φ this implies the assertion of the lemma.

(Instead of (5.5) we might also have used the Bochner type estimate (5.3).) □

Finally, we need a regularity result:

5.11 Lemma. *If $Q = B_R \times]0, T[$ and $u \in C^\infty(Q; N)$ with $|\partial_t u|$, $|\nabla^2 u| \in L^2(Q)$ solves (5.1), then u extends smoothly to $B_R \times]0, T]$.*

Proof. In order to avoid using cut-off functions, consider $\Sigma = T^2$, $Q = \Sigma \times]0, T[$ for simplicity. Multiplying (5.1) by $-\nabla \cdot (\nabla u |\nabla u|^{s-2})$ at any (fixed) time $t < T$ and integrating by parts, on account of (5.5) we obtain the estimate

$$\int_\Sigma \left(\partial_t \left(\frac{|\nabla u|^s}{s} \right) + |\nabla^2 u|^2 |\nabla u|^{s-2} + \frac{s-2}{4} |\nabla(|\nabla u|^2)|^2 |\nabla u|^{s-2} \right) dx$$

$$\leq c \int_\Sigma |\nabla^2 u| |\nabla u|^s \, dx + c \int_\Sigma |\nabla u|^{s+2} \, dx$$

$$\leq \frac{1}{2} \int_\Sigma |\nabla^2 u|^2 |\nabla u|^{s-2} \, dx + c \int_\Sigma |\nabla u|^{s+2} \, dx \, ,$$

for all $t < T$, $s \geq 2$. That is, for $t_0 < t_1 < T$ we have

$$\sup_{t_0 < t < t_1} \int_\Sigma |\nabla u(t)|^s \, dx + \int_{t_0}^{t_1} \int_\Sigma |\nabla(|\nabla u|^{s/2})|^2 \, dx \, dt$$

$$\leq c \int_{t_0}^{t_1} \int_\Sigma |\nabla u|^{s+2} \, dx \, dt + \int_\Sigma |\nabla u(t_0)|^s \, dx$$

$$\leq c \left(\int_{t_0}^{t_1} \int_\Sigma |\nabla u|^{2s} \, dx \, dt \right)^{1/2} \left(\int_{t_0}^{t_1} \int_\Sigma |\nabla u|^4 \, dx \, dt \right)^{1/2}$$

$$+ \int_\Sigma |\nabla u(t_0)|^s \, dx$$

with constants depending only on Σ, N, and s.

Now note that by Lemma 5.7

$$\|v\|_{L^{2s}(Q_0)}^{2s} = \|v^{s/2}\|_{L^4(Q_0)}^4$$

$$\leq c_0 \sup_{t_0 \leq t \leq t_1} \|(v(t))^{s/2}\|_{L^2(\Sigma)}^2 \cdot \left(\|\nabla(v^{s/2})\|_{L^2(Q_0)}^2 + \|v^{s/2}\|_{L^2(Q_0)}^2 \right)$$

on $Q_0 = \Sigma \times [t_0, t_1]$, for any $s \geq 2$. Letting $v = \nabla u$, from the preceding inequality hence we obtain the estimate

$$\int_{t_0}^{t_1} \int_\Sigma |\nabla u|^{2s} dx \, dt \leq c \left(\sup_{t_0 < t < t_1} \int_\Sigma |\nabla u(t)|^s \, dx \right) \cdot$$

$$\cdot \left(\sup_{t_0 < t < t_1} \int_\Sigma |\nabla u(t)|^s \, dx + \int_{t_0}^{t_1} \int_\Sigma |\nabla(|\nabla u|^{s/2})|^2 dx \, dt \right)$$

$$\leq c \left(\int_{t_0}^{t_1} \int_\Sigma |\nabla u|^{2s} \, dx \, dt \right) \left(\int_{t_0}^{t_1} \int_\Sigma |\nabla u|^4 \, dx \, dt \right)$$

$$+ c \left(\int_\Sigma |\nabla u(t_0)|^s \, dx \right)^2 \, ,$$

for any $s \geq 2$. Given $\delta > 0$, by Lemma 5.7 we may choose $\gamma > 0$ such that for $0 < t_0 < t_1 < T$ with $t_1 - t_0 < \gamma$ there holds

$$\int_{t_0}^{t_1} \int_\Sigma |\nabla u|^4 \, dx \, dt \leq c_0 E(u_0) \int_{t_0}^{t_1} \int_\Sigma \left(|\nabla^2 u|^2 + |\nabla u|^2 \right) dx \, dt < \delta .$$

Choosing δ sufficiently small, the preceding inequality then gives the *a priori* estimate

$$\int_{t_0}^{t_1} \int_\Sigma |\nabla u|^{2s} \, dx \, dt \leq c \left(\int_\Sigma |\nabla u(t_0)|^s \, dx \right)^2 ,$$

for any $s \geq 2$, with a constant c depending only on the geometry and s. Hence it follows that $\nabla u \in L^q(\Sigma \times [t_0, T])$ for any $q < \infty$ and any $t_0 > 0$. But then the linear theory for (5.5) implies that $|\partial_t u|, |\nabla^2 u| \in L^q(\Sigma \times [t_0, T])$ for any $q < \infty$, and hence that ∇u is Hölder continuous on $\Sigma \times [t_0, T]$, for any $t_0 > 0$; see Ladyzhenskaya-Ural'ceva [1; Theorem IV.9.1 and Lemma II.3.3]. To obtain higher regularity we now use the explicit form (5.1') of the evolution problem (5.1) and the usual Schauder iteration argument.   ☐

Proof of Theorem 5.6. (1°) We first consider smooth initial data $u_0 \in C^\infty(\Sigma; N)$. By the local solvability of ordinary differential equations in Banach spaces, problem (5.1), (5.2) has a local solution $u \in C^\infty(\Sigma \times [0, T[; N)$; see also Hamilton [1; p. 122 ff.]. By Lemma 5.8 we have $\partial_t u \in L^2(\Sigma \times [0, T])$ and $E(u(t)) \leq E(u_0)$ uniformly in $t \in [0, T]$.

Moreover, if $R > 0$ can be chosen such that

$$\sup_{\substack{x \in \Sigma \\ 0 \leq t \leq T}} E(u(t); B_R(x)) < \varepsilon_1 ,$$

it follows from Lemma 5.10 that

$$\int_0^T \int_\Sigma |\nabla^2 u|^2 \, dx \, dt \leq c E(u_0) \left(1 + \frac{T}{R^2} \right) ,$$

and hence from Lemma 5.11 that u extends to a C^∞-solution of (5.1) on the closure $\Sigma \times [0, T]$. Thus, if $T > 0$ is maximal, that is, if u cannot be extended beyond T as a smooth solution of (5.1), there exist points $x_1, x_2, \ldots$ such that

$$\limsup_{t \nearrow T} E(u(t), B_R(x_k)) \geq \varepsilon_1 ,$$

for any $R > 0$ and any index k. Choose any finite collection x_k, $1 \leq k \leq K$, of such points and for any $R > 0$, $k = 1, \ldots, K$, let $t_k < T$ be chosen such that

$$E(u(t_k); B_R(x_k)) \geq \varepsilon_1/2 .$$

We may assume that $B_{2R}(x_k) \cap B_{2R}(x_j) = \emptyset$ $(j \neq k)$, $t_k \geq T - \frac{\varepsilon_1 R^2}{4 c_1 E_0} =: t_0$. By Lemma 5.9 therefore

$$E\big(u(t_0)\big) \geq \sum_{k=1}^{K} E\big(u(t_0); B_{2R}(x_k)\big)$$

(5.6)
$$\geq \sum_{k=1}^{K} \left(E\big(u(t_k); B_R(x_k)\big) - c_1 \frac{t_k - t_0}{R^2} E_0 \right)$$

$$\geq K \left(\varepsilon_1/2 - c_1 \frac{T - t_0}{R^2} E_0 \right) \geq K \varepsilon_1/4 \ .$$

Since by Lemma 5.8 we have $E\big(u(t_0)\big) \leq E(u_0)$ uniformly, this gives an upper bound

$$K \leq \frac{4E(u_0)}{\varepsilon_1}$$

for the number K of singular points $x_1, \ldots, x_K$ at $t = T$. Moreover for any $Q \subset\subset \Sigma \times [0, T] \setminus \{(x_1, T), \ldots, (x_K, T)\}$ there exists $R = R_Q > 0$ such that

$$\sup_{(x,t) \in Q} E\big(u(t); B_R(x)\big) < \varepsilon_1$$

and by Lemma 5.10, 5.11 our solution u extends to a C^∞-solution of (5.1) on $\Sigma \times [0, T] \setminus \{(x_1, T), \ldots (x_K, T)\}$.

(2°) For initial data $u_0 \in H^{1,2}(\Sigma; N)$ choose a sequence $u_{0m} \in C^\infty(\Sigma; N)$ approximating u_0 in $H^{1,2}(\Sigma; N)$. This is possible by Theorem 5.2. For each m let u_m be the associated solution of (5.1), (5.2), and $T_m > 0$ its maximal time of existence. Let $R_0 > 0$ be such that

$$\sup_{x \in \Sigma} E\big(u_0; B_{2R_0}(x)\big) \leq \varepsilon_1/4 \ .$$

Then this inequality will also hold with $\varepsilon_1/2$ instead of $\varepsilon_1/4$ for u_{0m}, $m \geq m_0$. Thus, by Lemma 5.9, for $T = \frac{\varepsilon_1 R_0^2}{4c_1 E(u_0)}$ we have

$$\sup_{\substack{x \in \Sigma \\ 0 \leq t \leq \min\{T_m, T\}}} E\big(u_m(t); B_{R_0}(x)\big) \leq \sup_{x \in \Sigma} E\big(u_{0m}, B_{2R_0}(x)\big) + c_1 \frac{T}{R_0^2} E(u_0)$$

$$\leq \varepsilon_1/2 + \varepsilon_1/4 < \varepsilon_1 \ ,$$

whence by Lemma 5.10 it follows that $\nabla^2 u_m$ is uniformly bounded in $L^2\big(\Sigma \times [0, t]\big)$ for $t \leq \min\{T, T_m\}$ in terms of $E(u_0)$ and R_0 only. By Lemma 5.11, therefore, the interval of existence of u_m is both open and closed in $[0, T]$; that is, $T_m \geq T > 0$. Moreover

$$\int_0^T \int_\Sigma |\nabla^2 u_m|^2 \ dx \ dt \leq c E(u_0) \left(1 + \frac{T}{R_0^2} \right)$$

uniformly, and we may assume that u_m converges weakly to a solution u of (5.1) with $|\partial_t u|, |\nabla^2 u|^2 \in L^2\big(\Sigma \times [0, T]\big)$, and such that $E\big(u(t)\big) \leq E(u_0)$, uniformly in $t \in [0, T]$. Since $|\partial_t u| \in L^2\big(\Sigma \times [0, T]\big)$, the solution u also attains its initial data u_0 continuously in $L^2(\Sigma; N)$; by the uniform energy bound $E(u(t)) \leq E(u_0)$, moreover, this is also true in the $H^{1,2}(\Sigma; N)$-topology.

By Lemma 5.11 we have $u \in C^\infty(\Sigma \times]0, T_1[, N)$ for some maximal $T_1 > T$ and by part i) of this proof u extends smoothly to $\Sigma \times]0, T_1] \setminus \{(x_1, T_1), \dots, (x_{K_1}, T_1)\}$ for some finite collection of singular points x_k, $1 \leq k \leq K_1$. Moreover, as $t \nearrow T_1$ we have $u(t) \to u_0^{(1)} \in H^{1,2}(\Sigma; N)$ weakly and strongly in $H^{1,2}_{loc}(\Sigma \setminus \{x_1, \dots, x_{K_1}\}; N)$. Thus by (5.6)

$$E(u_0^{(1)}) = \lim_{R \to 0} E\left(u_0^{(1)}; \Sigma \setminus \bigcup_{k=1}^{K_1} B_{2R}(x_k)\right)$$

$$\leq \lim_{R \to 0} \limsup_{t \nearrow T_1} \left(E(u(t); \Sigma) - \sum_{k=1}^{K} E(u(t); B_{2R}(x_k))\right)$$

$$\leq E(u_0) - K_1 \varepsilon_1/4 .$$

Let $u^{(0)} = u, T_0 = 0$. By iteration we now obtain a sequence $u^{(m)}$ of solutions to (5.1) on $\Sigma \times]T_m, T_{m+1}[$ with initial data $u_0^{(m)}$ and such that $u^{(m)}(t) \to u_0^{(m+1)}$ weakly in $H^{1,2}(\Sigma; N)$ as $t \nearrow T_{m+1}$. Moreover, $u^{(m)}$ has finitely many singularities $x_1^{(m)}, \dots, x_{K_{m+1}}^{(m)}$ at $t = T_{m+1}$ with

(5.7) $$\sum_{l=1}^{m+1} K_l \varepsilon_1/4 \leq E(u_0) ,$$

and $u^{(m)}(t) \to u_0^{(m+1)}$ smoothly away from $x_1^{(m)}, \dots, x_{K_{m+1}}^{(m)}$ as $t \nearrow T_{m+1}$, for any $m \in \mathbb{N}$. In particular, the total number of singularites of the flows $u^{(m)}$ is finite. Piecing the $u^{(m)}$ together, we obtain a weak solution u to (5.1), (5.2) which is smooth on $\Sigma \times]0, \infty[$ up to finitely many points, for any initial $u_0 \in H^{1,2}(\Sigma; N)$.

(3°) Asymptotics: If for some $T > 0$, $R > 0$

$$\sup_{\substack{x \in \Sigma \\ t > T}} E(u(t); B_R(x)) < \varepsilon_1 ,$$

then by Lemma 5.10 for any $t > T$

$$\int_t^{t+1} \int_\Sigma |\nabla^2 u|^2 \, dx \, dt \leq c \, E(u_0)(1 + R^{-2})$$

is uniformly bounded while by Lemma 5.8

$$\int_t^{t+1} \int_\Sigma |\partial_t u|^2 \, dx \, dt \to 0 .$$

Hence we may choose a sequence $t_m \to \infty$ such that $u_m = u(t_m) \to u_\infty$ weakly in $H^{2,2}(\Sigma; N)$, while $\partial_t u(t_m) \to 0$ in L^2. Moreover, by the Rellich-Kondrakov

theorem $u_m = u(t_m) \to u_\infty$ also strongly in $H^{1,p}(\Sigma; N)$ for any $p < \infty$. Testing
(5.1') with $\Delta(u_m - u_\infty)$ and integrating by parts, we obtain that

$$\int_\Sigma |\nabla^2(u_m - u_\infty)|^2 \, dx$$

$$\leq \left| \int_\Sigma \partial_t u \Delta(u_m - u_\infty) dx \right| + \left| \int_\Sigma \Gamma(u_m)(\nabla u_m, \nabla u_m)\Delta(u_m - u_\infty) dx \right| \to 0 \; ;$$

that is, $u(t_m) \to u_\infty$ also strongly in $H^{2,2}(\Sigma; N)$. Passing to the limit in (5.1),
it follows that u_∞ is harmonic.

To study the remaining case, let $t_m \to \infty$ such that $u(t_m) \to \tilde{u}_\infty$ weakly
in $H^{1,2}(\Sigma; N)$ and suppose there exist points $x_1, \ldots, x_K$ such that

(5.8) $$\liminf_{m \to \infty} \left(E\big(u(t_m); B_R(x_k)\big) \right) \geq \varepsilon_1/2 \; , \qquad 1 \leq k \leq K \; ,$$

for all $R > 0$. Choose $R > 0$ such that $B_{2R}(x_j) \cap B_{2R}(x_k) = \emptyset$ $(j \neq k)$. Then
we have

(5.9)
$$E\big(u(t_m)\big) \geq \sum_{k=1}^{K} E\big(u(t_m); B_R(x_k)\big)$$
$$\geq K\big(\varepsilon_1/4\big) \; ,$$

and $K \leq 4E(u_0)/\varepsilon_1$. Let $x_1, \ldots, x_K$ denote all concentration points of $(u(t_m))$
in the sense that (5.8) holds. Then, by Lemma 5.9, for any $x \notin \{x_1, \ldots, x_K\}$,
there exists $R > 0$ such that with $\tau = \frac{\varepsilon_1 R^2}{2c_1 E(u_0)}$ there holds

$$\liminf_{m \to \infty} \left(\sup_{t_m \leq t \leq t_m + \tau} E\big(u(t); B_R(x)\big) \right) \leq \liminf_{m \to \infty} E\big(u(t_m); B_{2R}(x)\big) + \varepsilon_1/2 \leq \varepsilon_1 \; .$$

By repeated selection of subsequences of (t_m) and in view of the uniform boundedness of the number K of concentration points (independent of the sequence
(t_m)), we can even achieve that for any such x there is $R > 0$ such that

$$\limsup_{m \to \infty} \left(\sup_{t_m \leq t \leq t_m + \tau} E\big(u(t); B_R(x)\big) \right) \leq \varepsilon_1 \; .$$

Choose $\Omega \subset\subset \Sigma \setminus \{x_1, \ldots, x_K\}$. By compactness, $\overline{\Omega}$ is covered by finitely many
such balls $B_R(x)$. Hence by Lemma 5.10

(5.10) $$\int_{t_m}^{t_m + \tau} \int_\Omega |\nabla^2 u|^2 \, dx \, dt \leq c\, E(u_0) \left(1 + \frac{\tau}{R^2}\right) = c\, E(u_0)$$

while

(5.11) $$\int_{t_m}^{t_m + \tau} \int_\Omega |\partial_t u|^2 \, dx \, dt \to 0 \; .$$

Exhausting $\Sigma \backslash \{x_1, \ldots, x_K\}$ by such domains Ω, thus we may choose a sequence $t'_m \in [t_m, t_m + \tau]$ such that

$$u_m = u(t'_m) \to u_\infty \qquad \text{weakly in } H^{2,2}_{loc}(\Sigma \backslash \{x_1, \ldots, x_K\}, N) \,,$$

where u_∞ is harmonic from $\Sigma \backslash \{x_1, \ldots, x_K\}$ into N.

Finally, by the regularity result of Sacks-Uhlenbeck [1; Theorem 3.6], u_∞ extends to a regular harmonic map $u_\infty \in C^\infty(\Sigma; N)$.

(4°) Singularites: Suppose $(\overline{x}, \overline{t})$ is singular in the sense that for any $R \in]0, \frac{1}{2} \iota_\Sigma[$ we have

$$\limsup_{t \nearrow \overline{t}} E\big(u(t); B_R(\overline{x})\big) \geq \varepsilon_1 \,,$$

if $\overline{t} < \infty$, respectively – with (t_m) as in iii) – that

$$\liminf_{m \to \infty} \big(E\big(u(t_m); B_R(\overline{x})\big)\big) \geq \varepsilon_{1/2} \,,$$

if $\overline{t} = \infty$. Since, by finiteness of the singular set, $\overline{x}$ is isolated among concentration points, if $R_m \to 0$ we may choose $\overline{x}_m \to \overline{x}$, $\overline{t}_m \nearrow \overline{t}$ such that for some $R_0 > 0$ we have

$$E\big(u(\overline{t}_m); B_{R_m}(\overline{x}_m)\big) = \sup_{\substack{x \in B_{2R_0}(\overline{x}) \\ \overline{t}_m - \tau_m \leq t \leq \overline{t}_m}} E\big(u(t); B_{R_m}(x)\big) = \varepsilon_1 / 4 \,,$$

where $\tau_m = \frac{\varepsilon_1 R_m^2}{16 c_1 E(u_0)}$. We may assume $\overline{x}_m \in B_{R_0}(\overline{x})$. Rescale

$$u_m(x, t) := u(\overline{x}_m + R_m x, \overline{t}_m + R_m^2 t)$$

and note that $u_m : B_{R_0/R_m} \times [t_0, 0] \to N$, with $t_0 = -\frac{\varepsilon_1}{16 c_1 E(u_0)}$, solves (5.1) classically with

$$\sup_{\substack{R_m|x| \leq R_0 \\ t_0 \leq t \leq 0}} E\big(u_m(t); B_1(x)\big) \leq E\big(u_m(0), B_1\big) = \varepsilon_1 / 4 \,,$$

$$\int_{t_0}^0 \int_{B_{R_0/R_m}} |\partial_t u_m|^2 \, dx \, dt \leq \int_{\overline{t}_m - \tau_m}^{\overline{t}_m} \int_\Sigma |\partial_t u|^2 \, dx \, dt \to 0 \,,$$

as $m \to \infty$. From Lemma 5.10 now it follows that

$$\int_{t_0}^0 \int_{B_{R_0/2R_m}} |\nabla^2 u_m|^2 \, dx \, dt \leq c$$

uniformly, whence for a sequence $s_m \in [t_0, 0]$, if we let

$$\tilde{u}_m(x) := u_m(x, s_m) = u\big(\overline{x}_m + R_m x, \overline{t}_m + R_m^2 s_m\big) \,,$$

we can achieve that $\partial_t u_m(\cdot, s_m) \to 0$ in L^2 and $\tilde{u}_m \to \tilde{u}$ weakly in $H^{2,2}_{loc}(\mathbb{R}^2; N)$ and strongly in $H^{1,2}_{loc}(\mathbb{R}^2; N)$. (In fact, as above we can even show that $\tilde{u}_m \to \tilde{u}$

strongly in $H^{2,2}_{loc}(\mathbb{R}^2; N)$.) Upon passing to the limit in (5.1) we see that $\tilde{u}$ is harmonic. Moreover, with error $o(1) \to 0$ as $m \to \infty$,

$$
\begin{aligned}
E(\tilde{u}; B_2) &= E\big(u_m(s_m), B_2\big) - o(1) \\
&\geq E\big(u_m(0), B_1\big) - c_1 s_m E(u_0) - o(1) \\
&\geq \varepsilon_1/4 - \varepsilon_1/16 - o(1) > 0 \qquad (m \geq m_0) ,
\end{aligned}
$$

and $\tilde{u} \not\equiv const$. Finally, $E(\tilde{u}) \leq \liminf_{m \to \infty} E(\tilde{u}_m) \leq E(u_0)$, and by the result of Sacks-Uhlenbeck [1; Theorem 3.6] $\tilde{u}$ extends to a harmonic map $\bar{u} \colon S^2 \to N$, as claimed.

Thus also $E(\bar{u}) = E(\tilde{u}) \geq \varepsilon_0$, and therefore for large m and any $R > 0$ we obtain that

$$
E\big(\tilde{u}_m; B_{R/R_m}\big) = E\big(u(\bar{t}_m + R_m^2 s_m); B_R(\bar{x}_m)\big) \geq \varepsilon_0 - o(1) ,
$$

where $o(1) \to 0$ as $m \to \infty$. It follows that estimates (5.6), (5.7), (5.9) may be improved, yielding the upper bound $K \leq E(u_0)/\varepsilon_0$ for the total number of singularities.

(5°) See Struwe [10; Lemma 3.12] for a proof of the uniqueness of the solution constructed in Theorem 5.6. $\qquad\qquad\qquad\qquad\qquad\qquad\qquad\qquad\qquad\qquad$ □

5.12 Remarks. (1°) While by Theorem 5.6 the class of functions $u \colon \Sigma \times \mathbb{R}_+ \to N$ with $\partial_t u \in L^2$, $E\big(u(t)\big) \leq E_0$, and which are regular on $\Sigma \times]0, \infty[$ up to a finite set of points, is a uniqueness class for (5.1), (5.2), it would be highly interesting if uniqueness held without this last condition. In particular, such a uniqueness result would imply the regularity of weakly harmonic maps $u \in H^{1,2}(\Sigma; N)$.

(2°) Chang [6] has recently obtained the analogue of Theorem 5.6 for the evolution problem (5.1), (5.2) on manifolds Σ with boundary $\partial \Sigma \neq \emptyset$ and with Dirichlet boundary data. His result – in the same way as we used Theorem 5.6 to prove Theorem 5.5 – can be used to prove for instance the existence and multiplicity results of Brezis-Coron [1] and Jost [1] for harmonic maps with boundary.

(3°) Results like Theorem 5.6 may be applied to prove the existence of minimal surfaces (or, more generally, of surfaces of constant mean curvature) with free boundaries; see Struwe [16].

(4°) An intriguing problem is whether singularities in the evolution problem for harmonic maps of surfaces (or for minimal surfaces with free boundaries supported by smooth hypersurfaces) actually may occur in finite time. Recall that by Theorem 5.3 for certain geometries and initial data the flow (5.1) cannot stay regular *and* converge asymptotically as $t \to \infty$. Generally, it is believed that it takes infinite time for singularities to develop. Recent results by Chang-Ding [1] and Grayson-Hamilton [1; Theorem C.1] lend some support to this conjecture.

(5°) See Schoen-Uhlenbeck [1], [2], Struwe [14], Chen-Struwe [1], Coron [3], Coron-Ghidaglia [1], Chen-Ding [1] for results on harmonic maps and the evolution problem (5.1), (5.2) in higher dimensions $\dim(\Sigma) > 2$.

Appendix A

Here, we collect without proof a few basic results about Sobolev spaces. A general reference to this topic is Gilbarg-Trudinger [1], or Adams [1].

Sobolev Spaces

Let Ω be a domain in $\mathbb{R}^n$. For $u \in L^1_{loc}(\Omega)$ and any multi-index $\alpha = (\alpha_1, ..., \alpha_n) \in \mathbb{N}_0^n$, with $|\alpha| = \sum_{j=1}^n \alpha_j$, define the distibutional derivative $D^\alpha u = \frac{\partial^{\alpha_1}}{\partial x_1^{\alpha_1}} \cdots \frac{\partial^{\alpha_1}}{\partial x_n^{\alpha_1}} u$ by letting

$$(A.1) \qquad < \varphi, D^\alpha u > = \int_\Omega (-1)^{|\alpha|} u\, D^\alpha \varphi \; dx,$$

for all $\varphi \in C_0^\infty(\Omega)$. We say $D^\alpha u \in L^p(\Omega)$, if there is a function $g_\alpha \in L^p(\Omega)$ satisfying

$$< \varphi, D^\alpha u > = < \varphi, g_\alpha > = \int_\Omega \varphi g_\alpha \; dx,$$

for all $\varphi \in C_0^\infty(\Omega)$. In this case we identify $D^\alpha u$ with $g_\alpha \in L^p(\Omega)$.

Given this, for $k \in \mathbb{N}_0$, $1 \le p \le \infty$, define the space

$$W^{k,p}(\Omega) = \left\{ u \in L^p(\Omega); D^\alpha u \in L^p(\Omega) \quad \text{for all} \quad \alpha\colon |\alpha| \le k \right\},$$

with norm

$$\|u\|^p_{W^{k,p}} = \sum_{|\alpha| \le k} \|D^\alpha u\|^p_{L^p}, \quad \text{if} \ 1 \le p < \infty,$$

respectively, with norm

$$\|u\|_{W^{k,\infty}} = \max_{|\alpha| \le k} \|D^\alpha u\|_{L^\infty}.$$

Note that the distributional derivative (A.1) is continuous with respect to convergence in $L^p(\Omega)$. Thus, many properties of $L^p(\Omega)$ carry over to $W^{k,p}(\Omega)$.

A.1 Theorem. *For any $k \in \mathbb{N}_0$, $1 \le p \le \infty$, $W^{k,p}(\Omega)$ is a Banach space. $W^{k,p}(\Omega)$ is reflexive if and only if $1 < p < \infty$. Moreover, $W^{k,2}(\Omega)$ is a Hilbert space with scalar product*

$$(u, v)_{W^{k,2}} = \sum_{|\alpha| \le k} \int_\Omega D^\alpha u\, D^\alpha v \; dx \ ,$$

inducing the norm above.

For $1 \le p < \infty$, $W^{k,p}(\Omega)$ also is separable. In fact, we have the following result due to Meyers and Serrin, see Adams [1; Theorem 3.16]:

A.2 Theorem. *For any $k \in \mathbb{N}_0, 1 \leq p < \infty$, the subspace $W^{k,p} \cap C^\infty(\Omega)$ is dense in $W^{k,p}(\Omega)$.*

The completion of $W^{k,p} \cap C^\infty(\Omega)$ in $W^{k,p}(\Omega)$ is denoted by $H^{k,p}(\Omega)$. By Theorem A.2, $W^{k,p}(\Omega) = H^{k,p}(\Omega)$. In particular, if $p = 2$ it is customary to use the latter notation.

Finally, $W_0^{k,p}(\Omega)$ is the closure of $C_0^\infty(\Omega)$ in $W^{k,p}(\Omega)$; $\mathcal{D}^{k,p}(\Omega)$ is the closure of $C_0^\infty(\Omega)$ in the norm

$$\|u\|_{\mathcal{D}^{k,p}}^p = \sum_{|\alpha|=k} \|D^\alpha u\|_{L^p}^p .$$

Hölder Spaces

A function $u\colon \Omega \subset \mathbb{R}^n \to \mathbb{R}$ is Hölder continuous with exponent $\beta > 0$ if

$$[u]^{(\beta)} = \sup_{x \neq y \in \Omega} \frac{|u(x) - u(y)|}{|x - y|^\beta} < \infty.$$

For $m \in \mathbb{N}_0$, $0 < \beta \leq 1$, denote

$$C^{m,\beta}(\Omega) = \left\{ u \in C^m(\Omega); D^\alpha u \text{ is Hölder continuous} \right.$$
$$\left. \text{with exponent } \beta \text{ for all } \alpha: |\alpha| = m \right\} .$$

If Ω is relatively compact, $C^{m,\beta}(\bar\Omega)$ becomes a Banach space in the norm

$$\|u\|_{C^{m,\beta}} = \sum_{|\alpha| \leq m} \|D^\alpha u\|_{L^\infty} + \sum_{|\alpha|=m} [D^\alpha u]^{(\beta)}.$$

The space $C^{m,\beta}(\Omega)$ on an open domain $\Omega \subset \mathbb{R}^n$ carries a Fréchet space topology, induced by the $C^{m,\beta}$-norms on compact sets exhausting Ω. Finally, we may set $C^{m,0}(\Omega) := C^m(\Omega)$.

Imbedding Theorems

Let $(X, \|\cdot\|_X), (Y, \|\cdot\|_Y)$ be Banach spaces. X is (continuously) embedded into Y (denoted $X \hookrightarrow Y$) if there exists an injective linear map $i\colon X \to Y$ and a constant C such that

$$\|i(x)\|_Y \leq C\|x\|_X, \text{ for all } x \in X.$$

In this case we will often simply identify X with the subspace $i(X) \subset Y$.

X is compactly embedded into Y if i maps bounded subsets of X into relatively compact subsets of Y.

For the spaces that we are primarily interested in we have the following results. First, from Hölder's inequality we obtain:

A.3 Theorem. *For $\Omega \in \mathbb{R}^n$ with Lebesgue measure $\mathcal{L}^n(\Omega) < \infty$, $1 \le p < q \le \infty$, we have $L^q(\Omega) \hookrightarrow L^p(\Omega)$. This ceases to be true if $\mathcal{L}^n(\Omega) = \infty$.*

For Hölder spaces, by the theorem of Arzéla-Ascoli there holds (see Adams [1; Theorems 1.30, 1.31]):

A.4 Theorem. *Suppose Ω is a relatively compact domain in $\mathbb{R}^n$, and let $m \in \mathbb{N}_0$, $0 \le \alpha < \beta \le 1$. Then $C^{m,\beta}(\bar{\Omega}) \hookrightarrow C^{m,\alpha}(\bar{\Omega})$ compactly.*

Finally, for Sobolev spaces we have (see Adams [1; Theorem 5.4]):

A.5 Theorem (Sobolev embedding theorem). *Let $\Omega \in \mathbb{R}^n$ be a bounded domain with Lipschitz boundary, $k \in \mathbb{N}$, $1 \le p \le \infty$. Then the following holds:*
(1°) If $kp < n$, we have $W^{k,p}(\Omega) \hookrightarrow L^q(\Omega)$ for $1 \le q \le \frac{np}{n-kp}$; the embedding is compact, if $q < \frac{np}{n-kp}$.
(2°) If $0 \le m < k - \frac{n}{p} < m + 1$, we have $W^{k,p}(\Omega) \hookrightarrow C^{m,\alpha}(\bar{\Omega})$, for $0 \le \alpha \le k - m - \frac{n}{p}$; the embedding is compact, if $\alpha < k - m - \frac{n}{p}$.

Compactness of the embedding $W^{k,p}(\Omega) \hookrightarrow L^q(\Omega)$ for $q < \frac{np}{n-kp}$ is a consequence of the Rellich-Kondrakov theorem, see Adams [1; Theorem 6.2].
Theorem A.5 is valid for $W_0^{k,p}(\Omega)$-spaces on arbitrary bounded domains Ω.

Density Theorem

By Theorem A.2 Sobolev functions can be approximated by functions enjoying any degree of smoothness in the interior of Ω. Some regularity condition on the boundary $\partial\Omega$ is necessary if smoothness up to the boundary is required:

A.6 Theorem. *Let $\Omega \subset \mathbb{R}^n$ be a bounded domain of class C^1, and let $k \in \mathbb{N}$, $1 \le p < \infty$. Then $C^\infty(\bar{\Omega})$ is dense in $W^{k,p}(\Omega)$.*

More generally, it suffices that Ω has the segment property, see for instance Adams [1; Theorem 3.18].

Trace and Extension Theorems

For a domain Ω with C^k-boundary $\partial\Omega = \Gamma$, $k \in \mathbb{N}$, $1 < p < \infty$, denote by $W^{k-\frac{1}{p},p}(\Gamma)$ the space of "traces" $u|_\Gamma$ of functions $u \in W^{k,p}(\Omega)$. We think of $W^{k-\frac{1}{p},p}(\Gamma)$ as the set of equivalence classes $\{\{u\} + W_0^{k,p}(\Omega); u \in W^{k,p}(\Omega)\}$, endowed with the trace norm

$$\|u|_\Gamma\|_{W^{k-\frac{1}{p},p}(\Gamma)} = \inf\{\|v\|_{W^{k,p}(\Omega)}; u - v \in W_0^{k,p}(\Omega)\}.$$

By this definition, $W^{k-\frac{1}{p},p}(\Gamma)$ is a Banach space. Moreover, in case $k = 1$, $p = 2$ the trace operator $u \mapsto u|_{\partial\Omega}$ is a linear isometry of the (closed) orthogonal complement of $H_0^{1,2}(\Omega)$ in $H^{1,2}(\Omega)$ onto $H^{\frac{1}{2},2}(\Gamma)$. By the open mapping theorem this provides a bounded "extension operator" $H^{\frac{1}{2},2}(\Gamma) \to H^{1,2}(\Omega)$. In general, we have:

A.7 Theorem. *For any Ω with C^k-boundary Γ, $k \in \mathbb{N}$, $1 < p < \infty$ there exists a continuous linear extension operator* ext: $W^{k-\frac{1}{p},p}(\Gamma) \to W^{k,p}(\Omega)$ *such that $\big(\mathrm{ext}(u)\big)\big|_\Gamma = u$, for all $u \in W^{k-\frac{1}{p},p}(\Gamma)$.*

See Adams [1; Theorem 7.53 and 7.55].

Covering $\partial\Omega = \Gamma$ by coordinate patches and defining Sobolev spaces $W^{k,p}(\Gamma)$ as before via such charts (see Adams [1; 7.51]) an equivalent norm for $W^{s,p}(\Gamma)$, where $k < s < k+1$, $s = k + \sigma$, is given by

$$\|u\|_{\tilde{W}^{s,p}} = \left\{ \|u\|_{W^{k,p}}^p + \sum_{|\alpha|=k} \int_\Gamma \int_\Gamma \frac{|D^\alpha u(x) - D^\alpha u(y)|^p}{|x-y|^{n-1+\sigma p}} \, dx \, dy \right\}^{1/p} ;$$

see Adams [1; Theorem 7.48].

From this, the following may be deduced:

A.8 Theorem. *Suppose Ω is a bounded domain with C^k-boundary Γ, $k \in \mathbb{N}$, $1 < p < \infty$. Then $W^{k,p}(\Gamma) \hookrightarrow W^{k-\frac{1}{p},p}(\Gamma) \hookrightarrow W^{k-1,p}(\Gamma)$ and both embeddings are compact.*

In particular, we have

$$(\text{A.2}) \qquad\qquad H^{1,2}(\Omega) \hookrightarrow L^2(\partial\Omega)$$

compactly, for any bounded domain of class C^1.

Poincaré Inequality

For a bounded domain Ω of diameter d and $u \in H_0^{1,2}(\Omega)$ there holds

$$(\text{A.3}) \qquad\qquad \int_\Omega |u|^2 dx \le d^2 \int_\Omega |\nabla u|^2 \, dx .$$

This follows immediately from Hölder's inequality and the mean value theorem. (It suffices to consider $\Omega \subset [0,d] \times \mathbb{R}^{n-1} = S$, $u \in C_0^\infty(\Omega) \subset C_0^\infty(S)$.) More generally, we state:

A.9 Theorem. *For any bounded domain Ω of class C^1 there exists a constant $c = c(\Omega)$ such that for any $u \in H^{1,2}(\Omega)$ we have*

$$\int_\Omega |u|^2 \, dx \le c \int_\Omega |\nabla u|^2 \, dx + c \int_{\partial\Omega} |u|^2 \, do .$$

Proof. The argument is modelled on Nečas [1; p. 18 f.]. Suppose by contradiction that for a sequence (u_m) in $H^{1,2}(\Omega)$ there holds

$$(\text{A.4}) \qquad \|u_m\|_{L^2(\Omega)}^2 \ge m \left(\|\nabla u_m\|_{L^2(\Omega)}^2 + \|u_m\|_{L^2(\partial\Omega)}^2 \right) .$$

By homogeneity, we may normalize

$$\|u_m\|^2_{L^2(\Omega)} = 1 \ .$$

But then (u_m) is bounded in $H^{1,2}(\Omega)$, and we may assume that $u_m \to u$ weakly. Moreover, by Theorem A.5.$(1°)$ it follows that $u_m \to u$ strongly in $L^2(\Omega)$ and by (A.2) also $u_m|_{\partial\Omega} \to u|_{\partial\Omega}$ in $L^2(\partial\Omega)$.

But (A.4) also implies that $\nabla u_m \to 0$ in $L^2(\Omega)$, and $u_m|_{\partial\Omega} \to 0$ in $L^2(\partial\Omega)$. Hence $u \in H^{1,2}_0(\Omega)$ and satisfies $\nabla u = 0$. By (A.3) therefore $u \equiv \text{const.} = 0$, while $\|u\|_{L^2(\Omega)} = 1$. Contradiction. □

In the same spirit the following variant of Poincaré's inequality may be derived.

A.10 Theorem. *Let $A_R = B_{2R}(0) \setminus B_R(0) \subset \mathbb{R}^n$ denote the annulus of size R in $\mathbb{R}^n$. There exists a constant $c = c(n,p)$ such that for any $R > 0$, any $u \in H^{1,p}(A_R)$ there holds*

$$\int_{A_R} |u - \bar{u}_R|^p \, dx \le c \, R^p \int_{A_R} |\nabla u|^p \, dx \ ,$$

where $\bar{u}_R$ denotes the mean of u over the annulus A_R.

Proof. Scaling with R we may assume that $R = 1$, $A_R = A_1 =: A$. Moreover, it suffices to consider $\bar{u} = 0$. If for a sequence (u_m) in $H^{1,p}(A)$ with $\bar{u}_m = 0$ we have

$$1 = \int_A |u_m|^p \, dx \ge m \int_A |\nabla u_m|^p \, dx \ ,$$

by Theorem A.5 we conclude that $u_m \to u \equiv \text{const.} = \bar{u} = 0$ in $L^p(A)$. Contradiction. □

Appendix B

In this appendix we recall some fundamental estimates for elliptic equations. A basic reference is Gilbarg-Trudinger [1].

On a domain $\Omega \subset \mathbb{R}^n$ we consider second order elliptic differential operators of the form

$$(B.1) \qquad L\,u = -a_{ij}\frac{\partial^2}{\partial x_j \partial x_j}u + b_i\frac{\partial}{\partial x_i}u + cu,$$

or in divergence form

$$(B.2) \qquad L\,u = -\frac{\partial}{\partial x_i}\left(a_{ij}\frac{\partial}{\partial x_j}u\right) + c\,u\ ,$$

with bounded coefficients $a_{ij} = a_{ji}, b_i$, and c satisfying the ellipticity condition

$$a_{ij}\xi_i\xi_j \geq \lambda|\xi|^2$$

with a uniform constant $\lambda > 0$, for all $\xi \in \mathbb{R}^n$. Repeated indeces are summed 1 to n. The standard example is the operator $L = -\Delta$. If $a_{ij} \in C^1$, then any operator of type (B.2) also falls into category (B.1) with $b_j = -\frac{\partial}{\partial x_i}a_{ij}$.

Schauder Estimates

Let us first consider the (classical) C^α-setting; see Gilbarg-Trudinger [1; Theorems 6.2, 6.6].

B.1 Theorem. *Let L be an elliptic operator of type (B.1), with coefficients of class C^α, and let $u \in C^2(\Omega)$. Suppose $L\,u = f \in C^\alpha(\bar{\Omega})$. Then $u \in C^{2,\alpha}(\Omega)$, and for any $\Omega' \subset\subset \Omega$ we have*

$$(B.3) \qquad \|u\|_{C^{2,\alpha}(\Omega')} \leq C\left(\|u\|_{L^\infty(\Omega)} + \|f\|_{C^\alpha(\bar{\Omega})}\right).$$

If in addition Ω is of class $C^{2+\alpha}$, and if $u \in C^o(\bar{\Omega})$ coincides with a function $u_o \in C^{2+\alpha}(\bar{\Omega})$ on $\partial\Omega$, then $u \in C^{2,\alpha}(\bar{\Omega})$ and

$$(B.4) \qquad \|u\|_{C^{2,\alpha}(\bar{\Omega})} \leq C\left(\|u\|_{L^\infty(\Omega)} + \|f\|_{C^\alpha(\bar{\Omega})} + \|u_o\|_{C^{2,\alpha}(\bar{\Omega})}\right)$$

with constants C possibly depending on L, Ω, n, α, and – in the first case – on Ω'.

L^p-Theory

For solutions in Sobolev spaces the Calderón-Zygmund inequality is the counterpart of the Schauder estimates for classical solutions; see Gilbarg-Trudinger [1; Theorems 9.11, 9.13].

B.2 Theorem. *Let L be elliptic of type (B.1) with continuous coefficients a_{ij}. Suppose $u \in H^{2,p}_{loc}(\Omega)$ satisfies $L\,u = f$ in Ω with $f \in L^p(\Omega)$, $1 < p < \infty$. Then for any $\Omega' \subset\subset \Omega$ we have*

$$\text{(B.5)} \qquad \|u\|_{H^{2,p}(\Omega')} \leq C\left(\|u\|_{L^p(\Omega)} + \|f\|_{L^p(\Omega)}\right) .$$

If in addition Ω is of Class $C^{1,1}$, and if there exists a function $u_o \in H^{2,p}(\Omega)$ such that $u - u_o \in H^{1,p}_o(\Omega)$, then

$$\text{(B.6)} \qquad \|u\|_{H^{2,p}(\Omega)} \leq C\left(\|u\|_{L^p(\Omega)} + \|f\|_{L^p(\Omega)} + \|u_o\|_{H^{2,p}(\Omega)}\right)$$

The constants C may depend on L, Ω, n, p, and – in the first case – on Ω'.

Weak Solutions

Let L be elliptic of divergence type (B.2), $f \in H^{-1}(\Omega)$. A function $u \in H^{1,2}_0(\Omega)$ weakly solves the equation $L\,u = f$ if

$$< \varphi, L\,u > = \int_\Omega \left(a_{ij}\frac{\partial}{\partial x_i}u\,\frac{\partial}{\partial x_j}\varphi + c\,u\varphi\right)dx - \int_\Omega f\varphi dx = 0, \text{ for all } \varphi \in C^\infty_o(\Omega).$$

The integral

$$\mathcal{L}(u, \varphi) = \int_\Omega \left(a_{ij}\frac{\partial}{\partial x_i}u\,\frac{\partial}{\partial x_j}\varphi + c\,u\varphi\right)dx$$

continuously extends to a symmetric bilinear form $\mathcal{L}$ on $H^{1,2}_0(\Omega)$, the Dirichlet form associated with the operator L.

A Regularity Result

As an application we consider the equation

$$\text{(B.7)} \qquad -\Delta u = g(\cdot, u) \text{ in } \Omega,$$

on a domain $\Omega \subset \mathbb{R}^n$, with a Caratheodory function $g: \Omega \times \mathbb{R} \to \mathbb{R}$; that is, assuming $g(x, u)$ is measurable in $x \in \Omega$ and continuous in $u \in \mathbb{R}$. Moreover, we will assume that g satisfies the growth condition

$$\text{(B.8)} \qquad |g(x, u)| \leq C\left(1 + |u|^p\right)$$

where $p \leq \frac{n+2}{n-2}$, if $n \geq 3$. By (B.8) and Theorem A.5, for any $u \in H^{1,2}(\Omega)$ the composed function $g(\cdot, u(\cdot)) \in H^{-1}(\Omega)$; see Theorem C.2. The following estimate is essentially due to Brezis-Kato [1], based on Moser's [1] iteration technique.

B.3 Lemma. *Let Ω be a domain in $\mathbb{R}^n$ and let $g : \Omega \times \mathbb{R} \to \mathbb{R}$ be a Carathéodory function such that for almost every $x \in \Omega$ there holds*

$$(1°) \qquad\qquad |g(x,u)| \leq a(x)\bigl(1 + |u|\bigr)$$

with a function $a \in L_{loc}^{n/2}(\Omega)$. Also let $u \in H_{loc}^{1,2}(\Omega)$ be a weak solution of equation (B.7). Then $u \in L_{loc}^q(\Omega)$ for any $q < \infty$. If $u \in H_0^{1,2}(\Omega)$, and $a \in L^{n/2}(\Omega)$, then $u \in L^q(\Omega)$ for any $q < \infty$.

Proof. Choose $\eta \in C_0^\infty(\Omega)$ and for $s \geq 0$, $L \geq 0$ let $\varphi = \varphi_{s,L} = u \min\{|u|^{2s}, L^2\}\eta^2 \in H_0^{1,2}(\Omega)$, with supp $\varphi \subset\subset \Omega$. Testing (B.7) with φ we obtain

$$\int_\Omega |\nabla u|^2 \min\{|u|^{2s}, L^2\}\eta^2\, dx + \frac{s}{2}\int_{\{x \in \Omega\,;\, |u(x)|^s \leq L\}} \bigl|\nabla(|u|^2)\bigr|^2 |u|^{2s-2}\eta^2\, dx$$

$$\leq -2\int_\Omega \nabla u\, u\, \min\{|u|^{2s}, L^2\}\nabla\eta\, \eta\, dx$$

$$+ \int_\Omega a\bigl(1 + 2|u|^2\bigr)\min\{|u|^{2s}, L^2\}\eta^2\, dx$$

$$\leq \frac{1}{2}\int_\Omega |\nabla u|^2 \min\{|u|^{2s}, L^2\}\eta^2\, dx$$

$$+ c\int_\Omega |u|^2 \min\{|u|^{2s}, L^2\}|\nabla\eta|^2\, dx$$

$$+ 3\int_\Omega |a|\,|u|^2 \min\{|u|^{2s}, L^2\}\eta^2\, dx + \int_\Omega |a|\eta^2\, dx\ .$$

Suppose $u \in L_{loc}^{2s+2}(\Omega)$. Then we may conclude that with constants depending on the L^{2s+2}-norm of u, restricted to supp(η), there holds

$$\int_\Omega \Bigl|\nabla\bigl(u \min\{|u|^s, L\}\eta\bigr)\Bigr|^2\, dx \leq$$

$$\leq c + c\cdot\int_\Omega |a|\,|u|^2 \min\{|u|^{2s}, L^2\}\eta^2\, dx$$

$$\leq c + cK\int_\Omega |u|^2 \min\{|u|^{2s}, L^2\}\eta^2\, dx$$

$$+ c\int_{\{x \in \Omega\,;\, |a(x)| \geq K\}} |a|\,|u|^2 \min\{|u|^{2s}, L^2\}\eta^2\, dx$$

$$\leq c(1+K) + \left(c\cdot\int_{\{x \in \Omega\,;\, |a(x)| \geq K\}} |a|^{n/2}\, dx\right)^{2/n}\cdot$$

$$\cdot\left(\int_\Omega \bigl|u \min\{|u|^s, L\}\eta\bigr|^{\frac{2n}{n-2}}\, dx\right)^{\frac{n-2}{n}}$$

$$\leq c(1+K) + \varepsilon(K)\cdot\int_\Omega \Bigl|\nabla\bigl(u \min\{|u|^s, L\}\eta\bigr)\Bigr|^2\, dx\ ,$$

where

$$\varepsilon(K) = \left(\int_{\{x \in \Omega \ ; \ |a(x)| \geq K\}} |a|^{n/2} \, dx \right)^{2/n} \to 0 \qquad (K \to \infty) \ .$$

Fix K such that $\varepsilon(K) = \frac{1}{2}$ and observe that for this choice of K (and s as above) we now may conclude that

$$\int_{\{x \in \Omega \ ; \ |u(x)|^s \leq L\}} \left| \nabla \left(|u|^{s+1} \eta \right) \right|^2 \, dx \leq c \int_\Omega \left| \nabla \left(u \min \{ |u|^s, L \} \eta \right) \right|^2 \, dx$$

$$\leq c(1 + K)$$

remains uniformly bounded in L. Hence we may let $L \to \infty$ to derive that

$$|u|^{s+1} \eta \in H_0^{1,2}(\Omega) \hookrightarrow L^{2^*}(\Omega) \ ;$$

that is, $u \in L_{loc}^{\frac{(2s+2)n}{n-2}}(\Omega)$.

Now iterate, letting $s_0 = 0$, $s_i + 1 = (s_{i-1} + 1)\frac{n}{n-2}$, if $i \geq 1$, to obtain the conclusion of the lemma. If $u \in H_0^{1,2}(\Omega)$, we may let $\eta = 1$ to obtain that $u \in L^q(\Omega)$ for all $q < \infty$. $\qquad \square$

To apply Lemma B.3 note that if $u \in H_{loc}^{1,2}(\Omega)$ weakly solves (B.7) with a Carathéodory function g with polynomial growth

$$g(x, u) \leq C \left(1 + |u|^{p-1} \right) \ ,$$

and if $p \leq \frac{2n}{n-2}$ for $n > 2$, then u weakly solves the equation

$$-\Delta u = a(x) \left(1 + |u| \right)$$

with

$$a(x) = \frac{g(x, u(x))}{1 + |u(x)|} \in L_{loc}^{n/2}(\Omega) \ .$$

By Lemma B.3 therefore $u \in L_{loc}^q(\Omega)$, for any $q < \infty$. In view of our growth condition for g this implies that $\Delta u = -g(u) \in L_{loc}^q(\Omega)$ for any $q < \infty$. Thus, by the Caldéron-Zygmund inequality, Theorem B.2, $u \in H_{loc}^{2,q}(\Omega)$, for any $q < \infty$, whence also $u \in C_{loc}^{1,\alpha}(\Omega)$ by the Sobolev embedding theorem, Theorem A.5, for any $\alpha < 1$. Moreover, if $u \in H_0^{1,2}(\Omega)$, and if $\partial\Omega \in C^2$, by the same token it follows that $u \in H^{2,q} \cap H_0^{1,2}(\Omega) \hookrightarrow C^{1,\alpha}(\overline{\Omega})$. Now we may proceed using Schauder theory. In particular, if g is Hölder continuous, then $u \in C^2(\Omega)$ and is a non-constant, classical C^2-solution of equation (B.7). Finally, if g and $\partial\Omega$ are smooth, higher regularity (up to the boundary) can be obtained by iterating the Schauder estimates.

Maximum Principle

A basic tool for proving existence of solutions to elliptic boundary value problems in Hölder spaces is the maximum principle.

We state this in a form due to Walter [1; Theorem 2], allowing for more general coefficients c in the operator L than in classical versions.

B.4 Theorem. *Suppose L is elliptic of type (B.1) on a domain Ω and suppose $u \in C^2(\Omega) \cap C^1(\bar{\Omega})$ satisfies*

$$L\,u \geq 0 \ \text{in} \ \Omega, \ \text{and} \ u \geq 0 \ \text{on} \ \partial\Omega.$$

Moreover, suppose there exists $h \in C^2(\Omega) \cap C^0(\bar{\Omega})$ such that

$$L\,h \geq 0 \ \text{in} \ \Omega, \ \text{and} \ h > 0 \ \text{on} \ \Omega.$$

Then either $u > 0$ in Ω, or $u = \beta h$ for some $\beta \leq 0$.

In particular, let L be given by (B.2) with coefficients $a_{ij} \in C^{1,\alpha}(\bar{\Omega}), c \in C^{\alpha}(\bar{\Omega})$. Then L is self-adjoint and possesses a complete set of eigenfunctions (φ_j) in $H_0^{1,2}(\Omega) \cap C^{2,\alpha}(\bar{\Omega})$ with eigenvalues $\lambda_1 < \lambda_2 \leq \lambda_3 \leq \dots$. Moreover, $\varphi_1 > 0$ in Ω. Suppose that the first Dirichlet eigenvalue

$$\lambda_1 = \inf_{u \neq 0} \frac{(L\,u, u)_{L^2}}{(u, u)_{L^2}} > 0.$$

Then in Theorem B.4 we may choose $h = \varphi_1$, and the theorem implies that any solution $u \in C^2(\Omega) \cap C^1(\bar{\Omega})$ of $L\,u \geq 0$ in Ω, $u \geq 0$ on Ω either is positive throughout Ω or vanishes identically.

The strong maximum principle is based on the Hopf boundary maximum principle; see Walter [1; p. 294]:

B.5 Theorem. *Let L be elliptic of type (B.1) on the ball $B = B_R(0) \subset \mathbb{R}^n$, with $c \geq 0$. Suppose $u \in C^2(B) \cap C^1(\bar{B})$ satisfies $Lu \geq 0$ in B, $u \geq 0$ on ∂B, and $u \geq \gamma > 0$ in $B_\rho(0)$ for some $\rho < R, \gamma > 0$. Then there exists $\delta = \delta(L, \gamma, \rho, R) > 0$ such that*

$$u(x) \geq \delta\left(R - |x|\right) \qquad \text{in} \ B\ .$$

In particular, if $u(x_0) = 0$ for some $x_0 \in \partial B$, then the interior normal derivative of u is strictly positive.

Weak Maximum Principle

For weak solutions of elliptic equations we have the following analogue of Theorem B.4.

B.6 Theorem. *Suppose L is elliptic of type (B.2) and suppose the Dirichlet form of $\mathcal{L}$ is positive definite on $H_0^{1,2}(\Omega)$ in the sense that*

$$\mathcal{L}(u, u) > 0 \ \text{for all} \ u \in H_0^{1,2}(\Omega), \ u \neq 0.$$

Then, if $u \in H^{1,2}(\Omega)$ weakly satisfies $L\,u \geq 0$ in the sense that

$$\mathcal{L}(u, \varphi) \geq 0 \quad \text{for all non-negative} \ \varphi \in H_0^{1,2}(\Omega),$$

and $u \geq 0$ on ∂B, it follows that $u \geq 0$ in Ω.

Proof. Choose $\varphi = u_- = \max\{-u, 0\} \in H_0^{1,2}(\Omega)$. Then

$$0 \leq \mathcal{L}(u, u_-) = -\mathcal{L}(u_-, u_-) \leq 0$$

with equality if and only if $u_- \equiv 0$; that is $u \geq 0$. □

Theorem B.6 can be used to strengthen the boundary maximum principle Theorem B.5:

B.7 Theorem. *Let L satisfy the hypotheses of Theorem B.6 in $\Omega = B_R(0) = B \subset \mathbb{R}^n$ with coefficients $a_{ij} \in C^1$. Suppose $u \in C^2(B) \cap C^1(\overline{B})$ satisfies $Lu \geq 0$ in B, $u \geq 0$ on ∂B and $u \geq \gamma > 0$ in $B_\rho(0)$ for some $\rho < R, \gamma > 0$. Then there exists $\delta = \delta(L, \gamma, \rho, R) > 0$ such that $u(x) \geq \delta(R - |x|)$ in B.*

Proof. We adapt the proof of Walter [1; p. 294]. For large $C > 0$ the function $v = \exp(C(R^2 - |x|^2)) - 1$ satisfies $Lv \leq 0$. Moreover, for small $\varepsilon > 0$ the function $w = \varepsilon v$ satisfies $w \leq u$ for $|x| \leq \rho$ and $|x| = R$. Hence, Theorem B.6 – applied to $u - w$ on $B \setminus B_\rho(0)$ – shows that $u \geq w$ in $B \setminus B_\rho(0)$. □

Application

As an application, consider the operator $L = -\Delta - \delta$, where $\delta < \lambda_1$, the first Dirichlet eigenvalue of $-\Delta$ on Ω. Let $u \in H_0^{1,2}(\Omega)$ or $u \in C^2(\Omega) \cap C^0(\bar{\Omega})$ weakly satisfy $Lu \leq C_o$ in $\Omega, u \leq 0$ on $\partial\Omega$, and choose $v(x) = C(C - |x - x_0|^2)$ with $x_0 \in \Omega$ and C sufficiently large to achieve that $v > 0$ on $\bar{\Omega}$ and $Lv \geq C_0$. Then $w = v - u$ satisfies

$$Lw \geq 0 \text{ in } \Omega, \quad w > 0 \text{ on } \partial\Omega ,$$

and hence w is non-negative throughout Ω. Thus

$$u \leq v \quad \text{ in } \Omega .$$

More generally, results like Theorem B.4 or B.5 can be used to obtain L^∞- or even Lipschitz a priori bounds of solution to elliptic boundary value problems by comparing with suitably constructed "barriers".

Appendix C

In this appendix we discuss the issue of (partial) differentiability of variational integrals of the type

$$(C.1) \qquad E(u) = \int_\Omega F\big(x, u(x), \nabla u(x)\big)\, dx,$$

where $u \in H^{1,2}(\Omega)$, for simplicity. Differentiability properties will crucially depend on growth conditions for F.

Fréchet Differentiability

A functional E on a Banach space X is Fréchet differentiable at a point $u \in X$ if there exists a bounded linear map $DE(u) \in X$, called the differential of E at u, such that

$$\frac{|E(u+v) - E(u) - DE(u)v|}{\|v\|_X} \to 0$$

as $\|v\|_X \to 0$. E is of class C^1, if the map $u \mapsto DE(u)$ is continuous.

C.1 Theorem. *Suppose $F\colon \Omega \times \mathbb{R} \times \mathbb{R}^n \to \mathbb{R}$ is measurable in $x \in \Omega$, continuously differentiable in $u \in \mathbb{R}$ and $p \in \mathbb{R}^n$, with $F_u = \frac{\partial}{\partial u} F$, $F_p = \frac{\partial}{\partial p} F$, and the following growth conditions are satisfied:*
(1°) $|F(x,u,p)| \le C(1 + |u|^{s_1} + |p|^2)$, where $s_1 \le \frac{2n}{n-2}$, if $n \ge 3$,
(2°) $|F_u(x,u,p)| \le C(1 + |u|^{s_2} + |p|^{t_2})$, where $t_2 \le 2$, if $n \le 2$, $s_2 \le \frac{n+2}{n-2}$, $t_2 \le \frac{n+2}{n}$, if $n \ge 3$,
(3°) $|F_p(x,u,p)| \le C(1 + |u|^{s_3} + |p|)$, where $s_3 \le \frac{n}{n-2}$, if $n \ge 3$.
Then (C.1) defines a C^1-functional E on $H^{1,2}(\Omega)$. Moreover, $DE(u)$ is given by

$$< v, DE(u) >= \int_\Omega \big(F_u(x, u, \nabla u)v + F_p(x, u, \nabla u) \cdot \nabla v\big)\, dx \ .$$

Theorem C.1 applies for example to the functional

$$G(u) = \int_\Omega |u|^p\, dx$$

with $p \le \frac{2n}{n-2}$, if $n \ge 3$, or Dirichlet's integral

$$E(u) = \frac{1}{2} \int_\Omega |\nabla u|^2\, dx.$$

Theorem C.1 rests on a result by Krasnoselskii [1; Theorem I.2.1]. We state this result for functions $g\colon \Omega \times \mathbb{R}^m \to \mathbb{R}$, which includes the special class of functions considered in Theorem C.1 if we let $U = (u, \nabla u)$, $g(x, U) = F(x, u, \nabla u)$. To ensure measurability of composed functions $g(x, u(x))$, with $u \in L^p$, we assume $g\colon \Omega \times \mathbb{R}^m \to \mathbb{R}$ is a Carathéodory function; that is, g is measurable in $x \in \Omega$ and continuous in $u \in \mathbb{R}^m$.

C.2 Theorem. *Suppose $g\colon \Omega \times \mathbb{R}^m \to \mathbb{R}$ is a Carathéodory function satisfying the growth condition*
(1°) $|g(x, u)| \leq C(1 + |u|^s)$ for some $s \geq 1$.
Then the operator
$$u \mapsto (g(\cdot, u(\cdot)))$$
is continuous from $L^{sp}(\Omega)$ into $L^p(\Omega)$ for any p, $1 \leq p < \infty$.

Theorem C.2 asserts that Nemitskii operators – that is, evaluation operators like (C.1) – are continuous if they are bounded. For nonlinear operators this is quite remarkable.

Using this result, Theorem C.1 follows quite naturally from the Sobolev embedding theorem, Theorem A.5. To get a flavor of the proof, we establish continuity of the derivative of a functional E as in Theorem C.1. For $u_0, u \in H_0^{1,2}(\Omega)$, we estimate

$$\|DE(u) - DE(u_0)\| = \sup_{\substack{v \in H_0^{1,2} \\ \|v\|_{H_0^{1,2}} \leq 1}} |< v, DE(u) - DE(u_0) >|$$

$$\leq \sup_v \int_\Omega \left(|F_u(x, u, \nabla u) - F_u(x, u_0, \nabla u_0)| \, |v| \right) dx$$

$$+ \sup_v \int_\Omega \left(|F_p(x, u, \nabla u) - F_p(x, u_0, \nabla u_0)| \, |\nabla v| \right) dx$$

$$\leq \sup_v \left(\int_\Omega |F_u(x, u, \nabla u) - F_u(x, u_0, \nabla u_0)|^{\frac{2n}{n+2}} \, dx \right)^{\frac{n+2}{2n}} \cdot \left(\int_\Omega |v|^{\frac{2n}{n-2}} \, dx \right)^{\frac{n-2}{2n}}$$

$$+ \sup_v \left(\int_\Omega |F_p(x, u, \nabla u) - F_p(x, u_0, \nabla u_0)|^2 \, dx \right)^{\frac{1}{2}} \cdot \left(\int_\Omega |\nabla v|^2 \, dx \right)^{\frac{1}{2}}$$

if $n \geq 3$ – which we will assume from now on for simplicity. Now, by Theorem A.5 the integrals involving v are uniformly bounded for $v \in H_0^{1,2}(\Omega)$ with $\|v\|_{H_0^{1,2}} \leq 1$. By our growth conditions (2°) and (3°), moreover, F_u (respectively F_p) can be estimated like

$$|F_u(x, u, \nabla u)|^{\frac{2n}{n+2}} \leq C \left(1 + |u|^{\frac{2n}{n-2}} + |\nabla u|^2 \right),$$

and by Theorem C.2 it follows that $DE(u) \to DE(u_0)$, if $u \to u_0$, as desired.

Natural Growth Conditions

Conditions (1°)–(3°) of Theorem C.1 require a special structure of the function F; for instance, terms involving $|\nabla u|^2$ cannot involve coefficients depending on u. Consider for example the functional in Section I.1.5 given by

$$F(x, u, p) = g_{ij}(u)p_i p_j.$$

Note that F_u has the same growth as F in p. More generally, for analytic functions F such that

(C.2) $$|p|^2 \leq F(x, u, p) \leq C(|u|)(1 + |p|^2)$$

one would expect the following growth conditions:

(C.3) $$F_u(x, u, p) \leq C(|u|)(1 + |p|^2)$$

(C.4) $$F_p(x, u, p) \leq C(|u|)(1 + |p|)$$

for $x \in \Omega$, $u \in \mathbb{R}$, and $p \in \mathbb{R}^n$.

Under these growth assumptions, in general a functional E given by (C.1) cannot be Fréchet differentiable in $H^{1,2}(\Omega)$ any more. However, minimizers (in $H_0^{1,2}(\Omega)$, say) still may exist, compare Theorem I.1.5. Is it still possible to derive necessary conditions in the form of Euler-Lagrange equations? – The answer to this question may be positive, if we only consider a restricted set of minimizers and a narrower class of "testing functions", that is of admissible variations:

C.3 Theorem. *Suppose E is given by (C.1) with a Carathéodory function F, of class C^1 in u and p, satisfying the natural growth conditions (C.2)–(C.4). Then, if $u, \varphi \in H^{1,2} \cap L^\infty(\Omega)$, the directional derivative of E at u in direction φ exists and is given by:*

$$\frac{d}{d\varepsilon} E(u + \varepsilon\varphi)\Big|_{\varepsilon=0} = \int_\Omega \left(F_u(x, u, \nabla u)\varphi + F_p(x, u, \nabla u) \cdot \nabla\varphi\right) dx \ .$$

In particular, at a minimizer $u \in H^{1,2} \cap L^\infty$ of E, with F satisfying (C.2)–(C.4), the Euler-Lagrange equations are weakly satisfied in the sense that

$$\int_\Omega \left(F_u(x, u, \nabla u) \cdot \varphi + F_p(x, u, \nabla u)\nabla\varphi\right) dx = 0$$

holds for all $\varphi \in H_0^{1,2} \cap L^\infty(\Omega)$.

Note that the assumption $u \in L^\infty$ often arises naturally, as in Theorem I.1.5. Sometimes, boundedness of minimizers may also be derived a posteriori. For further details, we refer to Giaquinta [1] or Morrey [4].

References

Acerbi, E., Fusco, N.:
[1] Semicontinuity problems in the calculus of variations. Arch. Rat. Mech. Anal.
86 (1984) 125–145

Adams, R.A.:
[1] Sobolev spaces. Academic Press, New York-San Francisco-London (1975)

Ahmad, S., Lazer, A.C.-Paul, J.L.:
[1] Elementary critical point theory and perturbations of elliptic boundary value
problems at resonance. Indiana Univ. Math. J. **25** (1976) 933–944

Almgren, F.J.:
[1] Plateau's problem. Benjamin, New York-Amsterdam (1966)

Almgren, F., Simon, L.:
[1] Existence of embedded solutions of Plateau's problem. Ann Sc. Norm. Sup.
Pisa, Cl. Sci. (4) **6** (1979) 447–495

Alt, H.W.:
[1] Verzweigungspunkte von H-Flächen. Part I: Math. Z. **127** (1972) 333–362 Part
II: Math. Ann. **201** (1973) 33–55

Amann, H.:
[1] Ljusternik-Schnirelman theory and nonlinear eigenvalue problems. Math. Ann.
199 (1972) 55–72
[2] On the number of solutions of nonlinear equations in ordered Banach spaces.
J. Funct. Analysis **14** (1973) 346–384
[3] Existence and multiplicity theorems for semi-linear elliptic boundary value
problems. Math. Z. **150** (1976) 281–295
[4] Saddle points and multiple solutions of differential equations. Math. Z. **169**
(1979) 127–166

Amann, H., Zehnder, E.:
[1] Nontrivial solutions for a class of nonresonance problems and applications to
nonlinear differential equations. Ann. Sc. Norm. Sup. Pisa Cl. Sci. (4) **7** (1980)
539–603

Ambrosetti, A.:
[1] On the existence of multiple solutions for a class of nonlinear boundary value
problems. Rend. Sem. Mat. Univ. Padova **49** (1973) 195–204
[2] Problemi variazionali in analisi non lineare. Boll. Unione Mat. Ital. (7) 2-A
(1988) 169–188

Ambrosetti, A., Lupo, D.:
[1] On a class of nonlinear Dirichlet problems with multiple solutions. Nonlin.
Anal. Theory Meth. Appl. **8** (1984) 1145–1150

Ambrosetti, A., Mancini, G.:
[1] Sharp non-uniqueness results for some nonlinear problems. Nonlin. Anal. Theory Meth. Appl. **3** (1979) 635–645
[2] On a theorem by Ekeland and Lasry concerning the number of periodic Hamiltonian trajectories. J. Diff. Eqs. **43** (1981) 1–6

Ambrosetti, A., Rabinowitz, P.H.:
[1] Dual variational methods in critical point theory and applications. J. Funct. Anal. **14** (1973) 349–381

Ambrosetti, A., Struwe, M.:
[1] A note on the problem $-\Delta u = \lambda u + u|u|^{2^*-2}$. Manusc. math. **54** (1986) 373–379
[2] Existence of steady vortex rings in an ideal fluid. Arch. Rat. Mech. Anal. (to appear)

Arzéla, C.:
[1] Il principio di Dirichlet. Rend. Regia Accad. Bologna (1897)

Atkinson, F.V., Brezis, H.-Peletier, L.A.:
[1] Solutions d'équations elliptiques avex exposant de Sobolev critique qui changent de signe. C.R. Acad. Sci. Paris **306** Ser.I (1988) 711–714

Aubin, Th.:
[1] Problèmes isopérimétriques et espace de Sobolev. J. Diff. Geom. **11** (1976) 573–598
[2] Equations différentielles nonlinéaires et problème de Yamabe concernant la courbure scalaire. J. Math. Pures Appl. **55** (1976) 269–293
[3] Nonlinear analysis on manifolds. Monge-Ampère equations, Grundlehren **252**, Springer, New York-Heidelberg-Berlin (1982)

Bahri, A.:
[1] Topological results on a certain class of functionals and applications. J. Funct. Anal. **41** (1981) 397–427
[2] Critical points at infinity in some variational problems. Pitman Research Notes Math. **182**, Longman House, Harlow (1989)

Bahri, A., Berestycki, H.:
[1] A perturbation method in critical point theory and applications. Trans. Amer. Math. Soc. **267** (1981) 1–32
[2] Forced vibrations of superquadratic Hamiltonian systems. Acta Math. **152** (1984) 143–197

Bahri, A., Coron, J.-M.:
[1] On a nonlinear elliptic equation involving the critical Sobolev exponent: the effect of the topology of the domain. Max-Planck-Inst., Bonn (1987)

Bahri, A., Lions, P.-L.:
[1] Morse-index of some min-max critical points. Comm. Pure Appl. Math. **41** (1988) 1027–1037

Ball, J.M.:
[1] Convexity conditions and existence theorems in nonlinear elasticity. Arch. Rat. Mech. Anal. **63** (1977) 337–403
[2] Constitutive inequalities and existence theorems in nonlinear elastostatics. In: Nonlinear Analysis and Mechanics, Research Notes Math. **17**, Pitman, London (1977) 187–241

Ball, J.M., Murat, F.:
[1] $W^{1,p}$-quasiconvexity and variational problems for multiple integrals. J. Funct. Anal. **58** (1984) 225–253

Bangert, V.:
[1] Geodätische Linien auf Riemannschen Mannigfaltigkeiten. Jber. Dt. Math.-Verein **87** (1985) 39–66

Bartolo, P., Benci, V.-Fortunato, D.:
[1] Abstract critical point theorems and applications to some nonlinear problems with "strong" resonance at infinity. Nonlin. Anal. Theory Meth. Appl. **7** (1983) 981–1012

Bass, H., Connell, E.H., Wright, D.:
[1] The Jacobian conjecture: Reduction of degree and formal expansion of the inverse. Bull. Amer. Math. Soc. **7** (1982) 287–330

Benci, V.:
[1]   Some critical point theorems and applications. Comm. Pure Appl. Math. **33** (1980) 147–172
[2]   A geometrical index for the group S^1 and some applications to the study of periodic solutions of ordinary differential equations. Comm. Pure Appl. Math. **34** (1981) 393–432
[3]   On critical point theory for indefinite functionals in the presence of symmetries. Trans. Amer. Math. Soc. **274** (1982) 533–572
[4]   A new approach to the Morse-Conley theory and some applications. Preprint (1988)

Benci, V., Hofer, H., Rabinowitz, P.H.:
[1]   A priori bounds for periodic solutions on hypersurfaces. In: Periodic solutions of Hamiltonian systems and related topics (eds. Rabinowitz, P.H. et al.), D. Reidel Publishing Co. (1987)

Benci, V., Rabinowitz, P.H.:
[1]   Critical point theorems for indefinite functionals. Invent. Math. **52** (1979) 241–273

Bensoussan, A., Boccardo, L., Murat, F.:
[1]   On a nonlinear partial differential equation having natural growth terms and unbounded solution. Ann. Inst. H. Poincaré **5** (1988) 347–364

Berestycki, H., Lasry, J.M., Mancini, G., Ruf, B.:
[1]   Existence of multiple periodic orbits on star-shaped Hamiltonian surfaces. Comm. Pure Appl. Math. **38** (1985) 253–290

Berestycki, H., Lions, P.-L.:
[1]   Nonlinear scalar field equations. I. Existence of ground state, Arch. Rat. Mech. Anal. **82** (1983) 313–345
[2]   Existence d'états multiples dans des equations de champs scalaire non linéaires dans le cas de masse nulle. C. R. Acad. Sci. Paris, **297** (1983), Ser. I 267–270

Berger, M.S.:
[1]   Nonlinearity and functional analysis. Academic Press, New York (1978)

Berger, M.S., Berger, M.S.:
[1]   Perspectives in nonlinearity. Benjamin, New York (1968)

Berkowitz, L.D.:
[1]   Lower semi-continuity of integral functionals. Trans. Amer. Math. Soc. **192** (1974) 51–57

Birkhoff, G.D.:
[1]   Dynamical systems with two degrees of freedom. Trans. Amer. Math. Soc. **18** (1917) 199–300
[2]   Dynamical systems. Amer. Math. Soc. Coll. Publ. **9**, Prividence (1927)

Boccardo, L., Murat, F., Puel, J.P.:
[1]   Existence de solutions faibles pour des équations elliptiques quasi-linéaires à croissance quadratique. In: Nonlinear partial differential equations and their applications. College de France Seminar IV (ed.: Brezis-Lions), Research Notes Math. **84**, Pitman, London (1983) 19–73

Böhme, R.:
[1]   Die Lösungen der Verzweigungsgleichungen für nichtlineare Eigenwertprobleme. Math. Z. **127** (1972) 105–126

Brezis, H.:
[1]   Operateurs maximaux monotone. Math. Studies, North-Holland, Amsterdam-London (1973)

Brezis, H., Coron, J.-M.:
[1]   Large solutions for harmonic maps in two dimensions. Comm. Math. Phys. **92** (1983) 203–215
[2]   Multiple solutions of H-systems and Rellich's conjecture. Comm. Pure Appl. Math. **37** (1984) 149–187

[3]   Convergence of solutions of H-systems or how to blow bubbles. Arch. Rat.
        Mech. Anal. **89** (1985) 21–56

Brezis, H., Coron, J.-M., Nirenberg, L.:
[1]   Free vibrations for a nonlinear wave equation and a theorem of P. Rabinowitz.
        Comm. Pure Appl. Math. **33** (1980) 667–689

Brezis, H., Kato, T.:
[1]   Remarks on the Schrödinger operator with singular complex potentials. J.
        Math. Pures Appl. **58** (1979) 137–151

Brezis, H., Nirenberg, L.:
[1]   Forced vibrations for a nonlinear wave equation. Comm. Pure Appl. Math. **31**
        (1978) 1–30
[2]   Positive solutions of nonlinear elliptic equations involving critical Sobolev ex-
        ponents. Comm. Pure Appl. Math. **36** (1983) 437–477

Browder, F.E.:
[1]   Infinite dimensional manifolds and nonlinear elliptic eigenvalue problems. Ann.
        of Math. (2) **82** (1965) 459–477
[2]   Existence theorems for nonlinear partial differential equations. Proc. Symp.
        Pure Math. **16**, Amer. Math. Soc. Prividence (1970) 1–60
[3]   Nonlinear eigenvalue problems and group invariance. In: Functional analysis
        and related fields (ed.: Browder, F.E.), Springer (1970) 1–58

Capozzi, A., Fortunato, D., Palmieri, G.:
[1]   An existence result for nonlinear elliptic problems involving critical Sobolev
        exponent. Ann. Inst. H. Poincaré Analyse Nonlinéaire **2** (1985) 463–470

Castro, A., Kurepa, A.:
[1]   Radially symmetric solutions to a Dirichlet problem involving critical expo-
        nents. Preprint (1989)

Cerami, G.:
[1]   Un criterio di esistenza per i punti critici su varietà illimitate. Rend. Acad.
        Sci. Let. Ist. Lombardo **112** (1978) 332–336
[2]   Sull' esistenza di autovalori per un problema al contorno non lineare. Ann. di
        Mat. (IV) **24** (1980) 161–179

Cerami, G., Fortunato, D., Struwe, M:
[1]   Bifurcation and multiplicity results for nonlinear elliptic problems involving
        critical Sobolev exponents. Ann. Inst. H. Poincaré, Analyse Nonlinéaire 1
        (1984) 341–350

Cerami, G., Solimini, S., Struwe, M.:
[1]   Some existence results for superlinear elliptic boundary value problems involv-
        ing critical exponents. J. Funct. Anal. **69** (1986) 289–306

Chang, K.C.:
[1]   Solutions of asymptotically linear operator equations via Morse theory. Comm.
        Pure Appl. Math. **34** (1981) 693–712
[2]   Variational methods for non-differentiable functionals and their applications
        to partial differential equations J. Math. Anal. Appl. **80** (1981) 102–129
[3]   Infinite dimensional Morse theory and its applications. Lect. Notes 22nd Ses-
        sion Sem. Math. Sup. Montreal (1983)
[4]   A variant of the mountain pass lemma. Scientia Sinica **26** (1983) 1241–1255
[5]   Variational methods and sub- and super-solutions. Sci. Sinica Ser. A **26** (1983)
        1256–1265
[6]   Heat flow and boundary value problem for harmonic maps. Preprint (1988)
[7]   Infinite dimensional Morse theory and multiple solutions, Birkhäuser (to ap-
        pear)

Chang, K.C., Ding, W.-Y.:
[1]   A result on the global existence for heat flows of harmonic maps from D^2 into
        S^2. Preprint (1989)

Chang, K.C., Eells, J.:
[1]   Unstable minimal surface coboundaries. Acta Math. Sinica (New Ser.) **2** (1986) 233–247

Chang, S.-Y. A., Yang, P.:
[1]   Prescribing Gaussian curvature on S^2. Acta Math. **159** (1987) 215–259

Chen, W.-X., Ding, W.-Y.:
[1]   Scalar curvatures on S^2. Trans. Amer. Math. Soc. **303** (1987) 365–382

Chen, Y., Ding, W.-Y.:
[1]   Blow-up and global existence for heat flows of harmonic maps. Preprint (1989)

Chen, Y., Struwe, M.:
[1]   Existence and partial regularity results for the heat flow for harmonic maps. Math. Z. **201** (1989) 83–103

Chow, S.-N., Hale, J.K:
[1]   Methods of bifurcation theory. Grundlehren **251**, Springer, New York-Heidelberg-Berlin (1982)

Cianchi, A.:
[1]   A sharp form of Poincaré-type inequalities on balls and spheres. ZAMP (to appear)

Ciarlet, P.G.:
[1]   Three-dimensional elasticity. Math. Elasticity **1**, North-Holland, Amsterdam (1987)

Clark, D.C.:
[1]   A variant of the Ljusternik-Schnirelman theory. Indiana Univ. Math. J. **22** (1972) 65–74

Clarke, F.H.:
[1]   A new approach to Lagrange multipliers. Math. Oper. Res. **1** (1976) 165–174
[2]   Solution périodique des équations hamiltoniennes. C. R. Acad. Sci. Paris **287** (1978) 951–952
[3]   A classical variational principle for Hamiltonian trajectories. Proc. Amer. Math. Soc. **76** (1979) 186–188
[4]   Periodic solutions to Hamiltonian inclusions. J. Diff. Eq. **40** (1981) 1–6
[5]   Optimization and nonsmooth analysis. Wiley Interscience, New York, (1983)

Clarke, F.H., Ekeland, I.:
[1]   Hamiltonian trajectories having prescribed minimal period. Comm. Pure Appl. Math. **33** (1980) 103–116

Coffman, C.V.:
[1]   A minimum-maximum principle for a class of nonlinear integral equations. J. Analyse Math. **22** (1969) 391–419

Conley, C.:
[1]   Isolated invariant sets and the Morse index. CBMS **38**, Amer. Math. Soc., Providence (1978)

Conley, C., Zehnder E.:
[1]   A Morse type index theory for flows and periodic solutions to Hamiltonian systems. Comm. Pure Appl. Math. **37** (1984) 207–253

Connor, E., Floyd, E.E.:
[1]   Fixed point free involutions and equivariant maps. Bull. Amer. Math. Soc. **66** (1960) 416–441

Coron, J.M.:
[1]   Periodic solutions of a nonlinear wave equation without assumption of monotonicity. Math. Ann. **262** (1983) 273–285
[2]   Topologie et cas limite des injections de Sobolev. C.R. Acad. Sc. Paris **299**, Ser. I (1984) 209–212
[3]   Nonuniqueness for the heat flow of harmonic maps. Ann. Inst. M. Poincaré Analyse Nonlinéaire, (to appear)

Coron, J.M., Ghidaglia, J.M.:
[1]   Explosion en temps fini pour le flot des applications harmoniques. Preprint (1988)

Courant, R.:
[1]   Dirichlet's principle, conformal mapping and minimal surfaces. Interscience, New York (1950). Reprinted: Springer, New York-Heidelberg-Berlin (1977)

Crandall, M.G., Rabinowitz, P.H.:
[1]   Continuation and variational methods for the existence of positive solutions of nonlinear elliptic eigenvalue problems. Arch. Rat. Mech. Anal. **58** (1975) 201–218

Dacorogna, B.:
[1]   Weak continuity and weak lower semicontinuity of non-linear functionals. Lect. Notes Math. **922**, Springer, Berlin-Heidelberg-New York (1982)
[2]   Direct methods in the calculus of variations. Springer, Berlin-Heidelberg-New York (1989)

Dancer, E.N.:
[1]   Degenerate critical points, homotopy indices, and Morse inequalities. J. Reine Angew. Math. **350** (1984) 1–22
[2]   Counterexamples to some conjectures on the number of solutions of nonlinear equations. Math. Ann. **272** (1985) 421–440

de Figueiredo, D.G.:
[1]   Lectures on the Ekeland variational principle with applications and detours. Lect. notes, College on Variational Problems in Analysis, Trieste (1988)

Deimling, K.:
[1]   Nonlinear functional analysis. Springer, Berlin-Heidelberg-New York-Tokyo (1985)

Di Benedetto, E.:
[1] $C^{1+\alpha}$ local regularity of weak solutions to degenerate elliptic equations. Nonlinear Analysis, T.M.A. **7** (1983) 827–850

Ding, W.-Y.:
[1]   On a conformally invariant elliptic equation on $\mathbb{R}^n$. Comm. Math. Phys. **107** (1986) 331–335

Di Perna, R.J.:
[1]   Convergence of approximate solutions to conservation laws. Arch. Rat. Mech. Anal. **82** (1983) 27–70

Di Perna, R.J., Majda, A.J.:
[1]   Oscillations and concentrations in weak solutions of the incompressible fluid equations. Comm. Math. Phys. **108** (1987) 667–689

Dirichlet, L.:
[1]   Vorlesungen über die im umgekehrten Verhältnis des Quadrats der Entfernung wirkenden Kräfte. Göttingen (1856/57)

Douglas, J.:
[1]   Solution of the problem of Plateau. Trans. Amer. Math. Soc. **33** (1931) 263–321

Duc, D. M.:
[1]   Nonlinear singular elliptic equations. Preprint (1988)

Dunford, N., Schwartz, J.T.:
[1]   Linear operators. Vol. I, Intescience, New York (1958)

Edmunds, D.E., Moscatelli, V.B.:
[1]   Sur la distribution asymptotique des valeurs propres pour une classe générale d'opérateurs différentiels. C.R. Acad. Sci. Paris **284** (1978) 1283–1285

Eells, J.:
[1]   A setting for global analysis. Bull. Amer. Math. Soc. **72** (1966) 751–809

Eells, J., Lemaire, L.:
[1]   Report on harmonic maps. Bull. London Math. Soc. **10** (1978) 1–68
[2]   Another report on harmonic maps. Bull. London Math. Soc. **20** (1988) 385–524

Eells, J., Sampson, J.H.:
[1]   Harmonic mappings of Riemannian manifolds. Amer. J. Math. **86** (1964) 109–160

Eells, J., Wood, J.C.:
[1]   Restrictions on harmonic maps of surfaces. Topology **15** (1976) 263–266

Egnell, H.:
[1]   Linear and nonlinear elliptic eigenvalue problems. Dissertation, Univ. Uppsala (1987)

Eisen, G.:
[1]   A counterexample for some lower semicontinuity results. Math. Z. **162** (1978) 241–243
[2]   A selection lemma for sequences of measurable sets, and lower semi-continuity of multiple integrals. Manusc. math. **27** (1979) 73–79

Ekeland, I.:
[1]   On the variational principle. J. Math. Anal. Appl. **47** (1974) 324–353
[2]   Convexity methods in Hamiltonian mechanics. Ergebnisse d. Math. (Ser. III) **19**, Springer (1990)

Ekeland, I., Hofer, H.:
[1]   Periodic solutions with prescribed minimal period for convex autonomous Hamiltonian systems. Inv. Math. **81** (1985) 155–188

Ekeland, I., Lasry, J.M.:
[1]   On the number of periodic trajectories for a Hamiltonian flow on a convex energy surface. Ann. of Math. **112** (1980) 283–319

Ekeland, I., Lassoued, L.:
[1]   Un flot hamiltonien a au moins deux trajectoires fermées sur toute surface d' énergie convexe et bornée. C. R. Acad. Sci. Paris **361** (1985) 161–164

Ekeland, I., Temam, R.:
[1]   Convex analysis and variational problems. North Holland, Amsterdam (1976)

Evans, L.C.:
[1]   Quasiconvexity and partial regularity in the calculus of variations. Arch. Rat. Mech. Anal. **95** (1986) 227–252

Evans, L.C., Gariepy, R.F.:
[1]   Blowup, compactness and partial regularity in the calculus of variations. Indiana Univ. Math. J. **36** (1987) 361–371

Euler, L.:
[1]   Methodus inveniendi lineas curvas maximi minimive proprietate gaudentes sive solutio problematis isoperimetrici latissimo sensu accepti. Lausanne-Genève (1744), Opera, Ser. I, Vol. 24 (ed. C. Carathéodory), Bern (1952)

Fadell, E.R., Husseini, S.:
[1]   Relative cohomological index theories. Advances Math. **64** (1987) 1–31

Fadell, E.R., Husseini, S., Rabinowitz, P.H.:
[1]   Borsuk-Ulam theorems for arbitrary S^1 actions and applications. Trans. Amer. Math. Soc. **274** (1982) 345–360

Fadell, E.R., Rabinowitz, P.H.:
[1]   Bifurcation for odd potential operators and an alternative topological index. J. Funct. Anal. **26** (1977) 48–67
[2]   Generalized cohomological index theories for Lie group actions with applications to bifurcation questions for Hamiltonian systems. Invent. Math. **45** (1978) 139–174

Federer, H.:
[1]   Geometric measure theory. Grundlehren **153**, Springer, Berlin-Heidelberg-New York (1969)

Fet, A.I.:
[1]   Variational problems on closed manifolds. Mat. Sb. (N.S.) **30** (1952) 271–316
      (Russian) Amer. Math. Soc. Transl. **90** (1953)

Floer, A.:
[1]   Morse theory for fixed points of symplectic diffeomorphisms. Bull. Amer. Math.
      Soc. **16** (1987) 279–281
[2]   Symplectic fixed points and holomorphic spheres. Comm. Math. Phys. (to
      appear)

Fortunato, D., Jannelli, E.:
[1]   Infinitely many solutions for some nonlinear elliptic problems in symmetrical
      domains. Proc. Royal Soc. Edinb. **105 A** (1987) 205–213

Fraenkel, L.E., Berger, M.S.:
[1]   A global theory of steady vortex rings in an ideal fluid. Acta Math. **132** (1974)
      13–51

Frehse, J.:
[1]   Capacity methods in the theory of partial differential equations. Jber. d. Dt.
      Math.-Verein. **84** (1982) 1–44
[2]   Existence and perturbation theorems for non-linear elliptic systems. In: Non-
      linear partial differential equations and their applications, College de France
      Seminar IV (ed.: Brezis-Lions), Research Notes Math. **84**, Pitman, London
      (1983)

Fusco, N.:
[1]   Quasi convessità e semicontinuità per integrali multipli di ordine superiore.
      Ricerche di Mat. **29** (1980) 307–323

Gauss, C.F.:
[1]   Allgemeine Lehrsätze in Beziehung auf die im verkehrten Verhältnisse des
      Quadrats der Entfernung wirkenden Anziehungs- und Abstossungskräfte. Leip-
      zig (1840); Werke V, Göttingen (1867) 195–242

Gehring, F.W.:
[1]   The L^p-integrability of the partial derivatives of a quasi-conformal mapping.
      Acta Math. **130** (1973) 265–277

Gelfand, I.M.:
[1]   Some problems in the theory of quasilinear equations. Amer. Math. Soc.
      Transl., Ser. 2, **29** (1963) 295–381

Ghoussoub, N.:
[1]   Location, multiplicity and Morse indices of min-max critical points. Preprint
      (1989)

Giaquinta, M.:
[1]   Multiple integrals in the calculus of variations and nonlinear elliptic systems.
      Ann. Math. Studies **105**, Princeton Univ. Press, Princeton (1983)

Giaquinta, M., Giusti, E.:
[1]   On the regularity of the minima of variational integrals. Acta Math. **148** (1982)
      31–46

Giaquinta, M., Modica, G.:
[1]   Regularity results for some classes of higher order nonlinear elliptic systems.
      J. reine angew. Math. **311/312** (1979) 145–169
[2]   Partial regularity of minimizers of quasiconvex integrals. Ann. Inst. H. Poin-
      caré, Anal. Non Linéaire **3** (1986) 185–208

Giaquinta, M., Modica, G., Souček, J.:
[1]   Cartesian currents, weak diffeomorphisms and nonlinear elasticity. Preprint,
      Univ. Florence (1988)

Gidas, B., Ni, W.-M., Nirenberg, L.:
[1]   Symmetry and related properties via the maximum principle. Comm. Math.
      Phys. **68** (1979) 209–243

Gilbarg, D., Trudinger, N.S.:
[1]   Elliptic partial differential equations of second order. 2nd edition, Grundlehren
       224, Springer, Berlin-Heidelberg-New York-Tokyo (1983)

Giusti, E.:
[1]   Minimal surfaces and functions of bounded variation. Monographs in Mathe-
       matics **80**, Birkhäuser, Boston-Basel-Stuttgart (1984)

Girardi, M., Matzeu, M.:
[1]   Solutions of minimal period for a class of nonconvex Hamiltonian systems and
       applications to the fixed energy problem. Nonlinear Analysis T.M.A. **10** (1986)
       371–382
[2]   Periodic solutions of convex autonomous Hamiltonian systems with a quadratic
       growth at the origin and superquadratic at infinity. Ann. di Mat. Pura Appl.
       (IV) **147** (1987) 21–72

Goldstine, H.H.:
[1]   A history of the calculus of variations from the 17th through the 19th century.
       Studies Hist. Math. Phys. Sci. **5**, Springer, New York-Heidelberg-Berlin (1980)

Grayson, M., Hamilton, R.S.:
[1]   The formation of singularities in the harmonic map heat flow. Preprint

Guedda, M., Veron, L.:
[1]   Quasilinear elliptic equations involving critical Sobolev exponents. Nonlinear
       Analysis, T.M.A. **13** (1989) 879–902

Gulliver, R.D.:
[1]   Regularity of minimizing surfaces of prescribed mean curvature. Ann. of Math.
       (2) **97** (1973) 275–305

Gulliver, R.D., Lesley, F.D.:
[1]   On boundary branch points of minimizing surfaces. Arch. Rat. Mech. Anal.
       52 (1973) 20–25

Gulliver, R.D., Osserman, R., Royden, H.L.:
[1]   A theory of branched immersions of surfaces. Amer. J. Math. **95** (1973) 750–
       812

Günther, M.:
[1]   On the perturbation problem associated to isometric embeddings of Rieman-
       nian manifolds. Ann. Global Anal. Geom. **7** (1989) 69–77

Hamilton, R.:
[1]   Harmonic maps of manifolds with boundary. Lect. Notes Math **471**, Springer,
       Berlin (1975)

Hartman, P., Wintner, A.:
[1]   On the local behavior of solutions of non-parabolic partial differential equa-
       tions. Amer. J. Math. **75** (1953) 449–476

Heinz, E.:
[1]   Über die Existenz einer Fläche konstanter mittlerer Krümmung bei vorgege-
       bener Berandung. Math. Ann. **127** (1954) 258–287

Hempel, J.A.:
[1]   Multiple solutions for a class of nonlinear elliptic boundary value problems.
       Indiana Univ. Math. J. **20** (1971) 983–996

Hilbert, D.:
[1]   Über das Dirichletsche Prinzip. Jber. Deutsch. Math. Vereinigung **8** (1900)
       184–188
[2]   Über das Dirichletsche Prinzip. Festschrift zur Feier des 150-jährigen Bestehens
       der Königlichen Gesellschaft der Wissenschaften zu Göttingen 1901, Math.
       Ann. **49** (1904) 161–186

Hildebrandt, S.:
[1]   Boundary behavior of minimal surfaces. Arch. Rat. Mech. Anal. **35** (1969)
       47–82

[2]   On the Plateau problem for surfaces of constant mean curvature. Comm. Pure
      Appl. Math. **23** (1970) 97–114
[3]   Nonlinear elliptic systems and harmonic mappings. Proc. Beijing Symp. Diff.
      Geom. and Diff. Eqs. (1980), Gordon and Breach (1983) 481–615
[4]   The calculus of variations today, as reflected in the Oberwolfach meetings. In
      Perspect. Math., Anniversary of Oberwolfach 1984, Birkhäuser (1985)

Hildebrandt, S., Tromba, A.:
[1]   Mathematics and optimal form. Scientific American Books, Inc. New York
      (1985)

Hildebrandt, S., Widman, K.-O.:
[1]   On the Hölder continuity of weak solutions of quasilinear elliptic systems of
      second order. Ann. Sc. Norm. Sup. Pisa, Ser. 4, **4** (1977) 146–178

Hofer, H.:
[1]   Variational and topological methods in partially ordered Hilbert spaces. Math.
      Ann. **261** (1982) 493–514
[2]   A geometric description of the neighborhood of a critical point given by the
      mountain pass theorem. J. London Math. Soc. (2) **31** (1985) 566–570

Hofer, H., Zehnder, E.:
[1]   Periodic solutions on hypersurfaces and a result by C. Viterbo. Invent. Math.
      90 (1987) 1–9

Jost, J.:
[1]   The Dirichlet problem for harmonic maps from a surface with boundary onto a
      2-sphere with non-constant boundary values. J. Diff. Geom. **19** (1984) 393–401
[2]   Harmonic maps between surfaces. Lect. Notes Math. **1062**, Springer, Berlin-
      Heidelberg-New York (1984)

Jost, J., Struwe, M.:
[1]   A Morse-Conley index theory for minimal surfaces of varying topological type.
      Preprint (1988)

Kazdan, J., Warner, F.:
[1]   Scalar curvature and conformal deformation of Riemannian structure. J. Diff.
      Geom. **10** (1975) 113–134
[2]   Existence and conformal deformations of metrics with prescribed Gaussian
      and scalar curvatures. Ann. of Math. **101** (1975) 317–331

Kelley, J.L.:
[1]   General topology. Graduate Texts Math. **27**, Springer, New York-Heidelberg-
      Berlin (1955)

Kinderlehrer, D., Stampacchia, G.:
[1]   An introduction to variational inequalities and their applications. Academic
      Press, New York-London (1980)

Klingenberg, W.:
[1]   Lectures on closed geodesics. Grundlehren **230**, Springer, Berlin-Heidelberg-
      New York (1978)

Krasnoselskii, M.A.:
[1]   Topological methods in the theory of nonlinear integral equations. Macmillan,
      New York (1964)

Krasnoselskii, M.A., Zabreiko, P.P.:
[1]   Geometrical methods in nonlinear analysis. Grundlehren **263**, Springer, Berlin-
      Heidelberg-New York-Tokyo (1984)

Ladyzhenskaya, O.A.:
[1]   The mathematical theory of viscous incompressible flow. 2nd edition, Gordon
      & Breach, New York-London-Paris (1969)

Ladyzhenskaya, O.A., Solonnikov, V.A., Ural'ceva, N.N.:
[1]   Linear and quasilinear equations of parabolic type. Amer. Math. Soc. Transl.
      Math. Monogr. **23**, Providence (1968)

Ladyzhenskaya, O.A., Ural'ceva, N.N.:
[1]   Linear and quasilinear elliptic equations. Academic Press, New York (1968)

Lazer, A., Solimini, S.:
[1]   Nontrivial solutions of operator equations and Morse indices of critical points of min-max type. Nonlinear Analysis, T.M.A. **12** (1988) 761–775

Lebesgue, H.:
[1]   Sur le problème de Dirichlet. Rend. Circ. Mat. Palermo **24** (1907) 371–402

Lee, J.M., Parker, T.H.:
[1]   The Yamabe problem. Bull. Amer. Math. Soc. **17** (1987) 37–91

Lemaire, L.:
[1]   Applications harmoniques de surfaces riemanniennes. J. Diff. Geom. **13** (1978) 51–78

Lévy, P.:
[1]   Théorie de l'addition des variables aléatoirs. Gauthier-Villars, Paris (1954)

Lions, J.L., Magenes, E.:
[1]   Non-homogeneous boundary value problems and applications I. Grundlehren **181**, Springer, Berlin-Heidelberg-New York (1972)

Lions, P.-L.:
[1]   The concentration-compactness principle in the calculus of variations. The locally compact case. Part 1, Ann. Inst. H. Poincaré **1** (1984) 109–145
[2]   The concentration-compactness principle in the calculus of variations. The locally compact case. Part 2, Ann. Inst. H. Poincaré **1** (1984) 223–283
[3]   The concentration-compactness principle in the calculus of variations. The limit case. Part 1, Rev. Mat. Iberoamericano **1.1** (1985) 145–201
[4]   The concentration-compactness principle in the calculus of variations. The limit case. Part 2, Rev. Mat. Iberoamericano **1.2** (1985) 45–121

Lions, P.-L., Pacella, F., Tricarico, M.:
[1]   Best constants in Sobolev inequalities for functions vanishing on some part of the boundary and related questions. Preprint

Ljusternik, L., Schnirelman, L.:
[1]   Sur le problème de trois géodésiques fermées sur les surfaces de genre 0. C.R. Acad. Sci. Paris **189** (1929) 269–271
[2]   Méthodes topologiques dans les problèmes variationelles. Actualites Sci. Industr. **188**, Paris (1934)

Long, Y.:
[1]   Multiple solutions of perturbed superquadratic second order Hamiltonian systems. Preprint (1988)
[2]   Thesis. Univ. Wisconsin (1988)

Lovicarová, H.:
[1]   Periodic solutions for a weakly nonlinear wave equation in one dimension. Czech. Math. J. **19** (1969) 324–342

Lyusternik, L.:
[1]   The topology of function spaces and the calculus of variations in the large. Trudy Mat. Inst. Steklov **19** (1947) (Russian); Transl. Math. Monographs **16**, Amer. Math. Soc., Prividence (1966)

Mancini, M., Musina, R.:
[1]   A free boundary problem involving limiting Sobolev exponents. Manusc. math. **58** (1987) 77–93

Marcellini, P., Sbordone, C.:
[1]   On the existence of minima of multiple integrals of the calculus of variations. J. Math Pures et Appl. **62** (1983) 1–9

Marino, A.:
[1]   La biforcazione nel caso variazionale. Confer. Sem. Mat. Univ. Bari **132** (1977)

236   References

Mawhin, J.:
  [1]   Problèmes de Dirichlet variationels non linéaires. Seminaire Math. Sup. **104**,
        Presses Univ. Montreal, Montreal (1987)
Mawhin, J., Willem, M.:
  [1]   Critical point theory and Hamiltonian systems. Appl. Math. Sci. **74**, Springer,
        New York-Berlin-Heidelberg-London-Paris-Tokyo (1989)
Maupertuis, P.L.M. de:
  [1]   Accord de différentes lois de la nature qui avaient jusqu'ici paru incompatibles.
        Mém. Acad. Sci. (1744) 417–426
Meeks, W.H., Yau, S.-T.:
  [1]   The classical Plateau problem and the topology of three-dimensional mani-
        folds. Top **21** (1982) 409–442
Miersemann, E.:
  [1]   Über höhere Verzweigungspunkte nichtlinearer Variationsungleichungen.
        Math. Nachrichten **85** (1978) 195–213
Milnor, J.:
  [1]   Morse theory. Ann. of Math. Studies 51, Princeton Univ. Press, Princeton
        (1963)
Morgan, F.:
  [1]   Geometric measure theory. A beginner's guide. Academic Press, London (1988)
Morrey, C.B.Jr.:
  [1]   Existence and differentiability theorems for the solutions of variational prob-
        lems for multiple integrals. Bull. Amer. Math. Soc. **46** (1940) 439–458
  [2]   The problem of Plateau on a Riemannian manifold. Ann. of Math. **49** (1948)
        807–851
  [3]   Quasiconvexity and the lower semicontinuity of multiple integrals. Pacific J.
        Math. **2** (1952) 25–53
  [4]   Multiple integrals in the calculus of variations. Springer Grundlehren **130**,
        New York (1966)
Morse, M.:
  [1]   The critical points of functions and the calculus of variations in the large. Bull.
        Amer. Math. Soc. **35** (1929) 38–54
  [2]   The calculus of variations in the large. Amer. Math. Soc. Coll. Publ. **18** (1934)
  [3]   Functional topology and abstract variational theory. Ann. of Math. **38** (1937)
        386–449
Morse, M., Tompkins, C.B.:
  [1]   The existence of minimal surfaces of general critical types. Ann. of Math. **40**
        (1939) 443–472
  [2]   Unstable minimal surfaces of higher topological structure. Duke Math. J. **8**
        (1941) 350–375
Moser, J.:
  [1]   A new proof of De Giorgi's theorem. Comm. Pure Appl. Math. **13** (1960)
        457–468
  [2]   A Harnack inequality for parabolic differential equations. Comm. Pure Appl.
        Math. **17** (1964) 101–134
  [3]   Stable and random motions in dynamical systems. Ann. Math. Studies **77**,
        Princeton Univ. Press, Princeton (1973)
Müller, S.:
  [1]   Weak continuity of determinants and nonlinear elasticity. C.R. Acad. Sci. Paris
        307, sér. **1** (1988) 501–506
  [2]   Weak continuity of determinants and existence theorems in nonlinear elasticity.
        (To appear)
Murat, F.:
  [1]   Compacité par compensation. Ann. Sc. Norm. Sup. Pisa (IV) **5** (1978) 489–507
  [2]   Compacité par compensation II. in: Proc. intern. meet. recent methods in
        non-linear analysis, Rome, Pitagora ed. (1978) 245–256

Nash, J.:
[1] The imbedding problem for Riemannian manifolds. Ann. of Math. **63** (1956) 20–63

Nečas, J.:
[1] Les méthodes directes en théorie des équations elliptiques. Academia, Prague (1967)

Nehari, Z.:
[1] Characteristic values associated with a class of nonlinear second-order differential equations. Acta Math. **105** (1961) 141–175

Ni, W.-M.:
[1] Some minimax principles and their applications in nonlinear elliptic equations. J. Analyse Math. **37** (1980) 248–275

Nirenberg, L.:
[1] Variational and topological methods in nonlinear problems. Bull. Amer. Math. Soc. (New Ser.) **4** (1981) 267–302

Nitsche, J.C.C.:
[1] The boundary behavior of minimal surfaces, Kellog's theorem and branch points on the boundary. Invent. Math. **8** (1969) 313–333 Addendum, Invent. Math. 9 (1970) 270
[2] Vorlesungen über Minimalflächen. Grundlehren **199**, Springer, Berlin-Heidelberg-New York (1975)

Obata, M.:
[1] The conjectures on conformal transformations of Riemannian manifolds. J. Diff. Geom. **6** (1971) 247–258

Osserman, R.:
[1] A survey of minimal surfaces. Math. Studies **25**, Van Nostrand Reinhold Co., New York (1969)
[2] A proof of the regularity everywhere of the classical solution to Plateau's problem. Ann. of Math. **91** (1970) 550–569

Palais, R.S.:
[1] Morse theory on Hilbert manifolds. Topology **2** (1963) 299–340
[2] Lusternik-Schnirelman theory on Banach manifolds. Topology **5** (1966) 115–132
[3] Foundations of global non-linear analysis. Benjamin, New York (1968)
[4] Critical point theory and the minimax principle. Proc. Symp. Pure Math. **15** (1970) 185–212

Palais, R.S., Smale, S.:
[1] A generalized Morse theory. Bull. Amer. Math. Soc. **70** (1964) 165–171

Pohožaev, S.:
[1] Eigenfunctions of the equation $\Delta u + \lambda f(u) = 0$. Soviet Math. Dokl. **6** (1965) 1408–1411

Poincaré, H.:
[1] Les méthodes nouvelles de la méchanique céleste. Tome I, Dover Publications, New York (1957); Engl. transl.: National Aeronautics Space Administration, Springfield (1967)

Polya, G., Szegö, G.:
[1] Isoperimetric inequalities in mathematical physics. Ann. Math. Studies **27**, Princeton Univ. Press, Princeton (1951)

Pucci, P., Serrin, J.:
[1] Extensions of the mountain pass theorem. J. Funct. Anal. **59** (1984) 185–210
[2] The structure of the critical set in the mountain pass theorem. Trans. Amer. Math. Soc. **299** (1987) 115–132

Pugh, C., Robinson, C.:
[1] The C^1-closing lemma, including Hamiltonians. Ergodic Theory Dyn. Syst. **3** (1983) 261–313

238   References

Quittner, P. :
[1]   Solvability and multiplicity results for variational inequalities. Comm. Mat.
       Univ. Carolineae (to appear)

Rabinowitz, P.H.:
[1]   Some aspects of nonlinear eigenvalue problems. Rocky Mount. J. Math. **3**
       (1973) 161–202
[2]   Variational methods for nonlinear elliptic eigenvalue problems. Indiana Univ.
       Math. J. **23** (1974) 729–754
[3]   Variational methods for nonlinear eigenvalue problems. In: Eigenvalues of non-
       linear problems (ed. Prodi, G.), C.I.M.E., Edizioni Cremonese, Roma (1975)
       141–195
[4]   A variational method for finding periodic solutions of differential equations.
       Proc. Symp. Nonlinear Evol. Eq. (ed. Crandall, M.G.), Wisconsin (1977) 225–
       251
[5]   Periodic solutions of Hamiltonian systems. Comm. Pure Appl. Math. **31** (1978)
       157–184
[6]   Free vibrations for a semilinear wave equation. Comm. Pure appl. Math. **31**
       (1978) 31–68
[7]   Some minimax theorems and applications to nonlinear partial differential equa-
       tions. In : Nonlinear analysis; a collection of papers in honor of Erich Rothe
       (ed. Cesari, L. et al.), Academic Press, New York (1978) 161–177
[8]   Some critical point theorems and applications to semilinear elliptic partial
       differential equations. Ann. Sc. Norm. Sup. Pisa, Ser. 4, **5** (1978) 215–223
[9]   Periodic solutions of Hamiltonian systems: A survey. Siam J. Math. Anal. **13**
       (1982) 343–352
[10]   Multiple critical points of perturbed symmetric functionals. Trans. Amer.
       Math. Soc. **272** (1982) 753–770
[11]   Minimax methods in critical point theory with applications to differential
       equations. CBMS Regional Conference Series Math. **65**, Amer. Math. Soc.,
       Providence (1986)

Radó, T.:
[1]   On Plateau's problem. Ann. of Math. **31** (1930) 457–469
[2]   The isoperimetric inequality and the Lebesgue definition of surface area. Trans.
       Amer. Math. Socc. **61** (1947) 530–555

Reshetnyak, Y.:
[1]   General theorems on semicontinuity and on convergence with a functional.
       Sibir. Math. **8** (1967) 801–816
[2]   Stability theorems for mappings with bounded excursion. Sibir. Math. **9** (1968)
       667–684

Riemann, B.:
[1]   Grundlagen für eine allgemeine Theorie der Funktionen einer veränderlichen
       komplexen Grösse. Dissertation, Göttingen (1851); Bernhard Riemann's ge-
       sammelte mathematische Werke (ed. Weber, H.), Teubner, Leibzig (1876) 3–
       43

Rockafellar, R.T.:
[1]   Convex analysis. Princeton Univ. Press, Princeton (1970)

Rodemich, E.:
[1]   The Sobolev inequalities with best possible constants. In: Analysis Seminar,
       Cal. Inst. Tech. (1966)

Rothe, E.:
[1]   Critical points and the gradient fields of scalars in Hilbert space. Acta Math.
       85 (1951) 73–98

Rudin, W.:
[1]   Functional analysis. McGraw-Hill, New York (1973), 7th reprint, Tata McGraw-
       Hill, New Dehli (1982)

Rybakowski, K.P.:
[1]   On the homotopy index for infinite-dimensional semiflows. Trans. Amer. Math. Soc. **269** (1982) 351–382
[2]   The homotopy index on metric spaces with applications to partial differential equations. Freiburg (1985)

Rybakowski, K.P., Zehnder, E.:
[1]   On a Morse equation in Conley's index theory for semiflows on metric spaces. Ergodic Theory Dyn. Syst. **5** (1985) 123–143

Sacks, P., Uhlenbeck, K.:
[1]   On the existence of minimal immersions of 2-spheres. Ann. of Math. **113** (1981) 1–24

Salamon, D.:
[1]   Connected simple systems and the Conley index of isolated invariant sets. Trans. Amer. Math. Soc. **291** (1985) 1–41

Salvatore, A.:
[1]   Solutions of minimal period of a wave equation, via a generalization of Hofer's theorem. Preprint, Bari (1987)

Schoen, R.:
[1]   Conformal deformation of a Riemannian metric to constant scalar curvature. J. Diff. Geom. **20** (1984) 479–495

Schoen, R.M., Uhlenbeck, K.:
[1]   A regularity theory for harmonic maps. J. Diff. Geom. **17** (1982) 307–335, and J. Diff. Geom. **18** (1983) 329
[2]   Boundary regularity and miscellaneous results on harmonic maps. J. Diff. Geom. **18** (1983) 253–268

Schwartz, J.T.:
[1]   Generalizing the Ljusternik-Schnirelmann theory of critical points. Comm. Pure Appl. Math. **17** (1964) 307–315
[2]   Nonlinear functional analysis. Gordon & Breach, New York (1969)

Sedlacek, S.:
[1]   A direct method for minimizing the Yang-Mills functional over 4-manifolds. Comm. Math. Phys. **86** (1982) 515–527

Seifert, H.:
[1]   Periodische Bewegungen mechanischer Systeme. Math. Z. **51** (1948) 197–216

Serrin, J.:
[1]   On a fundamental theorem of the calculus of variations. Acta Math. **102** (1959) 1–22
[2]   On the definition and properties of certain variational integrals. Trans. Amer. Math. Soc. **101** (1961) 139–167

Shiffman, M.:
[1]   The Plateau problem for minimal surfaces of arbitrary topological structure. Amer. J. Math. **61** (1939) 853–882
[2]   The Plateau problem for non-relative minima. Ann. of Math. **40** (1939) 834–854
[3]   Unstable minimal surfaces with several boundaries. Ann. of Math. **43** (1942) 197–222

Simader, C.G.:
[1]   On Dirichlet's boundary value problem. Lect. Notes Math. **268**, Springer, Berlin-Heidelberg-New York (1972)

Simon, L.:
[1]   Lectures on geometric measure theory. Centre Math. Analysis, Australian Nat. Univ. **3** (1983)

Smale, S.:
[1]   Generalized Poincaré's conjecture in dimensions greater than four. Ann. of Math. **74** (1961) 391–406

[2]   Morse theory and a nonlinear generalization of the Dirichlet problem. Ann. Math. **80** (1964) 382–396

Solimini, S.:
[1]   On the existence of infinitely many radial solutions for some elliptic problems. Revista Mat. Aplicadas, Univ. Chile (to appear)

Spanier, E.H.:
[1]   Algebraic topology. McGraw Hill, New York (1966)

Steffen, K.:
[1]   On the existence of surfaces with prescribed mean curvature and boundary. Math. Z. **146** (1976) 113–135
[2]   On the nonuniqueness of surfaces with prescribed constant mean curvature spanning a given contour. Arch. Rat. Mech. Anal. **94** (1986) 101–122

Struwe, M.:
[1]   Infinitely many critical points for functionals which are not even and applications to superlinear boundary value problems. Manusc. math. **32** (1980) 335–364
[2]   Multiple solutions of anticoercive boundary value problems for a class of ordinary differential equations of second order. J. Diff. Eq. **37** (1980) 285–295
[3]   Infinitely many solutions of superlinear boundary value problems with rotational symmetry. Archiv d. Math. **36** (1981) 360–369
[4]   Superlinear elliptic boundary value problems with rotational symmetry. Archiv d. Math. **39** (1982) 233–240
[5]   A note on a result of Ambrosetti and Mancini. Ann. di Mat. **81** (1982) 107–115
[6]   Multiple solutions of differential equations without the Palais-Smale condition. Math. Ann. **261** (1982) 399–412
[7]   Quasilinear elliptic eigenvalue problems. Comm. Math. Helv. **58** (1983) 509–527
[8]   A global compactness result for elliptic boundary value problems involving limiting nonlinearities. Math. Z. **187** (1984) 511–517
[9]   On a critical point theory for minimal surfaces spanning a wire in $\mathbb{R}^n$. J. Reine Angew. Math. **349** (1984) 1–23
[10]   On the evolution of harmonic maps of Riemannian surfaces. Comm. Math. Helv. **60** (1985) 558–581
[11]   Large H-surfaces via the mountain-pass-lemma. Math. Ann. **270** (1985) 441–459
[12]   Nonuniqueness in the Plateau problem for surfaces of constant mean curvature. Arch. Rat. Mech. Anal. **93** (1986) 135–157
[13]   A Morse theory for annulus-type minimal surfaces. J. Reine Angew. Math. **368** (1986) 1–27
[14]   On the evolution of harmonic maps in higher dimensions. J. Diff. Geom. **28** (1988) 485–502
[15]   Heat flow methods for harmonic maps of surfaces and applications to free boundary problems. In: Partial Differential Equations (eds.: Cardoso, F., de Figueiredo, D.G., Iório, R., Lopes O.), Lect. Notes Math. 1324, Springer, Berlin (1988)
[16]   The existence of surfaces of constant mean curvature with free boundaries. Acta Math. **160** (1988) 19–64
[17]   Plateau's problem and the calculus of variations. Math. Notes **35**, Princeton University Press, Princeton (1989)
[18]   Multiple solutions to the Dirichlet problem for the equation of prescribed mean curvature. Moser-Festschrift

Szulkin, A.:
[1]   Minimax principles for lower semicontinuous functions and applications to nonlinear boundary value problems. Ann. Inst. H. Poincaré, Analyse Non Linéaire, **3** (1986) 77–109

Talenti, G.:
[1]   Best constant in Sobolev inequality. Ann. Mat. Pura Appl. **110** (1976) 353–372

[2]   Some inequalities of Sobolev-type on two-dimensional spheres. In: General in-
       equalities **5** (ed.:Walter, W.), Intern. Ser. Numer. Math. **80**, Birkhäuser (1987)
       401–408

Tanaka, K.:
[1]   Infinitely many periodic solutions for a superlinear forced wave equation. Non-
       linear Analysis, T.M.A. **11** (1987) 85–104

Tartar, L.C.:
[1]   Compensated compactness and applications to partial differential equations.
       In: Nonlinear Analysis and Mechanics, Pitman Research Notes Math. **39**, Lon-
       don (1979) 136–212
[2]   The compensated compactness method applied to systems of conservation
       laws. In: Systems of nonlinear partial differential equations, Proc. NATO Ad-
       vanced Study Inst. 111, Oxford 1982 (ed.: J.M. Ball), Reidel, Dordrecht (1983)
       263–285

Thews, K.:
[1]   Non-trivial solutions of elliptic equations at resonance. Proc. Royal Soc. Edinb.
       85A (1980) 119–129

Tolksdorf, P.:
[1]   On the Dirichlet problem for quasilinear equations in domains with conical
       boundary points. Comm. P.D.E. **8** (1983) 773–817
[2]   Regularity for a more general class of quasilinear elliptic equations. J. Diff.
       Eq. **51** (1984) 126–150

Tomi, F., Tromba, A.J.:
[1]   Extreme curves bound embedded minimal surfaces of the type of the disc.
       Math. Z. **158** (1978) 137–145

Tonelli, L.:
[1]   Fondamenti di calcolo delle variazioni. Zanichelli, Bologna (1921–23)

Trudinger, N.S.:
[1]   Remarks concertning the conformal deformation of Riemannian structures on
       compact manifolds. Annali Sc. Norm. supp. Pisa **22** (1968) 265–274

Turner, R.:
[1]   Superlinear Sturm-Liouville problems. J. Diff. Eq. **13** (1973) 157–171

Uhlenbeck, K.:
[1]   Regularity for a class of nonlinear elliptic systems. Acta Math. **138** (1977)
       219–240

Vainberg, M.M.:
[1]   Variational methods for the study of non-linear operators. Holden Day, San
       Francisco (1964)

Viterbo, C.:
[1]   A proof of the Weinstein conjecture in $\mathbb{R}^{2n}$. Ann. Inst. H. Poincaré: Analyse
       non linéaire **4** (1987) 337–356
[2]   Indice de Morse des points critiques obtenus par minimax. Preprint

Walter, W.:
[1]   A theorem on elliptic differential inequalities with an application to gradient
       bounds. Math. Z. **200** (1989) 293–299

Wang, Z.Q.:
[1]   On a superlinear elliptic equation. Preprint (1989)

Weierstrass, K.:
[1]   Über das sogenannte Dirichlet'sche Princip. Mathematische Werke II, Mayer
       & Müller, Berlin (1895) 49–54

Weinstein, A.:
[1]   Normal modes for nonlinear Hamiltonian systems. Inv. Math. **20** (1973) 47–57
[2]   Periodic orbits for convex Hamiltonian systems. Ann. Math. **108** (1978) 507–
       518

[3]   On the hypotheses of Rabinowitz' periodic orbit theorem. J. Diff. Eq. **33** (1979) 353–358

Wente, H.C.:
[1]   An existence theorem for surfaces of constant mean curvature. J. Math. Anal. Appl. **26** (1969) 318–344
[2]   A general existence theorem for surfaces of constant mean curvature. Math. Z. **120** (1971) 277–288
[3]   The Dirichlet problem with a volume constraint. Manusc. math. **11** (1974) 141–157
[4]   The differential equation $\Delta x = 2H x_u \wedge x_v$ with vanishing boundary values. Proc. Amer. Math. Soc. **50** (1975) 59–77
[5]   Large solutions to the volume constrained Plateau problem. Arch. Rat. Mech. Anal. **75** (1980) 59–77

Werner, H.:
[1]   Das Problem von Douglas für Flächen konstanter mittlerer Krümmung. Math. Ann. **133** (1947) 303–319

Weyl, H.:
[1]   Über die asymptotische Verteilung der Eigenwerte. Gesammelte Abhandlungen I (ed. Chandrasekharan, K.), Springer, Berlin-Heidelberg-New York (1968) 368–375

Widman, K.O.:
[1]   Hölder continuity of solutions of elliptic systems. Manusc. math. **5** (1971) 299–308

Willem, M.:
[1]   Remarks on the dual least action principle. Z. Anal. Anwendungen **1** (1982) 85–90
[2]   Subharmonic oscillations of a semilinear wave equation. Nonlinear Analysis, T.M.A. **9** (1985) 503–514

Yamabe, H.:
[1]   On the deformation of Riemannian structures on compact manifolds. Osaka Math. J. **12** (1960) 21–37

Yang, C.T.:
[1]   On the theorems of Borsuk-Ulam. Kakutani-Yamabe-Yujobŏ and Dysin. Part I: Ann. of Math. (2) **60** (1954) 262–282, part II: Ann. of Math. (2) **62** (1955) 271–280

Zehnder, E.:
[1]   Periodische Lösungen von Hamiltonschen Systemen. Jber. d. Dt. Math.-Verein. **89** (1987) 33–59
[2]   Remarks on periodic solutions on hypersurfaces. In: Periodic solutions of Hamiltonian systems and related topics (ed. Rabinowitz, P.H.), Reidel, Dordrecht (1987) 267–279

Zeidler, E.:
[1]   Vorlesungen über nichtlineare Funktionalanalysis I–IV. Texte zur Math. **11**, Teubner, Leipzig (1977)
[2]   Nonlinear functional analysis and its applications I–IV. Springer, New York (1988) (Translation of [1])

Index